# Recent Trends in
# AGRICULTURE

(Volume - 8)



**Editors**

**Dr. Gaurang Chhangani**

**Dr. Tara Yadav**

**Dr. Subhra Sahoo**

**Sakshi Sharma**

**Sushma Sannidi**

**Integrated Publications ™**
**New Delhi**

***Published By:*** *Integrated Publications ™*

*Integrated Publications*
*H. No. - 3 Pocket - H34, Sector - 3,*
*Rohini, Delhi-110085, India*
*Email – info@integratedpublications.in*

***Editors:*** *Dr. Gaurang Chhangani, Dr. Tara Yadav, Dr. Subhra Sahoo, Sakshi Sharma and Sushma Sannidi*



***Edition:*** *1st*

***Publication Year:*** *2023*

***Pages:*** *421*

***ISBN:*** *978-93-5834-009-9*

***Book DOI:*** *https://doi.org/10.22271/int.book.302*

***Price:*** *₹ 1628/-*

# Contents

# Chapter - 1

## The Prospects of Biochar in Agriculture


**Authors**

**Sushma Sannidi**

PhD Scholar, Department of Agronomy, College of Agriculture, Professor Jayashankar Telangana State Agricultural University, Rajendranagar, Hyderabad, Telangana, India

**Bommidi Rajyalakshmi**

PhD Scholar, Department of Agronomy, College of Agriculture, Acharya N.G. Ranga Agricultural University, Bapatla, Andhra Pradesh, India

**B.V.S. Kiran**

Assistant Professor, K.L. University, Andhra Pradesh, India

**V. Prasad**

PhD Scholar, Department of Soil Science and Agricultural Chemistry, College of Agriculture, Professor Jayashankar, Telangana State Agricultural University, Rajendranagar, Hyderabad, Telangana, India

# Chapter - 1

## The Prospects of Biochar in Agriculture

**Sushma Sannidi, Bommidi Rajyalakshmi, B.V.S. Kiran and V. Prasad**


### Abstract

Biochar has emerged as an extraordinary catalyst for agricultural advancement, bearing a plethora of profound benefits that bestow upon the soil an array of improved physical, chemical, and biological properties, fostering an environment conducive to exponential plant growth, enhanced yield, and elevated productivity. Distinguished by its unparalleled physical attributes, including an expansive surface area, abundant pore space, and remarkable moisture retention capacity, biochar assumes the role of a superlative soil amendment, an instrumental agent in soil remediation and reclamation endeavors. Beyond its manifold provisions, the employment of biochar in agriculture finds its apex in the realm of sustainability, owing to its remarkable capacity for carbon sequestration and its potential to curtail pernicious greenhouse gas emissions. Bearing these myriad qualities in mind, we embark upon an exhaustive exploration of the distinctive properties of biochar, its intricate production and application methodologies, and its profound impact on soil, crops, and the environment, while delving into the daunting challenges encountered throughout its production and usage journey.


**Keywords:** Biochar, residue management, slow pyrolysis, fast pyrolysis, nutrient retention, carbon sequestration.

### 1. Introduction

The word "biochar" came from a combination of "bio-" means "biomass" and "char" means "charcoal" (Schulz and Glaser 2012). Biochar is a variable charged organic material and highly porous fine-grained charcoal, which has been produced through thermal decomposition under limited oxygen condition using organic biomass that optimizes certain special characteristics like large surface area and porosity and ability to preserve in soils for a long time with very little biological deterioration (Robertson 2014). Biochar is a sustainable way of recycling the huge amount of crop residues used for producing the effective bio-amendment for soil restoration and reduce carbon emissions into the atmosphere and thus helps in reducing environmental risks (Abdelhafez *et al*, 2014; Ahmad *et al*, 2014). Biochar provides multiple benefits to the ecosystem through soil quality improvement, climate change mitigation, green energy production and waste management (Joseph *et al*, 2009).

### 1.1 The history of biochar as soil amendment

The first evidence of biochar in the history was in the form of "Terra Preta" which is also known as Indian black earth and was first discovered in the Amazon rain forest as soil influencer which demonstrated the transformation of less fertile degraded soils into enriched form (Glaser and Birk, 2012). Terra preta illustrated itself as the best way to sustain and flourish as nutrient rich, well structured and biome rich productive soil (Alling *et al,* 2014).

### 1.2 Origin of the concept of biochar

Biochar use is a millenia's old concept used in many cultures over the years. In the Brazilian part of the amazon basin, man-made soils made by charcoal burial of over 2000 years or more known as *terra preta* ("black earth" in Portugese) are found at several sites in the region. These soils are more productive than anywhere else on earth. These soils were formed by the application of stable organic matter in the form of biochar (charcoal). The "Terra Preta" soils which were regularly amended with biochar and other organic materials exhibit higher pH, greater nutrient availability; larger microbial populations more diversified microbial community (Liang *et al*, 2008; Germano *et al*, 2012; Taketani *et al*, 2013).

Carbon, ash, volatile matter, and moisture are the major constituents of biochar (Antal and Gronli, 2003; DeLuca *et al*, 2006).

## 2. Properties of biochar

It serves as a catalyst that enhances uptake of nutrient and water by the plant compared to other soil amendments, pertaining to high surface area and porosity of biochar enabling it to adsorb and retain nutrients and water and also provides a habitat for beneficial microorganisms to flourish (Glaser *et al*, 2002). Diffusion of gases increases with the application of biochar which in turn facilitates better soil aeration (Sun *et al*, 2013). The greater surface area of biochar leads to more opportunity for microbial colonization and its black color, attracts more heat and thus accelerating the microbial growth and enzyme activity (Gul *et al,* 2015). The presence of negatively charged phenolic, carboxyl and hydroxyl groups along with silicates, carbonates and bicarbonates originating from biochar can bind to $H^+$ ions and thereby removing them from soil solution and contributing to an increase in soil pH because of which biochar have been used in controlling soil pH, especially as liming material in acidic soils and soils with low organic matter content (Brewer and Brown, 2012; Chintala *et al*, 2014; Rodriguez *et al*, 2009; Lakaria *et al*, 2012). The slightly alkaline pH of biochar might attribute to the formation of oxides of alkali and alkaline earth metals during the incineration of plant tissues under limited oxygen supply. The EC of biochar in a study was found to be BC (0.0.097 dS $m^{-1}$) (Sanders and Bloomfield, 1980). The biochar contains C concentration of 475.8 g C $kg^{-1}$.

Biochar can act as effective organic manure for longer duration as the C present in it can be locked even for years together due to slower decomposition rate which may be influenced by the nature of feedstock and pyrolysis temperature and also due to reduced soil microbial activity (Li *et al*, 2018), or the sorption of soil organic matter (SOM) to biochar (Wang *et al*, 2016).

## 2.2 Nutrient release pattern of biochar

Both macro-and micro-nutrients in organic amendments were readily subjected to mineralization for ultimate release for subsequent plant uptake or chemical transformations, if these organic amendments were rich in water-soluble, fulvic acid- or humic acid-C. Whereas the biochar being recalcitrant to microbial decomposition with exceptionally condensed aromatic characteristics, ensures longer nutrient supply to the crop (Steiner *et al*, 2007; Hood-Nowotny *et al*, 2018).

The nutrient release pattern from biochar is the slowest when compared to other organic supplements like FYM, vermicompost, poultry manure, cattle manure etc. due to high C: N ratio and slow decomposition rate of biochar however among all the nutrients, K was quickly mineralized at a rate of 83.3% within 120 days of incubation. The nutrient release rate of Zn from biochar was higher due to lesser irreversible incorporation of Zn in humic acid- and humin fractions of C in it (Hattab *et al* 2014). The release pattern and rate of Fe and Mn from the biochar could be attributed to the presence of potential reducing agents like quinine and condensed aromatic structures in it (Klüpfel *et al* 2014). Release of Mn might delay due to its alkaline pH while the low release rate of B from biochar is due to the presence of insoluble boron compounds in it. Organic amendments with higher N content which could sustain faster decomposition rate might also release higher amounts of Fe as compared to organic amendments of lower N concentration (Dey *et al*, 2019).

## 3. Status of organic carbon content in the soil

Repeated annual removal of crop residues from the fields leads to reduced organic carbon stocks in the soil and stimulates erosion of fertile top soil (Laird *et al*, 2009).

In tropical and subtropical climates, high soil and air temperatures, along with high moisture content, stimulate microbial degradation of organic fertilizers such as compost, manure, crop residues, and other organic fertilizers put to soils, resulting in rapid organic matter loss. Lower carbon content in soils came from faster microbial degradation of residual agricultural residues and additional organic wastes (Agehara and Warncke 2005). As a result, the current carbon levels and carbon stocks in the soils of many nations in tropical and subtropical climate zones are rapidly reducing. In such case, biochar can be sustainable option for replenishing and conservation of soil organic carbon content in the soil. Also, biochar is highly resistant to decomposition when applied to soil and its residence time ranges from tens of years to millennia because of which it can act as a promising source of soil organic matter for a great deal of time (Preston and Schmidt, 2006; Verheijen *et al*, 2010).

It also helps to improve soil structure by encircling soil particles, helping biochar to not only capture its own carbon but also safeguard inherent and applied carbon from a variety of sources. Biochar mineralization is far slower than that of any other organic material that is applied to soil

## 4. Material used for production of biochar

Different types of sun dried organic materials (rice straw, rice husk, any other crop residues, wood and bark, bagasse, shrimp waste grass clippings, plant prunings, manure, and sewage sludge) can be used to produce biochar through pyrolysis process at 400–550°C in the absence of oxygen or limited oxygen. Wood chips, crop residues, nut shells, seed mill screenings, algae, animal manure and sewage sludge can also be used.

Sweet sorghum bagasse known to have a rich chemical composition of cellulose, hemicellulose, lignin and C/H ratio can product to be converted into biochar

## 4.1 Approaches for production

Pyrolysis is more acceptable method of thermos-chemical conversion of feed stock into biochar. Pyrolysis process is further distinguished into various types depending on operating conditions used such as the feedstock's heating rate, residence time and pyrolysis temperature (Czernik and Bridgwater, 2004; Mohan *et al*, 2006).

Slow pyrolysis biochar is a product of traditional heating of feedstocks under oxygen-limiting conditions, for cooking and house-warming purposes and it is achieved by heating the feedstocks at temperatures from 300 to 800°C at atmospheric pressure for hours to days (Brewer and Brown, 2012). Fast pyrolysis aims to maximize the production of bio-oil by rapid quenching of vapor produced from burning biomass at higher temperatures (400–1000°C) with a fast-heating rates i.e., >300°C for few hours (i.e., 1–2 h; Brewer and Brown, 2012; Mohanty *et al*, 2013).

## 4.2 Factors affecting the properties of biochar

### 4.2.1 Biochar properties as function of

#### i) Feedstock

The nutrient availability and carbon stocks in the biochar vary with the feed stock material (Gaskin *et al*, 2008). For instance, the biochar produced from the extracts of crop residues, sea weeds and organic manures are rich in nutrients with low pH. The wood feed stock biochar will lower ash content, soluble elemental components and high pH when compared to crop residue-derived and grass-derived biochar which possess higher EC, CEC, greater thermal stability and larger surface area (Kloss *et al,* 2012). In general, we can say that plant-based biochar prepared from agricultural residues has high pH and the animal-based biochar has low pH or is comparatively acidic (Solaiman and Anawar, 2015) and more specifically, leguminous plant-based biochar have strong alkaline properties that helps to neutralize the acidic soils (Yuan *et al*, 2011).

#### ii) Production temperature

The pyrolysis temperatures play a major role in determining the characteristic features of biochar. The properties of biochar such as pH, surface area, cation exchange capacity, nutrient availability and C: N & C:O ratios are highly dependent on the pyrolysis temperature. Biochar having higher soil nourishing properties can be obtained at lower PT and THT (Barrow, 2012) and research results demonstrate that biochar yield is inversely proportional to the pyrolysis temperature. Low pyrolysis temperature (<550°C) ensures controlled release of nutrients from fertilizers which is attributed to high amorphous carbon except the low ash content (Gruss *et al,* 2019) whereas the biochar prepared at high pyrolysis temperature (>550°C) has high EC, greater availability of macro and micronutrients excepts N and is biologically resistant ensuring longer residence in the soil with greater moisture holding capacity. The surface area of biochar is also dependent on the pyrolysis temperature where surface area (SA) of biochar increases with increase in PT due to loss of volatile matter, thus, in turn leading to increase in porosity and decrease in bulk density. Hence it can be admitted that biochar produced at low temperatures tend to have greater reactivity in soils and support higher yields (Chan *et al*, 2008; Steinbeiss *et al*, 2009). When compared to the high temperature biochars However, research results suggest that 550°C temperature will be appropriate for the production of biochar from crop residues (Baidoo *et al,* 2016).

Slow pyrolysis is characterized by relatively extended solids and gas residence times and slow heating (400°C) of the organic material (minutes to hours) in the absence of oxygen (typically several minutes to hours) (Mohan *et al*, 2006).

Under fast pyrolysis systems, fine grained organic matter is blown into a hot reactor (450-600°C) and exposed to heat transfer for very short duration (just a few milliseconds to seconds) (Bridgwater *et al*, 2007; Laird *et al*, 2009).

**iii) Fresh- versus aged-biochar-amended soils**

Aged biochar contains high C stocks, ash content and volatile matter when compared to freshly produced biochar which shows that incubated biochar has better results on the properties of biochar.

**iv) Temperature hold time (THT)**

Pyrolysis processes are generally classified on the basis of vapor residence time or peak temperature hold time (THT). CRB yield has been reported to decrease with increased THT (Peng *et al*, 2011).

**v) Heating rate**

Various studies have been performed at different heating rate ranging from 10°C/min (Wu *et al*, 2012; Lee *et al*, 2013) to 45°C/min (Cao *et al*, 2014). Higher heating rate leads to rapid carbonization of biomass, thus, resulted into extensive cracking and micropores generation producing high surface area (Mukherjee and Lal, 2013)

### 4.3 Classification of biochar

Based on the carbon content, the biochar is categorized into three categories by the International Biochar Initiative (IBI):

Class 1: >60% °C

Class 2: 30–60% °C

Class 3: 10–30% °C

### 4.4 Recovery rate of biochar

The biochar yield obtained per unit of feedstock depends on the rate of pyrolysis. The studies revealed that higher recovery of biochar was obtained with slow pyrolysis (Zhang *et al,* 2018). The recovery rate of biochar varying with the feedstock material, pyrolysis temperature and the retention time is presented in Table 1.

**Table 1:** Recovery rate of biochar obtained from different feedstock materials

| Crop | Biochar yield (%) |
|---|---|
| Castor (*Ricinus communis*) | 17-24 |
| Pigeon pea *(Cajanus cajan)* | 30-35 |
| Sorghum (*Sorghum bicolor*) | 25-30 |
| Rice straw *(Oryza sativa)* | 28-33 |
| Rice husk *(Oryza sativa)* | 33-38 |
| Corn stalks *(Zea mays*) | 25-30 |
| Wheat straw *(Triticum aestivum)* | 25-20 |
| Cotton stalk (*Gossypium sps*) | 25-30 |
| Sugarcane bagasse (*Saccharum officinarum*) | 24-28 |

### 4.5 Application rate of biochar

The rate of application of biochar has wide variation ranging from a minimum of 1 t $ha^{-1}$ to even 100 t $ha^{-1}$ (Chan *et al,* 2007). But the common application rate of biochar ranges from 10- 20 t $ha^{-1}$(Sun *et al,* 2014).

Application of biochar deep into the soil at a depth of 0-25 cm obtained better results when compared to shallow application at surface soil (0-10 cm) as it reached the rooting depth of the crop (Pietola and Alakukku 2005) and was more effective.

#### 4.5.1 Frequency of application of biochar

Having the slowest decomposition rate, biochar need not be applied every year unlike the conventional organic manures.

## 4.6 Method of application

Biochar can be applied to soil in the form of coarse pieces or finely grounded powder and can be mixed with soil for better incorporation as shown in the figure 1. Fine textured biochar are more susceptible to microbial degradation (Zimmerman, 2010) with high surface area tend to be physically protected from degradation and encapsulation, yet emit more $CO_2$ when compared to coarser biochar (Brodowski *et al*, 2006).

Band application of biochar along with synthetic fertilizers is beneficial in low rainfall areas which are susceptible to wind erosion (Blackwell *et al*, 2010) and ensures closer interaction of biochar and fertilizer for maximum benefits (nutrient and water supply, reduction of drought stress) as well (Blackwell *et al*, 2009a, b).

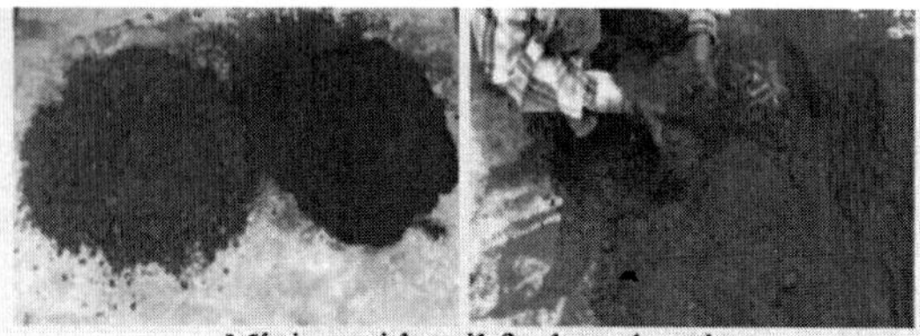

Mixing with soil for broadcasting

Incorporating biochar into soil

Band application

Spot placement

## 5. Rate of decomposition in the soil

Bird *et al* (1999) found a correlation between the granulometry of soil-charred particle and the time needed for biodegradation. The half-life of particles less than 2 mm is thought to be less than 50 years, whereas that of particles larger than 2 mm is thought to be less than 100 years, according to these authors' estimates. Research studies also revealed that slow pyrolyzed biochar has a slower rate of degradation comparatively and thus tends to retain for longer time in the soil.

### 5.1 Fate of biochar in the soil

On addition of biochar into the soil, the following changes are likely to take place in the soil to improve the soil physico-chemical properties (as mentioned in Figure 1).

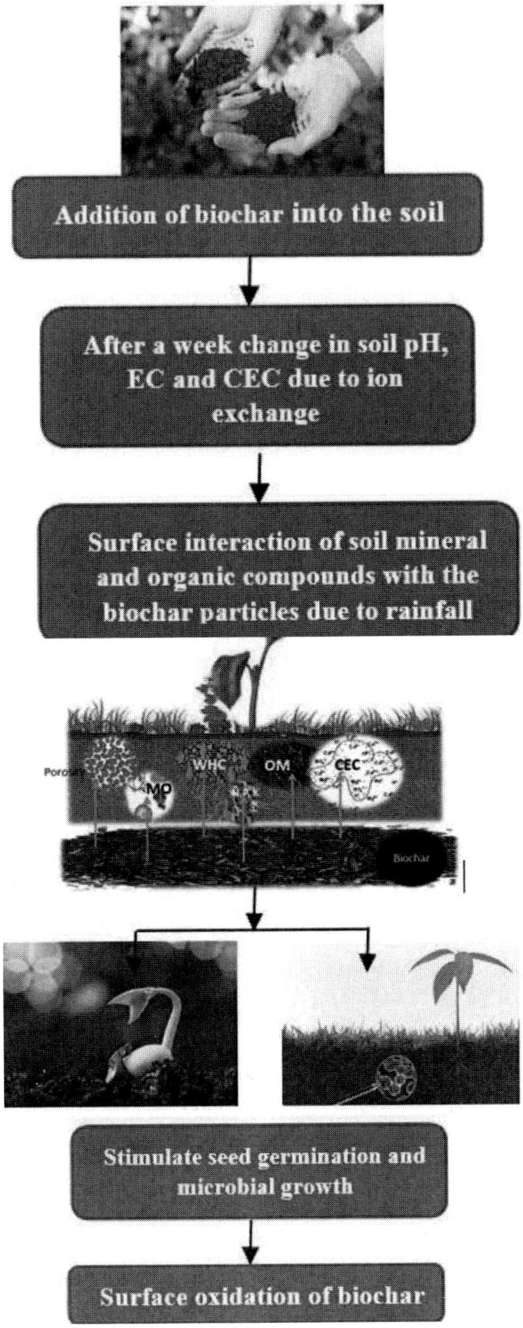

Addition of biochar into the soil
After a week change in soil pH, EC and CEC due to ion exchange
Surface interaction of soil mineral and organic compounds with the biochar particles due to rainfall
WHC
OM
CEC
MO
Biochar
Stimulate seed germination and microbial growth
Surface oxidation of biochar

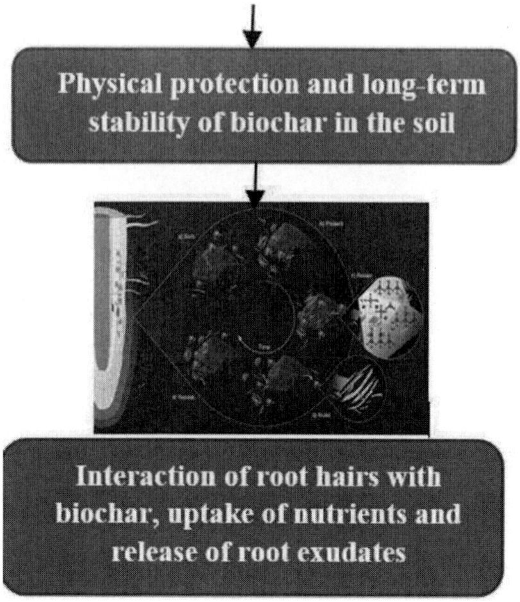


## 6. Benefits of biochar

### 6.1 Enhanced crop yields and productivity

The value of applying biochar to improve soil quality and boost the crop productivity appears to be promising due to the large surface area of sorghum biochars, the micropores which provide microhabitat for beneficial microorganism, and the rising soil organic carbon contents. The yield increase in biochar is not obvious in all conditions but always has a positive indirect effect on the yield and productivity of the crop which depends on rate and method of application of biochar. For example, the addition of sorghum biochars to winter wheat in a pot culture experiment, with or without supplementing additional inorganic phosphorus fertilizer in Ultisols of USA, enhanced total biomass of winter wheat plants by around 31% compared to control plants which was possible with the increase in above ground (mean: 30 g plot$^{-1}$; 26 g plot$^{-1}$) and below ground (mean: 26 g plot$^{-1}$; 20 g plot$^{-1}$) biomass of the plants due to the incraesed uptake of nutrients (Sigua *et al,* 2015).

### 6.2 Nutrient supply

In contrast to manure or straw, pyrogenic organic matter has an intrinsic stability that precludes the need for protection against aggregates and/or clay minerals facilitating accumulate higher concentrations of carbon (Smith 2016). Biochar application is more beneficial in the tropical and sub-tropical regions facilitating continuous nourishment and increased crop yields (Jeffery *et al* 2017) for longer period where the conventional organic manures fail due to faster decomposition rate at higher temperatures and also contribute to greenhouse gas emissions.

## 6.3 Improved soil physico-chemical properties

The biochar having gifted properties like reduced bulk density, higher porosity and large specific surface area provides greater aeration to the soil and also enhances the soil water holding capacity. The addition of biochar besides reducing the soil bulk density (Laird *et al* 2010 Oguntunde *et al* 2008), alters soil infiltration rate. As the surface area of biochar is often greater than that of sand and comparable to or greater than that of clay, it will enhance the total soil-specific surface area when added as an amendment (Downie *et al*, 2009). Higher rates of biochar application greatly enhance soil field capacity (Alburquerque *et al* 2014), and these effects can be more noticeable in non-irrigated regions with an increase in available water for crop development as well as a reduction in the occurrence of water stress in between rainfall episodes.

## 6.4 Enhanced microbial activity

In terrestrial ecosystems, biochar applied soil provides a favourable habitat for micro organisms. Improved bulk density as a result of greater soil porosity will improve soil aeration, which is important for root and microbial respiration. Biochar, in particular, stimulates the workings of mycorrhiza fungi by changing the soil physical/chemical structures and altering the environment favourable to mycorrhizae. It also interferes with plant-fungus signalling and allelochemical detoxification, in addition to providing shelter from fungal grazers (Ameloot *et al*, 2013). Biochars, as soil supplement, can potentially benefit diazotrophs (N2-fixing bacteria).

## 6.5 Soil conservation

Biochar has the ability to conserve the soil by reducing the soil erosion because biochar facilitates increased water holding capacity of the soil, fixing of nutrients and prevents the washing away of nutrients from the fertilizers.

## 6.6 Carbon sequestration potential

The use of biochar in soil represents a novel strategy to creating a major long-term sink of atmospheric carbon dioxide ($CO_2$) in terrestrial ecosystems due to due to slow microbial decomposition and chemical transformation. There is a possibility of sequestration of about 9.5 BT of carbon can in soils by the year 2100 with wide biochar application practices (Lehmann, *et al,* 2006). Approximately 50% of the carbon can be sequestered during the conversion of biomass carbon to biochar, compared to only 3% retention in soil by burning and less than 10-20% (after 5-10 years) through biological decomposition, resulting in higher yields of stable soil carbon in soil upon application (Lehmann, *et al,* 2006).

### 6.7 Reduced emissions of GHG

Biochar has the greatest potential in controlling the release pattern of $N_2O$ gas from the soil and thus extremely effective in reducing soil $N_2O$ emissions but many factors like variation within the soil, biochar material and its nutrient release pattern, fertilizer composition and application rate also influence the emissions of $N_2O$. (Cayuela *et al,* 2015)

### 6.8 Reclamation of problematic soils

#### 6.8.1 Salt-affected Soils

The biochar alleviated the adverse impact of abiotic stress both by decreasing plant exposure to stress or by enhancing the plant stress resistance (Buss *et al,* 2012). The mineral ash content of biochar attributes to its liming effect and thus is amenable for alleviating the adverse impact of salinity by three mechanisms: (i): reducing transient Na+ by adsorption, (ii): releasing mineral nutrients and decreasing osmotic stress by improving soil water availability (Akhtar *et al,* 2015). The strong adsorptive properties refer to its high porosity, large surface area and high cation exchange capacity which aids in adsorbing toxic ions or releasing more beneficial ions which help in reducing the adverse impact of salinity on plants (Novak *et al,* 2021). Also, the biochar application can increase the antioxidant enzymes like CAT< POD< SOD which would activate the plant resistance mechanisms under saline conditions. (Wang *et al*, 2016).

#### 6.8.2 Acidic soils

The pH of biochar ranges from 9-10.7 (alkaline range) based on the feedstock material and pyrolysis temperature (Marshall *et al,* 2019). Biochar has a potential liming effect in acidic soils by increasing base saturation and CEC thereby elevating the pH of the soil and alleviating the toxicity of Fe and Al and increasing the nutrient availability at neutral pH.

### 6.9 Phytoremediation

Biochar can be used as a practicable adsorbent for removing wide range of organic and inorganic contaminants like polyaliphatic and polyaromatic hydrocarbons, metals, toxic chemicals, and elements from soil, sediments and water (Hale *et al*, 2012; Mohan *et al*, 2014). It is also employed to stabilize the heavy metals in the soil which reduces the soil contamination and reduces the crop intake as well (Ippolito *et al*, 2012). A study conducted by Cheng *et al* (2006) reported the absorption of Cu, Ni, Pb and Hg by bamboo biochar from polluted soil and water.

### 6.9.1 Sorption and activity of agrochemicals

Biochar facilitates the sorption of agrochemical residues and metal toxicants in the soil benefitting the crops by immobilizing the residues

## 7. Challenges in biochar production and application

- The impediment to technology adoption has been the limited supply of commercial biochar available to farmers.
- Adsorption of agrochemicals like fertilizers, pesticides and herbicides by biochar reduces their efficacy and immobilizes thus increasing the rate of application which in turn increases the cost of cultivation to the farmers.
- Based on the nutrient release pattern from the biochar and from the synthetic fertilizers used in the crop, there may be rise in $N_2O$ emissions with the application of biochar as amendment (Cayuela *et al*, 2015).
- Biochar when applied to N deficit soils, further limits the N availability due to high C:N ratio thus reducing the crop productivity (Lehmann *et al,* 2002).

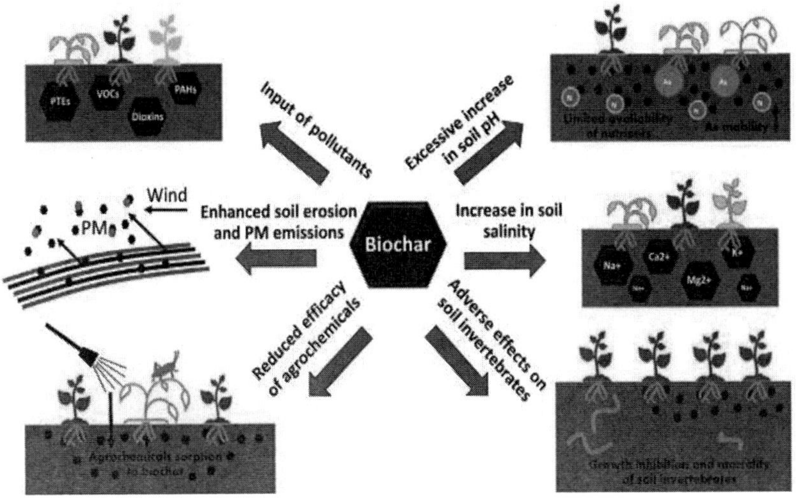

## Conclusion

Using biochar instead of chemical fertilisers has been found to provide a number of advantages for enhancing soil health, plant health, and crop productivity in a sustainable way. Through easier bioavailability of essential plant nutrients and enhanced soil physicochemical qualities, biochar has a significant ameliorating potential that may very well resolve soil concerns and raise agricultural production. Due to its unique physico-chemical characteristics, it is appropriate for organic farming, dryland farming, conservation agriculture, and land reclamation. A long-term soil carbon store can be created by adding biochar to agricultural soil in light of climate change and carbon trading. The feedstock and pyrolysis temperature have an impact on how heterogeneous and variable biochar is. Therefore, biochars should be carefully described before being applied to soil.

## References


1. Abdelhafez, A.A., Li, J. and Abbas, M.H., 2014. Feasibility of biochar manufactured from organic wastes on the stabilization of heavy metals in a metal smelter contaminated soil. *Chemosphere*, *117*, pp.66-71.
2. Agehara, S. and Warncke, D.D., 2005. Soil moisture and temperature effects on nitrogen release from organic nitrogen sources. *Soil Science Society of America Journal*, *69*(6), pp.1844-1855.
3. Ahmad, M., Rajapaksha, A.U., Lim, J.E., Zhang, M., Bolan, N., Mohan, D., Vithanage, M., Lee, S.S. and Ok, Y.S., 2014. Biochar as a sorbent for contaminant management in soil and water: a review. *Chemosphere*, *99*, pp.19-33.
4. Akhtar S.S., Andersen, M.N and Liu, F. 2015. Biochar mitigates salinity stress in potato. *Journal of Agronomy and Crop Science*. 201: 368-378.
5. Alburquerque, J.A.; Calero, J.M.; Barrón, V.; Torrent, J.; del Campillo, M.C.; Gallardo, A.; Villar, R. Effects of biochars produced from different feedstocks on soil properties and sunflower growth. J. Plant Nutr. Soil Sci. 2014, 177, 16–25.
6. Alling V; Hale, S.E.; Martinsen, V.; Mulder, J.; Smebye, A.; Breedveld, G.D.; Cornelissen, G. The role of biochar in retaining nutrients in amended tropical soils. J. Plant Nutr. Soil Sci. 2014, 177, 671–680.
7. Ameloot, N., Graber, E.R., Verheijen, F.G.A., De Neve, S., 2013. Interactions between biochar stability and soil organisms: review and research needs. Eur. J. Soil Sci. 64, 379–390.

8. Antal J and Grönli M (2003). The art, science and technology of charcoal production. Ind Eng Chem Res 42(8): 1619-1640.

9. Baidoo, Isaac, Sarpong, Bruce, D and Simon, B. 2016. Biochar amended soils and crop productivity: A critical and meta-analysis of literature. International Journal of Development and Sustainability. 5 (9): 414-432.

10. Barrow, C.J., 2012. Biochar: potential for countering land degradation and for improving agriculture. *Applied Geography*, *34*, pp.21-28.

11. Bird, M.I.; Moyo, C.; Veenendaal, E.M.; Lloyd, J.; Frost, P. 1999. Stability of elemental carbon in savanna soil. Global Biogeochemical Cycles 13: 923-932.

12. Blackwell P, Joseph S, McHenry M, Downie A, Park D, Bellamy P. 2009a. Concepts of dryland farming systems incorporating biochar and carbon-rich biological fertilisers. In Proceedings of the 1st Asia Pacific Biochar Conference. May 17–20, 2009, Gold Coast. p. 85.

13. Blackwell P, Krull E, Butler G, Herbert A, Solaiman Z. 2010. Effect of banded biochar on dryland wheat production and fertiliser use in south-western Australia: an agronomic and economic perspective. Aust J Soil Res. 48: 531–545.

14. Blackwell P, Riethmuller G, Collins M. 2009b. Biochar application to soil. In Lehmann J, Joseph S (eds.) Biochar for En vironmental Management: Science and Technology. Earthscan, London. pp. 207–226.

15. Brewer, C.E., Brown, R.C., 2012. Biochar. In: Sayigh, A. (Ed.), Comprehensive Renewable Energy. Elsevier, Oxford, pp. 357–384.

16. Bridgwater, A.V., Carson, P., Coulson, M., 2007. A comparison and fast and slow pyrolysis liquids from mallee. International Journal of Global Energy Issues 27, 204-216.

17. Brodowski, S., John, B., Flessa, H., Amelung, W., 2006. Aggregate-occluded black carbon in soil. European Journal of Soil Science 57, 539-546.

18. Buss, W., Kammann, C and Koyro, H.W. 2012. Biochar reduces copper toxicity in *Chenopodium quinoa* Wild. in a sandy soil. *Journal of Environmental Quality*. 41: 1157-1165.

19. Cao, X., Zhong, L., Peng, X., Sun, S., Li, S., Liu, S. and Sun, R., 2014. Comparative study of the pyrolysis of lignocellulose and its major components: Characterization and overall distribution of their biochars

and volatiles. *Bioresource technology*, *155*, pp.21-27.

20. Cayuela, M.L., Jeffery, S., Van Zwieten, L., 2015. The molar H:C org ratio of biochar is a key factor in mitigating N2O emissions from soil. Agric. Ecosyst. Environ. 202, 135–138.

21. Chan K Y, Van Zwieten L, Meszaros I, Downie A, Joseph S. 2008. Using poultry litter biochars as soil amendments. Aust J Soil Res. 46: 437–444.

22. Chan, K.Y., Van Zwieten, L., Meszaros, I., Downie, A., Joseph, S., 2007. Agronomic values of greenwaste biochar as a soil amendment. Australian Journal of Soil Research 45, 629-634.

23. Cheng CH, Lehmann J, Thies JE, Burton SD and Engelhard MH (2006). Oxidation of black carbon by biotic and abiotic processes. Org Geochem 37(11):1477-1488. DOI: 10.1016/j.oeggeochem.2006.06.022.

24. Chintala, R., Schumacher, T.E., Kumar, S., Malo, D.D., Rice, J.A., Bleakley, B., Chilom, G., Clay, D.E., Julson, J.L., Papiernik, S.K. and Gu, Z.R., 2014. Molecular characterization of biochars and their influence on microbiological properties of soil. *Journal of Hazardous Materials*, *279*, pp.244-256.

25. Czernik, S. and Bridgwater, A.V., 2004. Overview of applications of biomass fast pyrolysis oil. *Energy & fuels*, *18*(2), pp.590-598.

26. DeLuca TH, MacKenzie MD, Gundale MJ and Holben WE (2006). Wildfire-producedcharcoal directly influences nitrogen cycling in forest ecosystems. Soil Sci Soc Am J 70: 448-453.

27. Dey, A., Srivastava, P.C., Pachauri, S.P. and Shukla, A.K., 2019. Time-dependent release of some plant nutrients from different organic amendments in a laboratory study. *International Journal of Recycling of Organic Waste in Agriculture*, *8*, pp.173-188.

28. Downie, A., Crosky, A., Munroe, P., 2009. Physical properties of biochar. Biochar for Environmental Management. Science and Technology, p. 416.

29. Gaskin, J.W., Steiner, C., Harris, K., Das, K.C., Bibens, B., 2008. Effect of low-temperature pyrolysis conditions on biochar for agricultural use. Transactions of the ASABE 51, 2061-2069.

30. Germano M.G., Cannavan, F.D.S., Mendes, L.W., Lima, A.B., Teixeira, W.G., Pellizari, V.H. and Tsai, S.M., 2012. Functional diversity of bacterial genes associated with aromatic hydrocarbon degradation in

anthropogenic dark earth of Amazonia. *Pesquisa Agropecuária Brasileira*, *47*, pp.654-664.

31. Glaser, B., Lehmann, J. and Zech, W., 2002. Ameliorating physical and chemical properties of highly weathered soils in the tropics with charcoal–a review. *Biology and fertility of soils*, *35*, pp.219-230.

32. Glaser, B.; Birk, J.J. State of the scientific knowledge on properties and genesis of Anthropogenic Dark Earths in Central Amazonia (terra preta de Índio). Geochim. Cosmochim. Acta 2012, 82, 39–51.

33. Gruss, I.; Twardowski, J.; Latawiec, A.; Medy ´nska-Juraszek, A.; Królczyk, J. Risk assessment of low-temperature biochar used as soil amendment on soil mesofauna. Environ. Sci. Pollut. Res. 2019, 26, 18230–18239.

34. Gul, S., Whalen, J.K., Thomas, B.W., Sachdeva, V. and Deng, H., 2015. Physico-chemical properties and microbial responses in biochar-amended soils: mechanisms and future directions. *Agriculture, Ecosystems & Environment*, *206*, pp.46-59.

35. Hale, S.E., Lehmann, J., Rutherford, D., Zimmerman, A.R., Bachmann, R.T., Shitumbanuma, V., O'Toole, A., Sundqvist, K.L., Arp, H.P.H. and Cornelissen, G., 2012. Quantifying the total and bioavailable polycyclic aromatic hydrocarbons and dioxins in biochars. *Environmental science & technology*, *46*(5), pp.2830-2838.

36. Hattab, N., Soubrand, M., Guégan, R., Motelica-Heino, M., Bourrat, X., Faure, O. and Bouchardon, J.L., 2014. Effect of organic amendments on the mobility of trace elements in phytoremediated techno-soils: role of the humic substances. *Environmental Science and Pollution Research*, *21*, pp.10470-10480.

37. Hood-Nowotny, R., A. Watzinger, A. Wawra, and G. Soja. 2018. The impact of biochar incorporation on inorganic nitrogen fertilizer plant uptake; an opportunity for carbon sequestration in temperate agriculture. Geosciences 8:420.

38. Ippolito JA, Scheckel KG and Barbarick KA (2012). Selenium adsorption to aluminum-based water treatment residuals. J Colloid Interface Sci 338: 48–55.

39. Jeffery, S., Abalos, D., Prodana, M., Bastos, A.C., Van Groenigen, J.W., Hungate, B.A. and Verheijen, F., 2017. Biochar boosts tropical but not temperate crop yields. *Environmental Research Letters*, *12*(5),

p.053001.

40. Joseph, S., Peacocke, C., Lehmann, J. and Munroe, P., 2009. Developing a biochar classification and test methods. *Biochar for environmental management: science and technology*, *1*, pp.107-126.

41. Kloss, S., Zehetner, F., Dellantonio, A., Hamid, R., Ottner, F., Liedtke, V., Schwanninger, M., Gerzabek, M.H. and Soja, G., 2012. Characterization of slow pyrolysis biochars: effects of feedstocks and pyrolysis temperature on biochar properties. *Journal of environmental quality*, *41*(4), pp.990-1000.

42. Klupfel, L., Keiluweit, M., Kleber, M., Sander, M., 2014. Redox properties of plant biomass derived black carbon (biochar). Environ. Sci. Technol. 48, 5601–5611

43. Laird, D.A., Brown, R.C., Amonette, J.E., Lehmann, J., 2009. Review of the pyrolysis platform for coproducing bio-oil and biochar. Biofuels Bioproducts & Biorefining 3, 547-562.

44. Laird, D.A.; Fleming, P.; Davis, D.D.; Horton, R.; Wang, B.; Karlen, D.L. Impact of biochar amendments on the quality of a typical Midwestern agricultural soil. Geoderma 2010, 158, 443–449.

45. Lakaria, B.L., Sumitra G, Pramod J and Biswas, A.K. (2012). Biochar preparation and characterization. Ind J Soil Cons 40(2): 141-146.

46. Lee, Y., Park, J., Ryu, C., Gang, K.S., Yang, W., Park, Y.K., Jung, J. and Hyun, S., 2013. Comparison of biochar properties from biomass residues produced by slow pyrolysis at 500 C. *Bioresource technology*, *148*, pp.196-201.

47. Lehmann J, da Silva Jr., J.P., Rondon, M., Steiner, C., Nehls, T., Zech, W., Glaser, B., 2002. Nutrient availability and leaching in an archaeological Anthorosol and a Ferralsol of the Central Amazon basin: fertilizer, manure and charcoal amendments. Plant and Soil 249, 343–357.

48. Lehmann, J.; Gaunt, J.; Rondon, M. Bio-char sequestration in terrestrial ecosystems-A review. Mitig. Adapt. Strateg. Glob. Chang.2006, 11, 403–427.

49. Li, Y., Li, Y., Chang, S. X., Yang, Y., Fu, S., Jiang, P., Luo, Y., Yang, M., Chen, Z., Hu, S., Zhao, M., Liang, X., Xu, Q., Zhou, G., & Zhou, J. (2018). Biochar reduces soil heterotrophic respiration in a subtropical plantation through increasing soil organic carbon recalcitrancy and decreasing carbon-degrading microbial activity. Soil Biology and

Biochemistry, 122, 173–185. https://doi.org/10.1016/j.soilbio.2018.04.019

50. Liang, B., Lehmann, J., Solomon, D., Sohi, S., Thies, J.E., Skjemstad, J.O., Luizao, F.J., Engelhard, M.H., Neves, E.G. and Wirick, S., 2008. Stability of biomass-derived black carbon in soils. *Geochimica et Cosmochimica Acta*, *72*(24), pp.6069-6078.

51. Marshall, J., Muhlack, R., Morton, B.J., Dunnigan, L., Chittleborough, D. and Kwong, C.W., 2019. Pyrolysis temperature effects on biochar–Water interactions and application for improved water holding capacity in vineyard soils. *Soil Systems*, *3*(2), p.27.

52. Mohan D., Pittman Jr., C.U., Steele, P.H., 2006. Pyrolysis of wood/ biomass for bio-oil: a critical review. Energy & Fuels 20, 848-889.

53. Mohan, D., Sarswat, A., Ok, Y.S. and Pittman Jr, C.U., 2014. Organic and inorganic contaminants removal from water with biochar, a renewable, low cost and sustainable adsorbent–a critical review. *Bioresource technology*, *160*, pp.191-202.

54. Mohanty, P., Nanda, S., Pant, K.K., Naik, S., Kozinski, J.A. and Dalai, A.K., 2013. Evaluation of the physiochemical development of biochars obtained from pyrolysis of wheat straw, timothy grass and pinewood: effects of heating rate. *Journal of analytical and applied pyrolysis*, *104*, pp.485-493.

55. Mukherjee, A. and Lal, R., 2013. Biochar impacts on soil physical properties and greenhouse gas emissions. *Agronomy*, *3*(2), pp.313-339.

56. Novak, J.M., Busscher, W.J., Watts, D.W., Amonette, J.E., Ippolito, J.A and Lima. 2012. Biochars impact on soil moisture storage in an ultisol and two aridisols. *Soil Science.* 177: 310-320.

57. Oguntunde, P.G.; Abiodun, B.J.; Ajayi, A.E.; Van De Giesen, N. Effects of charcoal production on soil physical properties in Ghana. J. Plant Nutr. Soil Sci. 2008, 171, 591–596.

58. Peng, X.Y.L.L., Ye, L.L., Wang, C.H., Zhou, H. and Sun, B., 2011. Temperature-and duration-dependent rice straw-derived biochar: Characteristics and its effects on soil properties of an Ultisol in southern China. *Soil and tillage research*, *112*(2), pp.159-166.

59. Pietola, L. and Alakukku, L., 2005. Root growth dynamics and biomass input by Nordic annual field crops. *Agriculture, ecosystems & environment*, *108*(2), pp.135-144.

60. Preston, C.M. and Schmidt, M.W., 2006. Black (pyrogenic) carbon: a synthesis of current knowledge and uncertainties with special consideration of boreal regions. *Biogeosciences*, *3*(4), pp.397-420.

61. Robertson M (2014) Sustainability principles and practice. Routledge, New York

62. Rodríguez, L., Salazar, P. and Preston, T.R., 2009. Effect of biochar and biodigester effluent on growth of maize in acid soils. *Integrated Farming Systems for Food and Energy in a Warming, Resource-depleting World*, p.84.

63. Sanders JR, Bloomfield C (1980) The influence of pH, ionic strength, and reactant concentration on copper complexing by humifed organic matter. J Soil Sci 31:53–63. https://doi. org/10.1111/j.1365-2389.1980.tb02064.x

64. Schulz H, Glaser B (2012) Effects of biochar compared to organic and inorganic fertilizers on soil quality and plant growth in a greenhouse experiment. J Plant Nutr Soil Sci 175(3):410–422.

65. Sigua, G.C., Stone, K.C., Hunt, P.G., Cantrell, K.B. and Novak, J.M., 2015. Increasing biomass of winter wheat using sorghum biochars. *Agronomy for Sustainable Development*, *35*, pp.739-748.

66. Smith, P., 2016. Soil carbon sequestration and biochar as negative emission technologies. *Global change biology*, *22*(3), pp.1315-1324.

67. Solaiman, Z.M.; Anawar, H.M. Application of Biochars for Soil Constraints: Challenges and Solutions. Pedosphere 2015, 25, 631–638.

68. Steinbeiss S, Gleixner G, Antonietti M. 2009. Effect of biochar amendment on soil carbon balance and soil microbial activity. Soil Biol Biochem. 41: 1301–1310.

69. Steiner, C., Teixeira, W.G., Lehmann, J., Nehls, T., de Macedo, J.L.V., Blum, W.E.H., Zech, W., 2007. Long term effects of manure, charcoal and mineral fertilization on crop production and fertility on a highly weathered Central Amazonian upland soil. Plant and Soil 291, 275–290.

70. Sun Z, Moldrup P, Elsgaard L, Arthur E, Bruun EW, Hauggaard-Nielsen H, de Jonge LW (2013) Direct and indirect short-term effects of biochar on physical characteristics of an arable sandy loam. Soil Sci 178:465–473.

71. Sun Z, Bruun, E.W., Arthur, E., de Jonge, L.W., Moldrup, P., Hauggaard-Nielsen, H. and Elsgaard, L., 2014. Effect of biochar on aerobic processes, enzyme activity, and crop yields in two sandy loam soils. *Biology and Fertility of Soils*, *50*, pp.1087-1097.

72. Verheijen, F., Jeffery, S., Bastos, A.C., Van der Velde, M. and Diafas, I., 2010. Biochar application to soils. *A critical scientific review of effects on soil properties, processes, and functions. EUR*, *24099*(162), pp.2183-2207.

73. Wang, J., Z. Xiong, and Y. Kuzyakov. 2016. Biochar stability in soil: meta-analysis of decomposition and priming effects. GCB Bioenergy 8:512–523.

74. Wu, W., Yang, M., Feng, Q., McGrouther, K., Wang, H., Lu, H. and Chen, Y., 2012. Chemical characterization of rice straw-derived biochar for soil amendment. *Biomass and bioenergy*, *47*, pp.268-276.

75. Yuan, J.H., Xu, R.K., Wang, N. and Li, J.Y. 2011. Amendment of acid soils with crop residues and biochars. *Pedosphere.* 21 (3): 302–308.

76. Zhang, H.; Yue, X.; Li, F.; Xiao, R.; Zhang, Y.; Gu, D. Preparation of rice straw-derived biochar for efficient cadmium removal by modification of oxygen-containing functional groups. Sci. Total Environ. 2018, 631-632, 795–802.

77. Zimmerman, A.R., 2010. Abiotic and microbial oxidation of laboratory-produced black carbon (biochar). Environmental Science & Technology 44, 1295-1301.

# Chapter - 2

## The Vital Role of Pollinators in Agriculture and Ecosystems


**Authors**

**S. Abdul Mujeeb**

Ph.D. Research Scholar, Department of Entomology, S.V. Agricultural College Tirupati, Acharya N.G. Ranga Agricultural University, Lam Guntur, Andhra Pradesh, India

**P. Divya**

Ph.D. Research Scholar, Department of Entomology, S.V. Agricultural College Tirupati, Acharya N.G. Ranga Agricultural University, Lam Guntur, Uttar Pradesh, India

**G. Chaitanya**

Department of Entomology and Agril. Zoology, BHU, Varanasi, Uttar Pradesh, India

**B.L. Manisha**

Ph.D. Research Scholar, Department of Entomology, S.V. Agricultural College Tirupati, Acharya N.G. Ranga Agricultural University, Lam Guntur, Andhra Pradesh, India

# Chapter - 2

## The Vital Role of Pollinators in Agriculture and Ecosystems

**S. Abdul Mujeeb, P. Divya, G. Chaitanya and B.L. Manisha**

### Abstract

The rapid expansion of the human population has led to an increased need for food. To meet this demand sustainably, our agricultural systems must produce more food in the future. Pollinators play a crucial role in these systems, both wild and domesticated. They provide essential pollination services, which are necessary for the growth of 87 out of the top 115 global crops that humans consume. Approximately 35 percent of the food we eat depends, to some extent, on animal pollination. The value of crop production directly attributed to animal pollination is estimated to be between 235 and 577 billion US dollars annually, accounting for five to eight percent of the world's crop production. There is a wide range of animal species acting as pollinators, estimated to be between 200,000 and 350,000. Vertebrates, including birds, bats, and certain marsupials like the Small Pygmy Possum, contribute to this diversity. However, insects, particularly bees, are the most significant group of pollinators. Despite their importance, pollinators and their services are currently facing numerous threats. Climate change has altered the distribution, density, and behavior of some wild pollinator species. Human-induced changes in land use and landscape have led to habitat loss for wild pollinators, while environmental contamination has further damaged their natural habitats. These factors have resulted in reduced pollinator diversity, abundance, and overall health. Consequently, the availability of food and nesting resources for pollinators has decreased due to habitat fragmentation, degradation, and destruction. It is everyone's responsibility to conserve pollinators by minimizing the harmful effects caused by human activities. By doing so, we can protect the future of agriculture and maintain the balance of ecosystems. Conserving pollinators involves actions such as preserving and restoring their habitats, adopting sustainable agricultural practices, raising awareness about their importance, fostering collaboration and policy support, and conducting research and monitoring to inform conservation efforts. By taking these measures collectively, we can contribute to the preservation of pollinators, promote sustainable agriculture, and safeguard the well-being of our ecosystems.

**Keywords:** Agriculture, bees, food, pollinators and pollination.

## Introduction

Pollination plays a vital role in crop growth, and its significance extends to economic benefits. Understanding the latest research on pollination and its effects is crucial for conserving pollinators. Pollination is the process of transferring pollen grains from a flower's male anther to its female stigma. It is an essential step for plants to produce offspring and create seeds. In India, insect pollination, primarily by bees, is crucial for many food crops. Oilseeds like sunflower and safflower, vegetables such as carrot, coriander, cucumber, onion, and fruits like apple, almond, apricot, and peach all rely on pollinators, according to a study published by the International Research Journal of Natural and Applied Sciences. Additionally, the Zoological Survey of India highlights the economic importance of fly pollination (myophily) in tropical areas, where flies are primary pollinators of crops like cacao, mango, cashew, and tea.

Bees, along with other pollinators like birds and bats, are vital for food production. They contribute to 35 percent of the world's crop production and increase yields for 87 major food crops worldwide, as reported by the Food and Agriculture Organization (FAO). Pollinator Week, observed from June 22-28, aims to raise awareness about how pollination affects the entire agricultural process. The FAO's Rapid Assessment of Pollinators' Status report emphasizes that well-pollinated crops tend to have better quality, which is valued by markets. Insect pollinators not only enable crop reproduction but also enhance yields.

Pollination is not just a fascinating aspect of natural history; it is an essential ecological function for survival. Without pollinators, the human race and terrestrial ecosystems would struggle to exist. Approximately 80% of the world's 1,400 crop plants, which provide our food and plant-based industrial products, require animal pollination. Visits from bees and other pollinators result in larger, more flavorful fruits and higher crop yields. In the United States alone, the value of pollination services for agricultural crops is estimated at 10 billion dollars annually. Globally, the services provided by pollinators are believed to be worth over 3 trillion dollars Overall, recognizing the importance of pollination, both for crop production and ecosystem sustainability, underscores the need to protect and conserve pollinators.

# Pollination

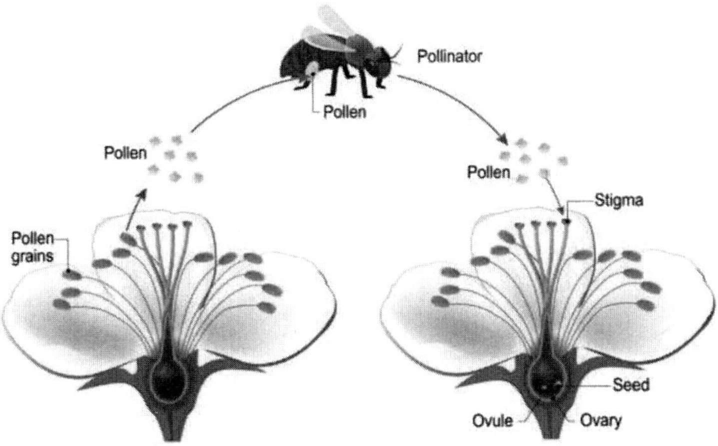


**Figure 1:** Pollination process by insects

Some global statistics illustrate the scale of pollinator contribution to agriculture and food security:

- Out of the 115 leading global crops consumed by humans, 87 rely on animal pollination, to some degree.
- 35 percent of the crops we eat, in terms of the volume produced globally, depend on animal pollination, to some extent.
- It is estimated that five to eight percent of global crop production, with an annual market value of 235 billion - 577 billion US dollars, is directly attributable to animal pollination.
- Globally nearly 90% of wild flowering plant species depend on animal pollination.
- More than 75% of leading global crop types benefit from animal pollination in production, yield and quality.
- Around 5-8% of current global crop production is directly ascribed to animal pollination, which equates to somewhere between 235 and 577 billion American dollars worldwide.
- Crop Dependence: Approximately 75% of the world's major food crops rely on animal pollination to some extent. This includes fruits, vegetables, nuts, and oilseeds.

- Economic Value: The value of pollination services for agricultural crops is estimated to be around 235 to 577 billion US dollars annually. This accounts for roughly 5 to 8 percent of the total global crop production.
- Crop Yield Increase: Pollination by bees and other pollinators can lead to higher crop yields. Studies have shown that well-pollinated crops often exhibit larger, more uniform fruits and increased overall productivity.
- Food Production Impact: The Food and Agriculture Organization (FAO) reports tha pollinators affect 35 percent of the world's crop production. They contribute to higher outputs for 87 of the leading food crops globally.
- Genetic Diversity: Pollinators play a crucial role in maintaining genetic diversity within plant populations. This diversity is essential for enhancing crop resilience to diseases, pests, and environmental changes.
- Nutritional Diversity: Pollinators contribute to the production of diverse and nutritious food. They are instrumental in the production of fruits, vegetables, and nuts, which are vital components of a healthy diet.
- Ecosystem Services: Pollinators provide essential ecosystem services beyond crop pollination. They contribute to the reproduction of wild plants, maintaining biodiversity, and supporting the functioning of ecosystems.
- Sustainable Agriculture: The presence of diverse pollinator communities is associated with sustainable agricultural practices. By promoting biodiversity and reducing reliance on chemical inputs, pollinators contribute to resilient and environmentally friendly farming systems.

These statistics emphasize the critical role that pollinators play in sustaining global agriculture, ensuring food security, and supporting ecosystem health. Protecting and conserving pollinators is crucial for the long-term sustainability of our food systems and the overall well-being of our planet.

Insects are important pollinators for mangoes but the specific type of insect varies across the globe. For instance, honey bees are the primary pollinators in parts of Brazil, whereas in Israel, bumble bees and flies are known to increase yields. In tropical countries like India, Malaysia, Colombia, Costa Rica and the Philippines, flies (Diptera) are the main mango pollinators.

The effectiveness of pollinators varies according to factors, such as their abundance; whether they reach individual plants of the same species and their ability to pick pollen up from, transfer and deposit it to the appropriate plant organs.56 For example, pollen collected on the specialized soft down on a mining bee's body is more easily transferred to stigma than from the small brush-like tufts of hair on the hind legs of honey bees

**Different crops have different dependencies**

The extent to which different plants rely on animal pollinators varies considerably, from being an essential service to unimportant. Many fruits, vegetables and stimulant crops (e.g. coffee) are highly dependent on animal pollinators, specifically insects. In contrast, some important staple crops like cereals, maize or potatoes do not require insect pollination at all.37 Figure three illustrates the range of dependencies for leading, global crops

A few plants that grow in the vicinity of cashew plantations produce oil, which is attractive for some oil-collecting bees. These bees are also good pollinators of cashew. Oil-collecting bees gather floral oil on hairs on their legs or abdomen and use it as food for their larvae, to build nests or to feed themselves.

Cocoa flowers are thought to attract their pollinators, including midges, in several ways. They have 'guidelines' on their petals and staminodia (sterile stamens), they emit a fragrance; have brightly colored leaf parts16; and can absorb and reflect ultraviolet light.

**Farming wouldn't be possible without pollinators**

Although small, pollinators are responsible for the growth and vitality of whole ecosystems and the foundation of entire food webs. The human food system would look dramatically different without the ecosystem services pollinators have been providing for millions of years. Pollinators of many different varieties play a critical role in the agriculture industry. From bees to bats, we depend on a diverse array of pollinators for food security, nutrition, and the rural economies producing our food.

Farmland investing relies on the work pollinators do behind the scenes. Most of the permanent crops Farm Together offers for farmland investments require insects and small animals for successful crop yields and sustainable returns. In recent decades, pollinator numbers have taken a downturn as a result of habitat loss, climate change, and exposure to chemicals. The recent surge in interest in regenerative farming practices could ultimately support pollinator habitat restoration, and has the potential to begin reversing the decline in pollinator numbers. Regenerative farming practices offer a safe harbor for the small animals and insects that are responsible for three out of four fruit and nut producing crops across the globe.

**What crops need pollinators?**

After crops bloom, they must be pollinated in order to bear fruit. Pollination occurs through the action of either insects or small animals, which feed on the nectar in the flowers and transfer pollen among the plants they visit, or through the action of the wind dispersing pollen between plants. Common grain crops like wheat, rice, barley, and oats are examples of plants that utilize the wind for pollination. Most permanent crops, on the other hand, cannot depend on wind for successful fertilization. Most varieties of permanent crops like almonds, apples, hazelnuts, and pistachios all need pollinators to grow a successful harvest. Pollinators are the highest contributors to crop yields worldwide. According to the Penn State Center for pollinator research, honeybees and wild bees alone contribute more than $20 billion to the U.S. agriculture industry each year based on the total value of the crops they pollinate.

**Types of pollinators**

Bees are the most famous of the pollinators but there are many that get far less publicity for their crucial work. Moths, flies, wasps, beetles, and butterflies are all on the list of critters we have to thank for successful crop yields. There are even many vertebrate pollinators including bats and other mammals like some species of monkeys, rodents, lemurs, and tree squirrels. A handful of birds such as hummingbirds, sunbirds, and few parrot species are on the list as well.

Each of these pollinator species contribute to their own unique niche in the food web and rely on different plants for survival, making it crucial that we support pollinators by growing a diverse array of plant species to accommodate pollinator needs. The abundance and diversity of pollinators ensure the resiliency of plant communities. Strong plant and pollinator communities lead to better food security and nutrition.

Pollinators affect 35 percent of agricultural production globally and are necessary for the successful harvest yields of 87 of the leading food crops worldwide. Crops like fruits and nuts that depend on pollinators are five times more valuable than crops like grains that do not need pollinators. The value of crops that require pollinators is estimated to be between \$235 and \$577 billion annually. The demand for healthy snack foods like fruits and nuts that come from food production systems reliant on pollinators is on the rise and the volume of crop production dependent on these wild insects and animals has increased by 300 percent in the last 50 years.

### Seven insect heroes of pollination

#### Bumblebees

Bumblebees can vibrate their bodies to dislodge pollen from a flower. The bee then combs the pollen off its hairy body into little baskets on its legs, called corbiculae. This way of gathering pollen is called buzz pollination. Most of the pollen collected will be taken back to the nest to feed young bumblebee larvae, but some grains will transfer to the next flowers the bumblebee visits. Bumblebees can forage several kilometres from their nest, but this varies depending on the species. Some bumblebee species can also get up to cracking speeds of 15 kilometres per hour.

#### Hoverflies

Hoverflies are prolific pollinators. They are known to visit at least 72% of global food crops and over 70% of animal-pollinated wildflowers. Highly migratory, hoverflies can travel hundreds of kilometres a day. They can also carry pollen over distances greater than 100 kilometres over open water. Some plants use deception to attract pollinators. The orchid *Cypripedium subtropicum* (which is native to southeast Asia) mimics aphids to lure hoverflies. The orchid's flower has hairy white tufts that look like an aphid colony covered with honey dew. It also emits an aphid-imitating smell. This leads certain hoverfly species whose young feed on aphids to wrongly believe the plant will offer their predatory brood plenty to eat. When the female lays her eggs, she picks up some pollen.

#### Solitary bees

Solitary bees don't travel very far, although foraging distances vary among species. Most forage up to 500 metres from their nest and some as little as 100 metres. Like honeybees and bumblebees, solitary bees gather pollen to feed to their offspring, but they don't collect it in pollen baskets. Instead, the grains stick to a pollen brush, or scopa, which can be on the legs or under the abdomen, depending on the species. Some solitary bees have an interesting relationship with a particular plant. The yellow loosestrife bee (*Macropis europaea*) forages exclusively from yellow loosestrife (*Lysimachia vulgaris*). Joe explains, '*M. europaea* collects both pollen and oil from the plant. It uses the pollen to feed its larvae and the oil provides a waterproof lining to the nest cells, allowing the bee to nest in wetland habitats.

### Wasps

Wasps are hungry mini beasts with high energy needs, so many of them search for flower nectar. Wasps are not as hairy as bees, so pollen does not stick to them as easily, but they are still important pollinators.

Some wasps (in the family Agaonidae) have a special relationship with wild fig plants. In an alliance that is around 60 million years old, the fig plant needs to be pollinated by fig wasps. In turn, these wasps have evolved to live inside the figs and feed exclusively on them.

### Moths

Most moths fly at night, so you might not notice as they flit from flower to flower under the cover of darkness. Some nocturnal moths prefer to feed from pale and tubular, scented flowers, such as honeysuckle. Pollen can get stuck to a moth's furry body and on its long tongue or face as it reaches down into the flower looking for nectar. Some species of moth can travel long distances, carrying the genetic information in pollen far and wide

### Butterflies

Due to this low species diversity, butterflies are not as important for pollination as some other insects, but their abundance in florally diverse meadows and chalk downlands, for example, can sometimes make them more important than moths for pollinating flowers during the day. 'For example, if you watch a hedgerow, you might see a brimstone (*Gonepteryx rhamni*) moving from flower to flower of the same species, such as bush vetch (*Vicia sepium*).

### Beetles

Beetles have been pollinating flowers for millions of years. Many different species of beetles eat pollen. For some, such as the aptly named pollen beetles, pollen is their main diet. For others, it is an additional tasty snack, including for some ladybirds that supplement their usual carnivorous diet of aphids with pollen. Beetle-pollinated flowers tend to make extra pollen so there is some left over to be carried to another flower on the beetle's body. The flowers tend to be smelly or fragrant, because beetles navigate by a strong sense of smell. Beetles are attracted to flat, open flowers, which allow them to graze, and flowers in clusters, such as cow parsley.

**Figure 2:** Different pollinators

## Current status of pollinators

In the last several decades, many pollinators have experienced significant population declines. In particular, wild and domestic bee populations and species count have seen a severe decline. Bee numbers have suffered for a variety of reasons including pesticide use, climate change, disease, and habitat loss. According to the USDA, the total number of managed honeybee colonies dropped from 5 million in the 1940s to only 2.66 million today.

Managed honeybee colonies are responsible for adding at least $15 billion to the value of U.S. agriculture annually through increased yields and successful harvests. Honeybee numbers from managed colonies are suffering from a variety of different stressors like drought, disease and habitat loss, resulting in beekeepers losing many colonies. Some permanent crops, like almonds for example, rely entirely on honeybees for pollination. The almond industry in California utilizes nearly three-quarters of all managed honeybee colonies in the United States. Colonies are brought from all over the country during the short almond bloom from January to February each year.

## Pollinators and regenerative agriculture

The recent decline in pollinator numbers has led to extensive research and development of mechanical pollination solutions like pollinator drones. While these new developments in Ag technology have promise, the cost of producing and using new methods of pollination are cost prohibitive. The cost barriers to engineered solutions for the loss in pollinator numbers has sparked renewed interest in farming practices, known as regenerative farming, that nurture wild pollinators.

Regenerative farming has the potential to build resilient farm systems by restoring biodiversity to the landscape. Adding biodiversity to our food production system provides habitat for the many different species of pollinators that all contribute to their niche in the food web. Regenerative agriculture helps ensure we will have natural pollination services for many different types of plants, providing long term food security and economic stability in our farming systems. Restoring biodiversity to the landscape helps the 20,000 bee species have the opportunity to complete their life cycles not to mention the thousands of other pollinator species. Bee colonies rely on a diverse food supply to produce honey, support the queen, and reproduce. Bees must be able to return to the hive and produce honey free of synthetic chemicals commonly used in conventional farming practices. Regenerative farming leaves out the harmful chemicals that disrupt the bee life cycle and increases flowering plant diversity.

Organic farming prohibits the use of chemical pesticides and herbicides which are a main cause of bee number and species loss. This offers bees and other pollinators safe havens from the stressors they experience elsewhere. Regenerative farming practices can go even further by providing underlying support to biodiversity in the farming landscape, giving bees and other pollinators an additional edge in keeping their populations healthy.

**Pollination is crucial to life on earth**

Pollination is crucial to life on earth as we know it. Every year, we owe many of our crop harvests to the insects and small animals that carry pollen from one plant to the next. Honeybees are responsible for one out of every three mouthfuls of food in the American diet yet the majority of farming practices in the US are contributing to the loss in pollinator populations and species. FarmTogether offers a variety of farmland investing opportunities that not only add valuable returns to your investment portfolio but also support farming practices that grow habitat for pollinators.

**Insects: The leading group of pollinators**

Plants are static organisms: That means they cannot move around. To reproduce, they must either self-pollinate or use external vectors to carry their male gametes (transported as pollen) from one flower to another, fertilizing the female parts in each flower. While wind and water carry some pollen, animals help to pollinate an estimated 87.5 percent of flowering plant species. These animals come in many shapes, sizes and numbers, as figure one demonstrates. As many as 200,000 different animal species are thought to act as pollinators, though some estimates put this figure as high as

350,000. Around 1,000 are vertebrates, including small mammals (such as bats or some marsupials like the Small Pygmy Possum from Oceania) and birds. Insects are by far the largest group and bees are the most important group of pollinating insects. The relationships between pollinators and plants range from generalist – pollinators (such as honey bees) that visit many different plants, and plants that have many pollinators – to highly specialized, one-to-one relationships. These relationships are not fixed: A plant species may be visited by different pollinators in different regions and relationships can also vary during the year depending on changing pollinator abundance.

Around 15,000 flowering plants, including important crops such as tomatoes and peppers, hide their pollen in pouches on their anthers that can only be reached via very small openings.27 To access this, the females of certain bee species, such as bumble bees with tomato plants, have evolved to use a technique known as sonication or 'buzz pollination'.

These specially adapted anthers prevent some insects from accessing the pollen (such as pollen-eating beetles, flies and non-buzzing bees), so that pollen is not lost to inefficient pollinators and 'pollen thieves' (visitors that remove pollen but make little or no contribution to pollination). The anthers may also increase the efficiency of pollen transfer by ejecting it onto specific areas of the pollinator's body which are more likely to come into contact with the stigma of other flowers.

How it works:

- ✓ A bee lands on the flower and curls its body around the anthers.
- ✓ It starts to 'buzz' with its body, causing the pollen to fall.
- ✓ The pollen lands on its body and is carried to the hive or another flower.

On average, wild bees contribute 3,251 US dollars per hectare to the production of crops worldwide – about the same as managed honey bees.

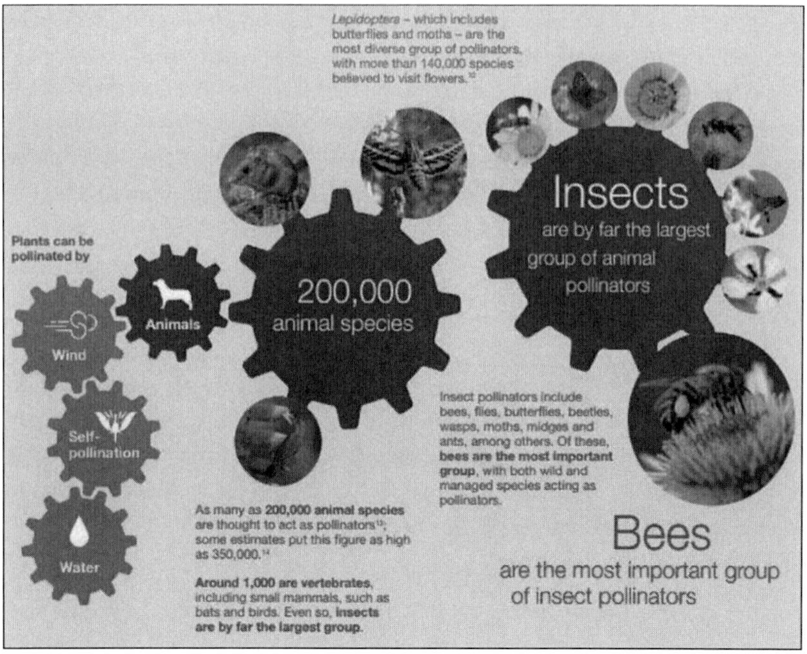


**Figure 3:** Pollinators in Agriculture The variety of pollinator groups and species (BEENOW, 2018)

**Bees: Agriculture's helpers**

Insects are the dominant pollinators in agricultural systems – and bees top this list, providing pollination services for many of our crops. Bees are a highly diverse group. Over 20,000 species have been described globally. Around 50 of these species are managed by humans for different purposes, for instance honey production, with twelve specifically managed for crop pollination. Some managed species have a widespread usage, notably the Western Honey Bee (*Apis mellifera*)

By contrast, the composition of wild bee species varies greatly around the world, with the highest number found in warmer temperate areas such as the Mediterranean basin, the Californian region and certain semi-desert areas. Among them are bumble bees, stingless bees, carpenter bees and orchid bees. Not all bee species are important for agriculture. When we look across crops, years and regions, a small number of common species dominate (see figure two): this means that 80 percent of crop pollination is provided by only two percent of bee species and threatened or rare species are rarely observed in crop fields.

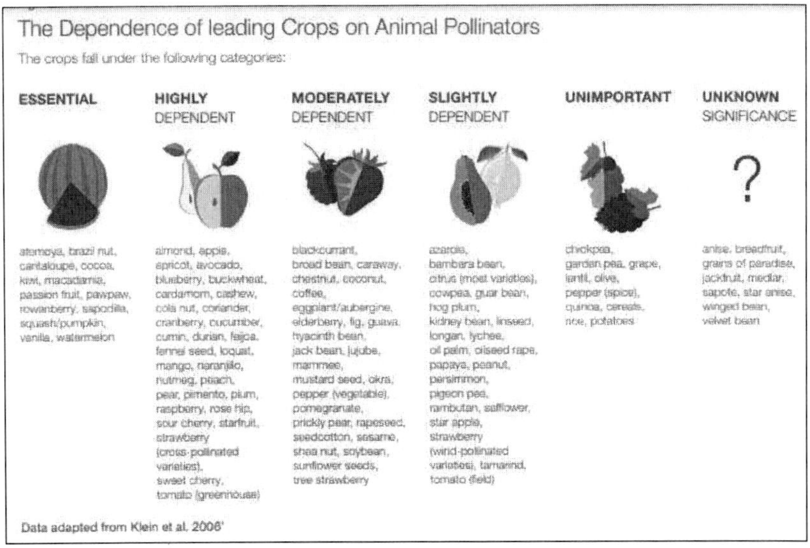


**Figure 4:** Dependence of crops on animal pollianators (Klein *et al.* (2006)

**What are the threats?**

Human activities threaten pollinators. Both the abundance, diversity and health of pollinators are threatened, as well as the provision of pollination. In recent decades, climate change has altered the range, abundance and seasonal activities of some wild pollinator species. Humans have altered the natural landscape and changed land use, reducing the natural habitat of wild pollinator species. Simultaneously, environmental pollution has degraded the natural habitats. Habitat destruction, fragmentation and degradation often reduce pollinators' food and nesting resources.

Contemporary forms of agriculture have changed the influence of pollinators with a shift to large-scale and artificial forms of pollination. Intensive agricultural management is a leading threat to pollinator species.

Controlled forms of pollination by managed pollinators have brought in risks. Mass breeding and large-scale transport of managed bees contributed to an enhanced transmission of parasites and pathogens. This spreads diseases across the entire community of both wild and managed pollinators. Additionally, it increases the occurrence of more dangerous pathogens, alien species invasions and regional extinction of native species.

**The need to protect all pollinators**

The right balance of wild and managed pollinators is instrumental in achieving the necessary pollination services for a good crop harvest. Yet, land use changes like urbanization, the intensification of agriculture and other factors have negatively affected local pollinator populations in different parts of the world. A range of conservation measures in intensively-farmed regions can help to maintain diversity, by preserving the resources that pollinators need (figure five). These measures protect the dominant crop pollinators in agro-ecosystems and contribute to maintaining the biodiversity of the ecosystem, including the less agriculturally-important pollinators.

**What do pollinators need?**

- ✓ A source of food: Many wild bee species are fussy eaters, preferring specific foraging plants.
- ✓ 'Stepping stones': Are needed to connect natural or semi-natural areas across landscapes fragmented by farming or other human activities. Steps can be flowering strips, hedgerows, small forest patches with surrounding vegetation or other landscape structures.
- ✓ Fragments of natural vegetation: Provide nesting sites and shelter for wild bees.
- ✓ Nesting sites: Different wild bee species prefer different environments and materials for nesting. Many nest in the ground, so creating soil mounds in a landscape invites them to nest there

Despite this, there are several reasons for conserving pollinator diversity beyond these key pollinator species in the agricultural landscape. One is to provide insurance over time and space, as species that are rarely observed on crops now may, in future, become important. Protecting pollinator diversity also helps to counter the risks associated with relying on a small number of species for pollination and may make agricultural systems more resilient in the long term. More immediately, pollinator diversity needs protecting because there are specific relationships between certain crops and certain pollinators. If a pollinator disappears from a landscape, it is not guaranteed that another will take its place and provide the same crop pollination service. This diversity needs protection by integrating conservation measures with sustainable agricultural practices, which may raise crop yields and protect bees (both wild and managed species) and other pollinators. But, while we know quite a lot about the biology and ecology of managed pollinators and their importance to different agricultural systems, our understanding and knowledge of most wild pollinators are less developed. Increasing this will support conservation efforts and help to maintain pollinator diversity and effectiveness.

## Conclusion

Absolutely! Bees and other pollinators undoubtedly hold great significance in global agricultural systems. Their contributions are invaluable, encompassing economic benefits, food security, nutrition, livelihood support, agricultural diversification, and the maintenance of natural ecosystems. Ongoing research and experimentation in crop-pollinator relationships are continuously expanding our knowledge and understanding of these remarkable creatures. With this growing understanding, management strategies aimed at enhancing pollination or conserving pollinators will become more precise and targeted. Farmers will have access to valuable information and tools to strike a harmonious balance between agriculture and nature conservation on their farms. This knowledge will ensure the continuity and potential amplification of the pollination services provided by bees and other animal pollinators, ultimately benefiting farmers, consumers, and the natural worldBy recognizing the importance of pollinators and implementing appropriate conservation and enhancement measures, we can strengthen the symbiotic relationship between agriculture and nature. This holistic approach will contribute to sustainable and resilient farming practices, promote food security, and safeguard the integrity of our ecosystems.

## References


1. BEENOW. (20180. Pollinator project in India: Delivering a honey-sweet message. Bayer Bee Care Center, Monheim, Germany
2. Brittain, C. (2013). Synergistic effects of non-Apis bees and honey bees for pollination service.
3. De Luca, P.A and M. Vallejo, Marín. (2013). What's the 'buzz' about? The ecology and evolutionary significance of buzz-pollination. *Current Opinion of Plant Biology*. 16(4): 429–435.
4. FAO. (2018). AGP - Integrated Pest Management. Food and Agriculture Organization of the United Nations, Rome, Italy.
5. Herrera, C.M. (1987). Components of pollinator quality: comparative analysis of a diverse insect assemblage. *Oikos*. 50(1): 79–90.
6. Hung, K.L.J. (2018). The worldwide importance of honey bees as pollinators in natural habitats. *Proceedings of Royal Society B: Biological Sciences*. 285(1870): 2017–2040.
7. https://www.nhm.ac.uk/discover/insect-pollination.html.

# Chapter - 3

## Precision Farming Technologies

**Authors**

**Rejith S.**

Research Scholar, Kerala Agricultural University, Kerala, India

# Chapter - 3

## Precision Farming Technologies

**Rejith S.**

**Abstract**

The origins of agriculture date back thousands of years, with its development occurring independently in various regions based on climate and geography. Agriculture revolutionized civilization by providing a sustainable food source, enabling rapid population growth compared to hunting and gathering. In the twentieth century, precision farming emerged as a response to increasing food demands and the need for resource optimization. This approach utilizes technology to measure and analyse individual fields and crops, optimizing management practices and resource allocation. Precision farming integrates tools such as remote sensing, GPS and sensors to gather extensive field data. By leveraging AI and IoT, farmers can make informed decisions, maximize productivity, reduce waste and minimize environmental impact. The book explores these applications, techniques, benefits and real-time devices in detail, revolutionizing agriculture into an analytics-driven industry.

**Keywords**; Precision farming, remote sensing, data analytics, resource optimization, IoT.

### Introduction

The origins of agriculture can be traced back to ancient times, spanning thousands of years. Its evolution was not uniform; rather, it developed independently in many parts of the planet, influenced by varying climatic conditions and geographical landscapes. Agriculture transformed civilization as a whole by offering a consistent and sustainable source of food production, allowing the population to grow rapidly in comparison to the restricted resources accessible through hunting and gathering. As the twentieth century began, advances in technology and scientific understanding gave rise to precision farming, a modern approach to agriculture. Precision farming arose in response to rising food production demands, as well as the need to optimize resource use and reduce environmental effect. This novel technique entails using technology to precisely measure, observe and analyze particular fields and crops in order to adapt to their specific needs. Farmers can acquire extensive information about their fields by using precision farming techniques, such as soil composition, nutrient levels, moisture content and crop health. This information is gathered using a variety of methods, including remote sensing, global positioning systems (GPS) and sensors embedded in the field. Farmers may make informed decisions on crop management practices, resource allocation and the use of inputs such as fertilizer, water and pesticides using this vast amount of information. Precision farming's major purpose is to maximize productivity and profitability while minimising waste and environmental damage. Farmers can optimize resource utilization, reduce input costs and limit the danger of over-application of fertilisers or pesticides by adapting agricultural practices to specific areas within a field. Furthermore, precision farming allows for the early detection of crop stress, illnesses and pests, enabling prompt interventions and focused treatments. The integration of technology and data analytics has changed agriculture into an analytics-driven industry in which farmers can make accurate decisions based on real-time data. This strategy improves the efficiency and sustainability of farming operations, resulting in higher crop yields, a smaller environmental footprint and increased profitability. In the present day, integrating Artificial Intelligence (AI) and the Internet of Things (IoT) has become critical across a wide range of applications. These technical advances have also made their way into modern agriculture, revolutionizing the industry and making it more precise and profitable. This chapter addresses the benefits and advancements brought about by AI and IoT in the context of smart precision farming. The chapter discusses how cutting-edge technologies are being used to produce more precise and efficient agricultural practices. Farmers may make data-driven decisions about crop management, resource allocation and yield optimization by using AI algorithms and machine learning approaches. The Internet of Things improves precision farming even further by connecting physical devices and sensors to collect real-time data on a variety of environmental and crop characteristics. This information is then analysed to provide useful insights for increasing productivity and sustainability. A comprehensive review of the tools and procedures used in precision farming is also carried out. It includes techniques such as remote sensing, satellite imaging, drones and sensor networks. Each of these technologies is critical in acquiring information about soil conditions, crop health, weather patterns and other pertinent elements. Farmers can get a holistic view of their fields and make informed decisions about irrigation, fertilisation, pest control and other critical practices by using these tools. Furthermore, the chapter emphasizes the numerous advantages of precision farming and the real-time instruments used in its execution. It investigates how precision farming helps to reduce waste, reduce environmental impact and optimize resource utilization such as water, fertilisers and pesticides. Precision farming technology, such as weather stations, soil moisture sensors and automatic irrigation systems, enable farmers to monitor and manage their fields in real time. This ensures that crops get the exact amount of water, nutrients and other inputs they require for optimal development and yield.

**Fundamentals of precision farming**

Precision Farming is generally defined as an information and technology based farm management system to identify, analyze and manage variability within fields for optimum profitability, sustainability and protection of the land resource. Robert *et al.* (1994) proposed three "R", the Right time, the Right amount and the Right place. Later, the International Plant Nutrition Institute added another "R" to that list, "the Right Source" and more recently, Khosla (2008) proposed an additional "R", the Right manner.

The use of developing information technologies in this farming technique enables farmers to make more informed decisions about multiple facets of crop production. Precision farming aims to enhance efficiencies by recognising and managing the inherent variability within a field. The goal is not to attain uniform yields throughout the entire field, but to manage and allocate inputs depending on unique site needs in order to maximize long-term cost-effectiveness. The traditional practice of applying inputs uniformly across the field may no longer be the best option. Farmers can use precision farming to optimize resource allocation, boost productivity and reduce wasteful costs by adopting technology. Precision farming is revolutionizing agricultural practices around the world by allowing farmers to maximize crop input efficiency. On closer examination, it is evident that knowledge is the primary driving force behind accurate farming. Farmers can use the targeted approach to make data-driven decisions and change their practices to match the particular characteristics and needs of distinct zones within a field. Farmers may maximize output, eliminate input waste and optimize their overall agricultural operations by utilizing the power of information and using precise management approaches.

(Searcy, 1997; Khosla 2010; Singh, 2010)

The global food system is currently facing considerable challenges, which are expected to worsen in the future decades. While existing technologies and expertise can help in the short term, solving future difficulties will require more radical changes in the food system as well as significant investment in research to discover creative solutions. The decline in overall productivity, depletion and degradation of natural resources, stagnant farm incomes, the lack of an ecoregional approach, land holding fragmentation, the impact of trade liberalization on agriculture, limited employment opportunities in non-farm sectors and the effects of global climate variations are now recognised as pressing concerns for agricultural growth and development. Given these limitations, the adoption of technological advances is seen as a critical aspect in increasing agricultural output in the future. Farmers can overcome the constraints of traditional ways and improve their efficiency and productivity by adopting innovative technology and practices. The integration of emerging technology such as precision farming, remote sensing, genetic engineering and data analytics offers promising options for addressing the agricultural sector's complicated difficulties. These developments have the potential to transform farming practices, optimize resource use, reduce environmental consequences and boost overall output. To address numerous issues, it is critical to invest in research and development to provide fresh solutions and strategies customized to changing agricultural demands. The agriculture sector can adapt to changing conditions and achieve sustainable growth by encouraging innovation and allowing the use of new technologies. Collaboration among academics, policymakers and stakeholders will be critical in fostering agricultural innovation and guaranteeing the global food system's long-term viability.

## Evolution of precision farming

Precision farming has evolved dramatically over the years, altering the agricultural landscape. Precision farming began with the capacity to accurately determine positions in the field and build detailed maps, which was initially pushed by the emergence of Global Positioning System (GPS) technology in the 1980s and 1990s. This paved the way for site-specific management. In the 1990s and 2000s, the development of Variable Rate Technology (VRT) advanced precision farming by allowing farmers to apply inputs at variable rates based on individual field conditions, optimizing resource utilization and increasing yields. With the advancement of remote sensing and imaging technologies in the 2000s, farmers were able to monitor crop health, soil moisture and vegetation indices using satellite imagery and drones. In recent years, the integration of data analytics and machine learning has provided farmers with useful insights into crop performance, disease detection and yield estimates, allowing for informed decision-making. Furthermore, the use of Internet of Things (IoT) devices and connectivity has enabled real-time data collecting and remote monitoring. The continued integration of Artificial Intelligence (AI) technology holds enormous potential for the future, as AI-powered systems may automate jobs, optimize resource allocation and boost overall efficiency. Precision farming's growth has resulted in major increases in productivity, resource management and sustainability, making it an essential component of modern agriculture.

(Mulla and Khosla, 2016; O'Connor *et al.*, 2019)

## Importance and benefits of implementing precision farming

Precision farming has the potential to generate significant economic and environmental benefits, especially through reducing the use of water, fertilizer, herbicide, pesticide and equipment. Precision farming in contrast to traditional farming approaches that focus on consistent management practices assuming an average condition throughout the entire field, recognises and accounts for the inherent heterogeneity within the field. It recognises that different parts of the same field have distinct characteristics and requirements. Precision farming allows farmers to adjust their management measures by using a site-specific strategy. This implies that inputs like water, fertilizer, herbicides and pesticides may be accurately distributed to specific locations, increasing their effectiveness and reducing waste. Furthermore, precision farming enables for more efficient equipment utilization, ensuring that resources are given precisely where and when they are required. Farmers may increase output, reduce costs and lessen the environmental effect of their agricultural practices by embracing precision farming techniques. Farmers are conscious that their fields provide varying yields across the landscape. These differences are caused by a combination of management practices, soil qualities and environmental conditions. Crop yields are influenced by soil qualities such as texture, structure, moisture, organic matter concentration, nutrient status and landscape position. Weather patterns, weed infestations, insect populations and the prevalence of illnesses are all examples of environmental influences. Farmers may recognise and account for these variances by using precision farming techniques, allowing them to employ customized management plans tailored to each individual area within their fields. To obtain a thorough grasp of how precision farming works, it is necessary to become acquainted with the many instruments and processes that serve as the cornerstone of this modern approach to agricultural management. Precision farming is based on a sophisticated infrastructure that combines a variety of technology and approaches to optimize agricultural practices. Global positioning systems (GPS) are used for precise location tracking, satellite imagery is used for remote sensing, drones are used for aerial surveillance, sensor networks are used for data collection and data analytics is used for analysis and decision-making. These instruments and strategies work together to offer farmers with up-to-date information about their fields, crops and environmental conditions. Farmers may make smart crop management, resource allocation and yield optimization decisions by leveraging the copious amount of data. Understanding the complexities of these equipment and procedures is critical for maximizing precision farming's potential and reaping its benefits in modern agriculture.

(Hakkim *et al.*, 2016; Finger *et al.*, 2019)

**Role of data collection and analysis in precision farming**

Data gathering and analysis are critical components of precision farming, enabling farmers to make sound choices and improve their farming practices. Farmers can collect substantial data on critical factors such as crop health, soil conditions, weather dynamics and more by utilizing a variety of technology such as sensors, drones, satellites and weather stations. This data is then subjected to sophisticated analytic tools, such as machine learning and data mining, in order to extract important insights and find trends that might guide agricultural plans. Farmers can use this real-time data to understand the variability in their fields and make site-specific decisions to maximize productivity. Farmers, for example, can estimate the precise irrigation requirements of different locations within a field by analysing soil moisture data, eliminating under or over-watering and guaranteeing optimal plant growth.

The interpretation of acquired data and the extraction of actionable insights rely heavily on data analysis. Farmers may find correlations between numerous factors, uncover patterns and even anticipate crop yields by using modern statistical models and algorithms. This analytical capability enables farmers to make appropriate adjustments to their farming practices, such as fine-tuning fertilizer application rates based on soil nutrient data or applying targeted pest management measures guided by pest population trends. Furthermore, data analysis helps to monitor crop health and detect early indicators of stress or disease. Farmers can discover small changes in vegetation indices or spot indicators of certain plant diseases by analysing images data received from drones or satellites. This early diagnosis allows for timely actions, limiting disease transmission and significant crop losses. Integrating data collecting and analysis into precision farming also makes it easier to construct decision support systems and automated processes. Farmers can gain access to user-friendly dashboards and mobile applications that provide real-time insights and recommendations based on data analysis. These methods aid in optimizing resource allocation, improving operational efficiency and reducing waste.

(Shankarnarayan and Ramakrishna, 2020)

**Data acquisition and sensing technologies**

The tools and systems used to collect, measure and record data from various sources are referred to as data acquisition and sensing technologies. These technologies are critical for gathering data from the physical environment and translating it into digital formats for further analysis and decision-making.

**Remote sensing technologies for field monitoring**

The gathering and analysing of information about items or places without physical contact is the goal of remote sensing technology. It facilitates data collecting from a distance, generally through the use of satellites, aircraft, or drones. The electromagnetic radiation reflected or emitted by objects or areas of interest is measured in remote sensing. Sensors aboard satellites or other platforms collect data in a variety of spectral bands, including visible, infrared, thermal and microwave. Scientists and analysts can gain vital information about the Earth's surface or atmosphere by analysing the patterns and features of reflected or emitted radiation.

Remote sensing (RS) technology provides several benefits for gathering information about Earth's properties. Because RS is a non-destructive approach, valuable data can be collected without causing harm. Furthermore, rather than relying on single-point observations, RS permits systematic data collecting over wide geographical areas, offering a full picture. It can gain access to and divulge information about inaccessible locations, so expanding our understanding. The systematic data gathering in RS avoids sample bias and provides fundamental biophysical knowledge useful to many scientific areas. Furthermore, unlike other mapping sciences such as cartography or GIS, RS is self-contained and generates its own data.

The origins of modern remote sensing (RS) can be traced back to the practice of taking black and white aerial images of landscapes. The practice of taking black and white aerial images of landscapes can be traced back to the beginnings of contemporary remote sensing (RS). These early photos were mostly used for military purposes. However, the Agriculture Stabilisation and Conservation Service quickly recognised the possibilities of aerial photography and began using the technique to collect indexed pictures for agricultural purposes. RS has evolved substantially throughout time, using new technologies to improve its efficiency, completeness, accuracy and timeliness in supplying information. RS has now developed as a powerful and practical management tool for site-specific crop management, enabling focused and precise agricultural practices based on detailed information gained through RS approaches.

(Liaghat and Balasundram, 2010)

**Satellite imagery and aerial photography**

A remote sensing satellite image is made up of Digital Numbers (DN), which are numerical values that demonstrate image qualities such as colour, brightness, wavelength, electromagnetic frequency, or the image's core components known as picture elements (pixels). Pixels are the smallest units that comprise an image. They are organised in a grid-like pattern, generating rows and columns to make what is known as a raster image. The resolution of the image determines the quantity of information stored in these pixels, as well as their size, shape and level of detail. In essence, the resolution determines the information content and dimensions of the pixels in a digital image.

Spatial Resolution: The level of detail collected by a sensor in a picture is referred to as spatial resolution. The spatial resolution of the sensor has a direct impact on the amount of detail present in an image. For example, if the spatial resolution is 10 metres, each pixel represents a 10 metre by 10 metre ground area. Higher resolution photos provide finer details, allowing for a clearer visualization of smaller objects or characteristics, but they cover a lesser area on the ground. Lower resolution photos, on the other hand, may lack clear clarity in terms of features, but they cover a bigger overall ground area. As a result, spatial resolution is important in establishing an image's level of detail and coverage area.

Spectral resolution: A sensor's spectral resolution refers to its capacity to distinguish and differentiate tiny intervals of wavelengths within the electromagnetic spectrum. The ability of the sensor to capture the reactions of electromagnetic radiation (EMR) received on objects across multiple wavelengths ranges also influences the level of detail visible in an image.

When distinguishing between vegetation and water, for example, a broader wavelength range such as visible and near-infrared is generally used. Differentiating between different sorts of rocks, on the other hand, may necessitate a shorter wavelength range within the same band. As a result, greater spectral resolution corresponds to a shorter range of wavelengths within a single band, allowing for more exact classification and analysis of distinct image aspects.

Radiometric resolution: Radiometric resolution refers to an image's ability to differentiate and identify objects with varying levels of contrast. It includes the ability to recognise and analyze an image's brightness, contrast, illumination variations and other subtle features. Radiometric resolution informs us about the range and precision of brightness values that can be collected and represented within an image. It is critical in accurately depicting tiny differences in intensity, enabling for the identification and distinction of objects or features with varied contrast levels. Higher radiometric resolution allows for better detail and fidelity, allowing for a more comprehensive interpretation and study of the image content.

Temporal resolution: In remote sensing, temporal resolution refers to the time necessary for a satellite to revisit the same area from the same viewing point. This is also known as absolute temporal resolution and it is measured by the time it takes a satellite to complete one entire orbit cycle. Temporal data is critical in remote sensing applications because it allows for the long-term monitoring of dynamic processes such as vegetation changes, flood occurrences, deforestation and urban expansion. The spectral resolution can change over time, which can be collected and analysed with multi-temporal photography. Researchers can notice and examine changes in spectral characteristics by reviewing data acquired at different time intervals, which can provide significant insights into the evolving conditions and transformations within an area of interest. The temporal resolution of satellite imagery is affected by a number of factors, including the satellite sensor's capabilities, the latitude of the observed area and the length of swath overlap. These parameters influence the frequency with which data can be collected and the intervals between trips to a certain place.

(Al-Wassai and Kalyankar, 2013; Sowmya *et al.*, 2017)

**Unmanned aerial vehicles (UAVs) and drones**

Remote sensing photographs are now obtained mostly through the use of piloted planes or satellites. However, when it comes to agricultural applications, these systems have severe limits. They have low geographical and temporal resolution, are limited in availability due to weather conditions and are expensive. Unmanned Aerial Vehicles (UAVs), on the other hand, have emerged as a viable solution for agricultural remote sensing data collection. Unmanned aerial vehicles (UAVs) are aircraft that function without the presence of a human pilot. They are outfitted with a variety of sensors, cameras and navigation systems that enable them to carry out a variety of jobs and operations. UAVs have grown in popularity and have applications in a wide range of businesses and areas. One of the primary benefits of UAVs is their capacity to explore regions that humans may find difficult or unsafe to approach. As a result, they are suitable for duties like aerial surveys, monitoring and inspections. UAVs are lightweight, low-cost aircraft that are ideal for data collection. Unlike larger planes and satellites, they may operate at low altitudes, below cloud cover and go unnoticed. UAV aerial photography bridges the gap between ground-based observations and remotely sensed imagery, providing various benefits. UAVs can be deployed rapidly and frequently, are less expensive and safer than piloted aircraft and offer greater flexibility in terms of flight height and mission scheduling. Furthermore, UAVs can capture high-resolution imagery, enabling for close examination of individual plants, patches and patterns across landscapes. Researchers created UAV-based agricultural remote sensing systems and shown its utility in mapping crop maturity, detecting irrigation difficulties, monitoring early-stage crop growth and collecting high-resolution georeferenced photos. It should be emphasised, however, that the UAVs utilised in some of these systems were fixed-wing aircraft, which require a runway for take-off and landing, causing difficulties in agricultural areas.

Although the terms "drones" and "UAVs" are sometimes used interchangeably, there is a distinction. According to industry analysts, UAVs have autonomous flight capabilities, whereas drones may or may not have this capability. Because of the lack of autonomous flight capability, all UAVs can be categorized as drones, but not all drones can be classified as UAVs. This distinction emphasizes the technological differences between these terminology in the field of unmanned aerial systems.

(Xiang and Tian, 2011; Kim *et al.*, 2019; Yavaşli, 2020; Velusamy *et al.*, 2021)

### Sensor technologies for crop and soil monitoring

Sensor technologies are critical in crop and soil monitoring because they allow farmers and agricultural professionals to collect reliable, real-time data for informed decision-making. Sensors of various types are used to evaluate crop health, soil conditions and environmental influences. Sensors such as multispectral and hyperspectral cameras provide data on plant health, chlorophyll content and nutrient levels for crop monitoring. Soil moisture sensors assess soil moisture content, allowing farmers to optimize irrigation and avoid water stress. Temperature and humidity sensors also provide information about microclimate conditions, which aids in pest management and illness prevention. These sensor technologies enable precise and focused interventions, thereby encouraging sustainable agricultural practices and increasing crop yields.

**Soil sensors for monitoring soil properties and conditions**

Soil sensors have become indispensable instruments in modern agriculture, allowing farmers and agronomists to precisely and efficiently monitor and control soil qualities and conditions. These sensors provide real-time data on critical soil properties, enabling for more informed decisions and better soil management practices.

One of the fundamental roles of soil sensors is to measure soil moisture content. Soil moisture sensors use several ways to properly estimate the quantity of water present in the soil, such as capacitance, time domain reflectometry (TDR), or neutron scattering. This data assists farms in scheduling irrigation, avoiding water stress or flooding and optimizing water usage, increasing water conservation and lowering irrigation expenses.

Temperature sensors offer information on soil temperature fluctuations, which affect plant growth, nutrient availability and microbial activity. Farmers can make judgements about planting schedules, crop choice and the placement of soil amendments or fertilisers by monitoring soil temperature. This aids in crop yield optimization and general plant health.

Soil pH sensors are essential for determining soil acidity or alkalinity. Because certain minerals are more accessible within specific pH ranges, soil pH has a direct impact on nutrient availability to plants. pH sensors monitor soil pH accurately, allowing farmers to change soil pH levels with appropriate additives or treatments. Farmers may promote healthy growth and maximize harvests by ensuring that plants have access to the nutrients they need by maintaining ideal pH levels.

Electrical conductivity (EC) sensors are used to measure soil electrical conductivity, which is a good indicator of soil salinity. High soil salinity can have a negative impact on plant growth and output. Farmers can utilise EC sensors to identify regions with excessive salt concentrations, allowing them to implement targeted management measures such as leaching or the use of salt-tolerant crops. Farmers can avoid crop losses and efficiently manage salt-affected soils by monitoring soil salinity.

Nutrient sensors provide essential information on soil nutrient levels, including key elements like nitrogen, phosphorous and potassium. Farmers can make data-driven fertilizer application decisions by analysing nutrient concentrations, ensuring that crops receive adequate nutrients. This not only encourages optimal plant growth and development, but it also aids in the prevention of nutrient deficits and decreasing fertilizer waste and minimising environmental implications.

Soil sensor integration with precision agriculture systems has transformed soil management practices. Sensor technology improvements have allowed these sensors to be integrated into automated systems, allowing for continuous monitoring and data collection across enormous agricultural areas. This real-time data is frequently wirelessly transferred to a central management system, providing farmers with rapid access to crucial data.

(Evett, 2000; Evett *et al.*, 2002; Gornall *et al.*, 2007; Kim *et al.*, 2009; Valdés *et al* 2015; Rossel and Bouma, 2016; Scheberl *et al.*, 2019; Bañón *et al.*, 2021; Yin *et al.*, 2021; Mary *et al.*, 2022)

**Weather sensors for monitoring environmental conditions**

Weather influences numerous aspects of our daily lives, including farming, transport and urban development. Environmental monitoring and awareness are critical for making well-informed choices and mitigating potential threats. Humidity, temperature, wind speed and direction, rainfall and other atmospheric factors are all measured by weather sensors. They are critical in forecasting the weather, climate surveillance and precision agriculture. Weather forecasting is one of the most common applications for weather sensors. Meteorological institutions rely on weather sensor data to forecast and monitor weather trends and severe weather events. Temperature sensors measure the ambient temperature, which aids in the prediction of heatwaves and cold spells. Humidity sensors measure the amount of moisture in the air and can help predict fog, dew, or drought conditions. Wind sensors assess wind speed and direction, which aids in storm, hurricane and wind pattern predictions. Precipitation sensors, such as rain indicators, monitor the amount of rainfall, allowing for precise flood or drought predictions. Weather sensors help to early warning systems by continuously monitoring environmental conditions, allowing officials as well as residents to take necessary steps to protect people and property.

Accurate and current weather data from sensors enables farmers to execute site-specific management practices tailored to their farms' individual conditions. Farmers can optimize irrigation, fertilisation and pest management practices by combining weather sensor information with other agricultural data such as moisture levels in the soil and nutrient content. If a temperature sensor predicts an impending heatwave, farmers can change irrigation schedules to ensure plants receive adequate water during hot periods. Similarly, wind sensors can aid in determining the best time to apply pesticides, minimising drift and optimizing efficacy. Farmers can maximize crop output, eliminate resource waste and minimize environmental consequences by making data-driven choices based on weather sensor information. Weather sensors can help detect and mitigate weather-related dangers and crop diseases early on. Farmers, for example, can recognise conditions favourable for the growth of fungal infections and take preventive actions to prevent their emergence by monitoring temperature and humidity levels. Real-time data from precipitation sensors helps farmers manage waterlogging or drought circumstances by allowing them to change irrigation practices as needed.

(Tenzin *et al.*, 2014; Kumar and Ambika, 2020; Dombrowski *et al.*, 2021)

**Crop sensors for monitoring plant health and growth**

Precision farming requires the capacity to correctly monitor and assess plant health and growth in order to maximize crop output and optimize resource management. Crop sensors have evolved as useful precision agriculture instruments, delivering vital insights into crops' physiological status and demands. Farmers may use these sensors to make data-driven decisions and perform targeted interventions, resulting in enhanced crop health and yields.

Crop sensors are meant to assess plant health factors such as chlorophyll content, leaf temperature, canopy framework and nutrient levels. Chlorophyll sensors, for example, offer data on plants' photosynthetic activity, which is an important indicator of their general health and production. Farmers can analyze the effectiveness of photosynthesis and discover any potential nutritional deficits or stresses impacting plant growth by measuring the percentage of chlorophyll present in leaves.

Plant stress produced by environmental variables such as heat, drought, or disease is monitored using leaf temperature sensors. Elevated leaf temperatures might signal a lack of water or heat stress, allowing farmers to change irrigation schedules or employ shade strategies to protect plants from the sun. These sensors also aid in the early diagnosis of diseases, as sick plants frequently have higher leaf temperatures than healthy plants.

Canopy sensors provide essential information about the crop canopy's structure and vigour. These sensors monitor factors such as the leaf area index (LAI), the vegetation index (VI) and the accumulation of biomass. Farmers can analyze crop canopy density and growth by monitoring these parameters, allowing them to optimize planting density, change fertilizer delivery rates and make educated judgements on crop trimming or pruning practices. Canopy sensors can also identify early indicators of nutritional deficits or imbalances, which will prompt corrective action.

The assessment of nutrient levels in plants is another important application of crop sensors in precision farming. Farmers can use nutrient sensors to monitor crop nutrient status, identifying deficits or excesses that may impede development and productivity. Farmers may customize fertilisation programmes to suit crop requirements by precisely evaluating plant nutrient needs, eliminating wasteful fertilizer application and minimising environmental impacts.

(Stamatiadis *et al.*, 2006; Yu *et al.*, 2016; Palazzi *et al.*, 2019; Jiang *et al.*, 2020; Kim and Lee, 2022; Ahmad and Sharma, 2023)

**Data management and integration**

Data management and integration are critical in precision farming, ensuring that the huge amount of data collected is used efficiently. Farmers use many technologies to collect data on soil conditions, weather trends, health of crops and machinery efficiency. Comprehensive evaluation and data-driven decision-making are made possible by proper gathering, storing and integration. Machine learning and other advanced analytics approaches can assist spot patterns, optimize resource allocation and boost production. Data interoperability, safety and privacy are also critical considerations. Effective data management and integration in precision farming boost agricultural production, sustainability and innovation.

**Geographic Information Systems (GIS)**

A GIS is a system of hardware, software and procedures to facilitate the management, manipulation, analysis, modelling, representation and display of georeferenced data to solve complex problems regarding planning and management of resources. (Escobar *et al.*, 2008)

Dr. John Snow's mapping investigation into an outbreak of cholera in London in 1854 marks the beginning of GIS. He identified a link between cholera cases and the water line by mapping outbreak areas and other topographical factors. Using paper maps as a tool, this event highlighted

GIS's problem-solving skills. The 1960s saw advances in computers that paved the door for GIS to move from paper to computer mapping. Roger Tomlinson was a crucial figure in the development of GIS, taking a layered approach to map processing. GIS began the commercialization phase between 1975 and 1990, with software such as ODYSSEY GIS and ARC/INFO serving as key milestones. Despite obstacles, GIS saw extensive acceptance in education, corporations and governments from 1990 to 2010. GIS (Geographic Information System) is made up of five major components: technology, software, data, people and methodologies. Computer systems such as servers and desktop PCs are examples of hardware. Software performs the functions required for storing, analysing and displaying geographic data. Data includes both geographical and tabular information gathered from various sources. People are essential in the management and application of GIS to real-world situations. Methods include well-thought-out plans and rules tailored to each organisation. GIS and remote sensing are inextricably linked. GIS receives spatial information from remote sensing and GIS improves remote sensing by integrating, analysing and visualizing data. This mutually beneficial partnership allows for thorough spatial evaluation and in decision-making.

(Ali, 2020)

**Data storage and cloud computing**

Data storage and cloud computing are revolutionizing the way agricultural data is managed, analysed and used in precision farming. Traditional methods of data storage and management have become inadequate with the arrival of new technologies and the exponential rise of data created in farming operations. As a result, cloud computing solutions that provide scalable, secure and accessible computing and storage capability for precision farming systems have gained popularity.

One of the primary benefits of cloud computing in precision farming is its capacity to efficiently manage massive amounts of data. Sensors, drones, satellites and weather stations generate massive volumes of data in modern farming practices. This information covers the condition of the soil, crop health, climate trends and machine performance. Local storage and management of this data can be difficult and costly. Cloud computing offers a scalable platform for farmers to securely store and analyze data without a requirement for extensive on-site gear and infrastructure. Cloud computing also offers real-time data access and collaboration across precision farming ecosystem stakeholders. Farmers, agronomists, researchers and consultants can now readily access and share data, making it easier for them to collaborate, analyze patterns and make informed decisions. This real-time data access allows for rapid interventions such as modifying irrigation schedules, optimizing fertilizer inputs and detecting early signs of crop disease. In addition, cloud-based platforms frequently include advanced analytics and visualization tools, allowing users to acquire important insights from collected data and make data-driven decisions.

Precision farming requires sensitive knowledge of farming operations as well as potentially valuable intellectual property, therefore data security is crucial. Cloud computing providers make significant investments in sophisticated security mechanisms to safeguard data against unauthorized access, data loss and other cyber dangers. To protect data integrity and confidentiality, they use encryption techniques, data backup processes and strict access controls. Furthermore, cloud-based solutions frequently adhere to industry norms and laws in order to ensure data privacy and suit the needs of farmers and agricultural organisations. Another benefit of cloud computing in precision farming is its flexibility. Farmers can choose from a variety of cloud-based services and just pay for what they use, avoiding the need for huge upfront hardware and software investments. They may scale their storage and processing capacity to meet their individual requirements, enabling affordable and effective data management. Cloud computing also allows for the seamless integration of other precision farming methods, such as data analytics channels, algorithms for machine learning and Internet of Things (IoT) sensors, allowing for a more holistic and seamless approach to agricultural operations.

(Tripathi and Mishra, 2011; Ferrández-Pastor *et al.*, 2016; López-Riquelme *et al.*, 2017)

**Decision support systems for precision farming**

Decision support systems (DSS) play an important role in precision farming because they provide farmers with useful insights and advice for making educated decisions. These systems analyze and understand complicated agricultural scenarios by integrating diverse data sources such as weather information, soil data, crop health data and historical records. DSS can produce predictions, discover patterns and optimize resource allocation by leveraging advanced analytics, machine learning and modelling techniques. Farmers may obtain real-time information, analyze multiple scenarios and receive customized advice for operations such as irrigation scheduling, fertilizer management, pest control and harvesting using user-friendly interfaces. Farmers can use decision support systems to make data-driven decisions, improve efficiency, cut costs and increase productivity while minimising environmental effect.

(Reddy and Rao, 1995; Mir and Quadri, 2009)

**Precision Farming Techniques**

Precision Farming Techniques includes the integration of technologies and automation to enable precise monitoring, analysis and management of various aspects of crop production.

**Variable rate technology (VRT) for input application**

Precision farming's variable rate application technology (VRT) transforms the way crop production inputs are applied within fields. Rather than employing a constant application rate, VRT modifies the input rate based on spatially variable parameters that influence the ideal rate. This method has the potential to improve input efficiency, field profitability and environmental stewardship dramatically. VRT tries to optimize application rates for specific areas by recognising that factors impacting crop production vary across fields. However, while VRT has many advantages, its economic rewards are not always consistent. VRT's effectiveness can be limited by a number of variables, including implementation costs, insufficient crop yield gain and insufficient input savings. Field characteristics, crop type, geographic location, pests, nutrients, soil conditions, variability and managerial capabilities have an impact on the decision to implement VRT practices.

To properly deploy VRT, reliable identification and interpretation of within-field variation are required. This necessitates the refinement of input-crop models as well as the development of cost-effective mapping technologies. Predictive economic modelling can also help discover fields that are ideal for VRT implementation. One frequent method for soil sampling is to collect samples from fields no larger than 20 acres in size. Soil cores are collected at random sites throughout the sampling area and transported to a laboratory for analysis. Crop advisors utilise soil test results to recommend fertilizer applications for the 20-acre area. Another method, known as grid soil sampling, increases the sample intensity. A 20-acre sampling region, for example, would contain 10 samples utilizing a 2-acre grid sampling system. These soil samples, gathered in a methodical grid, give location data for mapping and the production of nutrient demand maps, known as application maps. After that, the application maps are entered into a computer that is mounted on a variable-rate fertilizer spreader. Thc computer, which is equipped with a GPS receiver, leads a product-delivery controller, which modifies the amount and kind of fertilizer product based on the application map. Variable-rate technology (VRT) can be applied to a variety of farming operations, including fertilisation, seeding and pesticide application. It uses extrapolated information from the Geographic Information System (GIS) to regulate activities at a variable rate in the appropriate place and at the right time. In the United States, VRT has become one of the most extensively used precision agricultural technologies. VRT has the ability to increase specific crop quality in various cycles, such as obtaining specified protein content in cereals like wheat for baking flour or barley for malting. Pesticide, herbicide and seeding application rates can also be adjusted using VRT technology to fit soil potentials or problem regions. Except in fields with little spatial variability, studies have indicated that VRT can result in decreased nitrogen losses in the environment when compared to uniform rate application (URA). Variable rate fertilizer treatment has also been demonstrated to boost yields.

VRT employs two fundamental technologies: map-based and sensor-based. The application rate of map-based VRT is adjusted based on an electronic map, also known as a prescription map. The concentration of input is modified as the applicator passes through the field using GPS field position and a prescription map specifying the required rate. Sensor-based VRT, on the other hand, uses sensors on the applicator to monitor soil parameters or crop traits in real time. Based on this constant stream of information, a control system calculates the input requirements and delivers the input to the observed site. The most common VRT devices in use are huge chemical applicators that can handle many compounds at once, often up to 11 different substances. These VRT applicators are made up of three basic parts: a controller that adjusts the flow rate of the materials, a positioning system and a map that specifies the intended application rates for the field. Both map-based and sensor-based VRT have advantages and disadvantages. An site-specific crop management strategies have been implemented to take use of the benefits of both methods. These systems combine map-based and sensor-based VRT to maximize the benefits of variable rate application technology.

(Sawyer *et al.*, 1994; Searcy, 1997; Singh, 2010; Bakhtiari and Hematian, 2013; Fulton *et al.*, 2013; Hakkim *et al.*, 2016; Finger *et al.*, 2019

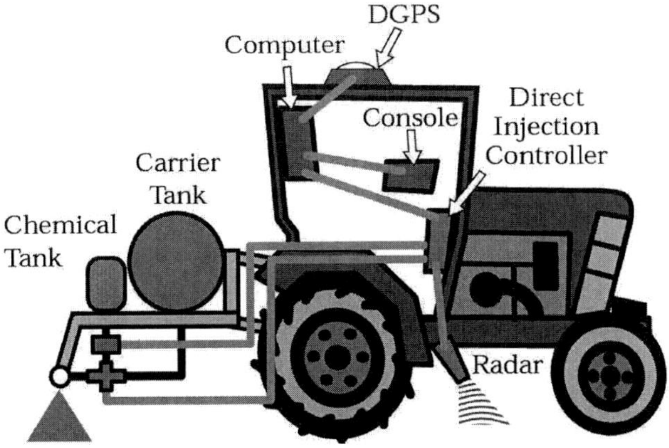


**Figure 1:** The elements of a VRT chemical applicator are depicted schematically (Searcy, 1997)

**Yield mapping**

Yield monitors and maps are critical components of precision farming systems, especially in site-specific crop management (SSCM). production monitors give farmers a precise insight of production changes within a field, allowing them to conduct on-farm trials and analyses without the need for extra equipment. These monitors are made up of different sensors, a data storage device, a user interface and a task computer in the combine cockpit that regulates their integration. The sensors measure grain flow, separator speed, ground speed, grain moisture and header height and use these characteristics to calculate yield. However, crop width, GPS inconsistencies, grain losses and sensor calibration can all cause yield measurement discrepancies. Yield mapping software is available for analysing and visualizing the acquired data, which is frequently used to build contour maps using interpolation techniques. GPS receivers are used to record position data and produce maps. Data acquired over multiple years using yield monitors can provide useful insights for regional and temporal analysis of trends, as well as decision-making. Forage crops, for example, have yield monitoring systems that monitor weight, moisture and other parameters each bale. Overall, yield monitors provide a substantial contribution to optimizing crop production practices and increasing agricultural productivity.

(Bakhtiari and Hematian, 2013; Hakkim *et al.*, 2016)

**Precision livestock farming**

Precision livestock farming (PLF) is the practice of applying precision agriculture principles and technology to livestock production. It includes activities such as animal development, milk and egg production, disease detection, animal behaviour monitoring and physical environment management. Milk monitoring is included in PLF systems to analyze lipid and microbial levels in order to identify potential illnesses. Stockmen can engage with animals more efficiently because to robotic feeding, weighing and cleaning equipment, as well as imaging devices. Data monitoring systems monitor feed and water use, which aids in the early detection of infections. Real-time growth measurement assists producers in monitoring feed conversion and growth rates. Coughing is detected by acoustic sensors in pigs as a sign of respiratory illnesses. Temperature, birthing readiness and fertility are all monitored by sensors on the animals' collars and farmers are notified by SMS. Farmers can arrange insemination using vaginal thermometers and sensor-equipped collars based on indicators of estrus and fertilisation readiness. PLF provides sophisticated technologies for efficient management of livestock and early diagnosis of health problems.

(Banhazi *et al.*, 2012; Hakkim *et al.*, 2016)

**Conclusion**

Precision farming uses geographical location as a primary technique to collect site-specific information regarding fields, crops, soils and yields. This study investigates various precision agriculture technologies, their advantages, disadvantages and limits. While the benefits of precision agriculture are widely acknowledged, barriers for complete acceptance and development remain. Farmers, equipment developers and academics are all concerned about profitability. The future of precision farming may involve a shift towards small-scale precision techniques, with autonomous drones replacing huge manned machines. Strategic assistance from both the public and corporate sectors is critical for driving rapid adoption, especially in underdeveloped nations where precision farming is only a theoretical concept. Exploration, analysis and implementation are the three stages of successful adoption. Precision agriculture provides solutions to modern agriculture's economic and environmental concerns by emphasizing the significance of completing the right actions at the right time and location. To accomplish a sustainable and evergreen agricultural revolution, efforts should be made to use new technological inputs. Precision farming's effectiveness is dependent on the availability and rapid acquisition of knowledge to successfully guide the adoption of new technology. It employs advanced information technology to predict soil and plant variety while including agricultural practices adapted to specific needs. The goal is to increase agricultural economic returns while decreasing energy input and environmental effect.

## References


1. Ahmad, U., & Sharma, L. (2023). A review of best management practices for potato crop using precision agricultural technologies. *Smart Agricultural Technology*, 100220.
2. Ali, E. (2020). Geographic information system (GIS): definition, development,applications & components. *Department of Geography, Ananda Chandra College. India.*
3. Al-Wassai, F. A., & Kalyankar, N. V. (2013). Major limitations of satellite images. *Journal of Global Research in Computer Science*, *4*(5).
4. Bakhtiari, A. A., & Hematian, A. (2013). Precision farming technology, opportunities and difficulty. *International Journal for Science and Emerging Technologies with Latest Trends*, *5*(1), 1-14.
5. Banhazi, T. M., Lehr, H., Black, J. L., Crabtree, H., Schofield, P., Tscharke, M., & Berckmans, D. (2012). Precision livestock farming: an international review of scientific and commercial aspects. *International Journal of Agricultural and Biological Engineering*, *5*(3), 1-9.
6. Bañón, S., Álvarez, S., Bañón, D., Ortuño, M. F., & Sánchez-Blanco, M. J. (2021). Assessment of soil salinity indexes using electrical conductivity sensors. *Scientia Horticulturae*, *285*, 110171.
7. Dombrowski, O., Hendricks Franssen, H. J., Brogi, C., & Bogena, H. R. (2021). Performance of the ATMOS41 all-in-one weather station for weather monitoring. *Sensors*, *21*(3), 741.
8. Escobar, F., Hunter, G., Bishop, I., & Zerger, A. (2008). Introduction to GIS. Department of Geomatics, the University of Melbourne.
9. Evett, S. (2000). Some aspects of time domain reflectometry, neutron scattering and capacitance methods for soil water content measurement. *Comparison of soil water measurement using the neutron scattering, time domain refiectometry and capacitance methods*, 5.
10. Evett, S., Laurent, J. P., Cepuder, P., & Hignett, C. (2002). Neutron scattering, capacitance and TDR soil water content measurements compared on four continents. In *17th World Congress of Soil Science* (pp. 14-21).
11. Ferrández-Pastor, F. J., García-Chamizo, J. M., Nieto-Hidalgo, M., Mora-Pascual, J., & Mora-Martínez, J. (2016). Developing ubiquitous sensor network platform using internet of things: Application in precision agriculture. *Sensors*, *16*(7), 1141.

12. Finger, R., Swinton, S. M., El Benni, N., & Walter, A. (2019). Precision farming at the nexus of agricultural production and the environment. *Annual Review of Resource Economics*, *11*, 313-335.

13. Fulton, J. P., Shearer, S. A., Higgins, S. F., & Mc Donald, T. P. (2013). A method to generate and use as-applied surfaces to evaluate variable-rate fertilizer applications. *Precision Agriculture*, *14*, 184-200.

14. Gornall, J. L., Jónsdóttir, I. S., Woodin, S. J., & Van der Wal, R. (2007). Arctic mosses govern below-ground environment and ecosystem processes. *Oecologia*, *153*, 931-941.

15. Hakkim, V. A., Joseph, E. A., Gokul, A. A., & Mufeedha, K. (2016). Precision farming: the future of Indian agriculture. *Journal of Applied Biology and Biotechnology*, *4*(6), 068-072.

16. Jiang, J., Zhang, Z., Cao, Q., Liang, Y., Krienke, B., Tian, Y., & Liu, X. (2020). Use of an active canopy sensor mounted on an unmanned aerial vehicle to monitor the growth and nitrogen status of winter wheat. *Remote Sensing*, *12*(22), 3684.

17. Khosla R (2008) The 9th International Conference on Precision Agriculture opening ceremony presentation. July 2023rd, 2008.

18. Khosla, R. (2010). Precision agriculture: challenges and opportunities in a flat world. In *Proceedings of the 19th World Congress of Soil Science: Soil solutions for a changing world, Brisbane, Australia, 1-6 August 2010. Division Symposium 3.1 Precision techniques for land use* (pp. 26-28). International Union of Soil Sciences (IUSS), c/o Institut für Bodenforschung, Universität für Bodenkultur.

19. Kim H.J., Sudduth K.A., & Hummel, J. W. (2009). Soil macronutrient sensing for precision agriculture. *Journal of Environmental Monitoring*, *11*(10), 1810-1824.

20. Kim, J., Kim, S., Ju, C., & Son, H. I. (2019). Unmanned aerial vehicles in agriculture: A review of perspective of platform, control and applications. *Ieee Access*, *7*, 105100-105115.

21. Kim, M. Y., & Lee, K. H. (2022). Electrochemical Sensors for Sustainable Precision Agriculture - A Review. *Frontiers in Chemistry*, *10*.

22. Kumar, A., & Ambika, R. (2020). Drone Integrated Weather Sensors for Agriculture Purpose. *Int. J. Electr. Eng. Technol*, *11*(5), 83-90.

23. Liaghat, S., & Balasundram, S. K. (2010). A review: The role of remote

sensing in precision agriculture. *American journal of agricultural and biological sciences*, *5*(1), 50-55.

24. López-Riquelme, J. A., Pavón-Pulido, N., Navarro-Hellín, H., Soto-Valles, F., & Torres-Sánchez, R. (2017). A software architecture based on FIWARE cloud for Precision Agriculture. *Agricultural water management*, *183*, 123-135.

25. Mary, A. X., Popov, V., Raimond, K., Johnson, I., & Vijay, S. J. (2022). Scope and recent trends of artificial intelligence in Indian agriculture. *The Digital Agricultural Revolution: Innovations and Challenges in Agriculture through Technology Disruptions*, 1-24.

26. Mir, S. A., & Quadri, S. M. K. (2009). Decision support systems: concepts, progress and issues–A review. *Climate Change, Intercropping, Pest Control and Beneficial Microorganisms: Climate change, intercropping, pest control and beneficial microorganisms*, 373-399.

27. Mulla, D., & Khosla, R. (2016). Historical evolution and recent advances in precision farming. *Soil-specific farming precision agriculture*, *1*, 35.

28. O'Connor, A.C., Gallaher, M.P., Clark-Sutton, K., Lapidus, D., Oliver, Z.T., Scott, T.J., Wood, D.W., Gonzalez, M.A., Brown, E.G. and Fletcher, J. (2019). Economic benefits of the global positioning system (GPS).

29. Palazzi, V., Gelati, F., Vaglioni, U., Alimenti, F., Mezzanotte, P., & Roselli, L. (2019, January). Leaf-compatible autonomous RFID-based wireless temperature sensors for precision agriculture. In *2019 IEEE topical conference on wireless sensors and sensor networks (WiSNet)* (pp. 1-4). IEEE.

30. Reddy, M. N., & Rao, N. H. (1995). GIS based decision support systems in agriculture. *National Academy of Agricultural Research Management Rajendranagar*, 1-11.

31. Robert P, Rust R, Larson W (1994) Sitespecific Management for Agricultural Systems, Proceedings of the 2nd International Conference on Precision Agriculture, 1994, Madison, WI. ASA/CSSA/SSSA.

32. Rossel, R. A. V., & Bouma, J. (2016). Soil sensing: A new paradigm for agriculture. *Agricultural Systems*, *148*, 71-74.

33. Sawyer, J. E. (1994). Concepts of variable rate technology with

considerations for fertilizer application. *Journal of Production Agriculture*, *7*(2), 195-201.

34. Scheberl, L., Scharenbroch, B. C., Werner, L. P., Prater, J. R., & Fite, K. L. (2019). Evaluation of soil pH and soil moisture with different field sensors: Case study urban soil. *Urban Forestry & Urban Greening*, *38*, 267-279.

35. Searcy, S. W. (1997). Precision Farming: A New Approach to Crop Management, Texas Agricultural Extension Service. *The Texas A&M University System, College Station, TX, USA.*

36. Shankarnarayan, V. K., & Ramakrishna, H. (2020). Paradigm change in Indian agricultural practices using Big Data: Challenges and opportunities from field to plate. *Information Processing in Agriculture*, *7*(3), 355-368.

37. Singh, A. K. (2010). Precision farming. *Water Technology Centre, IARI, New Delhi.*

38. Sowmya, D. R., Deepa Shenoy, P., & Venugopal, K. R. (2017). Remote sensing satellite image processing techniques for image classification: a comprehensive survey. *International Journal of Computer Applications*, *161*(11), 24-37.

39. Stamatiadis, S., Taskos, D., Tsadilas, C., Christofides, C., Tsadila, E., & Schepers, J. S. (2006). Relation of ground-sensor canopy reflectance to biomass production and grape color in two Merlot vineyards. *American Journal of Enology and viticulture*, *57*(4), 415-422.

40. Tenzin, S., Siyang, S., Pobkrut, T., & Kerdcharoen, T. (2017, February). Low-cost weather station for climate-smart agriculture. In *2017 9th international conference on knowledge and smart technology (KST)* (pp. 172-177). IEEE.

41. Tripathi, A., & Mishra, A. (2011, September). Cloud computing security considerations. In *2011 IEEE International Conference on Signal Processing, Communications and Computing (ICSPCC)* (pp. 1-5). IEEE.

42. Valdés, R., Ochoa, J., Franco, J. A., Sánchez-Blanco, M. J., & Bañón, S. (2015). Saline irrigation scheduling for potted geranium based on soil electrical conductivity and moisture sensors. *Agricultural Water Management*, *149*, 123-130.

43. Velusamy, P., Rajendran, S., Mahendran, R. K., Naseer, S., Shafiq, M.,

& Choi, J. G. (2021). Unmanned Aerial Vehicles (UAV) in precision agriculture: Applications and challenges. *Energies*, *15*(1), 217.

44. Xiang, H., & Tian, L. (2011). Development of a low-cost agricultural remote sensing system based on an autonomous unmanned aerial vehicle (UAV). *Biosystems engineering*, *108*(2), 174-190.

45. Yavaşli, D. D. (2020). *Drone applications in geography: Game of drones*. Current Studies in Social Sciences II; 117-131

46. Yin, H., Cao, Y., Marelli, B., Zeng, X., Mason, A. J., & Cao, C. (2021). Soil sensors and plant wearables for smart and precision agriculture. *Advanced Materials*, *33*(20), 2007764.

47. Yu, L., Wang, W., Zhang, X., & Zheng, W. (2016). A review on leaf temperature sensor: Measurement methods and application. In *Computer and Computing Technologies in Agriculture IX: 9th IFIP WG 5.14 International Conference, CCTA 2015, Beijing, China, September 27-30, 2015, Revised Selected Papers, Part I 9* (pp. 216-230).

# Chapter - 4

## Global Warming and Effect on Insect Pests


**Authors**

**Vijaya**

Assistant Professor (Entomology), School of Agricultural Sciences, Baddi University of Emerging Sciences and Technology, Baddi, Solan, Himachal Pradesh, India

**Diksha Abrol**

Assistant Professor (Entomology), School of Agricultural Sciences, Baddi University of Emerging Sciences and Technology, Baddi, Solan, Himachal Pradesh, India

# Chapter - 4

## Global Warming and Effect on Insect Pests

**Vijaya and Diksha Abrol**


### Abstract

Climate change and global warming are much more important to agriculture worldwide and are among the most discussed topics in today's era. Climatic parameters like increased temperature, rising $CO_2$ levels in the atmosphere, and changing precipitation patterns have significant impact on agricultural production and on insect pests of agricultural importance. Increased temperature can directly or indirectly influence the reproductive rate and survival rate of agricultural crop-pests. Elevated $CO_2$ can have its impact on the nutritional attributes of the crop plants which ultimately affect the insect pest feeding. Changes in the rainfall pattern can also directly influence the insect-pest population on the host crop plants. Introduction of an invasive insect pest species to a newer environment can also be benefitted due to the alterations in climatic factors. Ineffectiveness of bio-control agents may also be the impact of changed climatic factors like temperature or relative humidity. As a result, there is a great risk of economic production losses and a challenge to global food security in the long term. Climate change significantly affects pest population dynamics, adaptive management strategies are needed to deal with the changing status of insect-pests, therefore, several priorities have to be identified for conducting future research on the effects of climate change on insect pests of agricultural crop plants and for formulating their management strategies.


**Keywords:** Climate change, insect-pests, temperature, $CO_2$, management strategies.

### Introduction

Global warming and climate change are frequently utilized terms but have unmistakable association. Global warming is the gradual increase in Earth's surface temperature since the pre-industrial era as a result of human activity, mostly the burning of fossil fuels, which contributes to rising levels of greenhouse gases that trap heat. Climate change, on the other hand, refers to a long-term transformation of the usual climate patterns that have come to characterize Earth's local, territorial, and global climates, therefore this term cannot be used in place of it. Wide-ranging observed effects of the term are consistent with these modifications.

Insects are the most prevalent species on Earth, making up more than half of the 1.5 million species of organisms (Karthik *et al.*, 2021), which are the basic building blocks of ecosystem structure and function. They are cold-blooded organisms, which implies that variations in temperature have an impact on their ability to develop, reproduce, and survive. The size of the insect population, its velocity of dispersal, and its geographic range may all be directly impacted by changes in climatic variables. The interactions between insect pests, their surrounding habitat, natural enemies, predators, and other mutual creatures, among others, may be impacted by the indirect effects of climate change.

Increased $CO_2$ in the atmosphere, higher temperatures, and decreased soil moisture are most likely the primary causes of climate change. These changes may have a substantial impact on the population dynamics of insect pests and, consequently, the proportion of crop losses. (Skendzi *et al.,*2021). The interactions between agricultural crops and insect pests can be considerably impacted by the intricacy of the physiological effects of temperature and elevated $CO_2$ (Roth and Lindroth, 1995). As a result, growers should prepare for new and severe pest issues in the upcoming years as a result of climate change. Food security is under risk due to the worldwide issue of crop pests crossing geographical and political boundaries, which impacts all nations and areas. Therefore, an effort is being made to discuss each of these features individually in this chapter.

## 1. Effect of increased temperature on insect pests

According to Kocmankova *et al.* (2010), distribution patterns of insect-pests, their feeding behaviour, reproductive behaviour and growth & development are very much affected by the most important environmental factor i.e. temperature. Insect population dynamics can be impacted by elevated temperatures by impacting fecundity, survival, life cycle, size of the population, and geographic range (Bale *et al.*, 2002). Elevated temperatures also tend to increase insect consumption, development, and movement.

The metabolic rate of insects typically doubles with a 10 °C increase in temperature because of their very sensitive physiology to temperature changes (Dukes *et al.*, 2009). Temperature affects host availability, migration, and metamorphosis, which affects the likelihood of changes in pest population and dynamics (Shrestha, 2019).

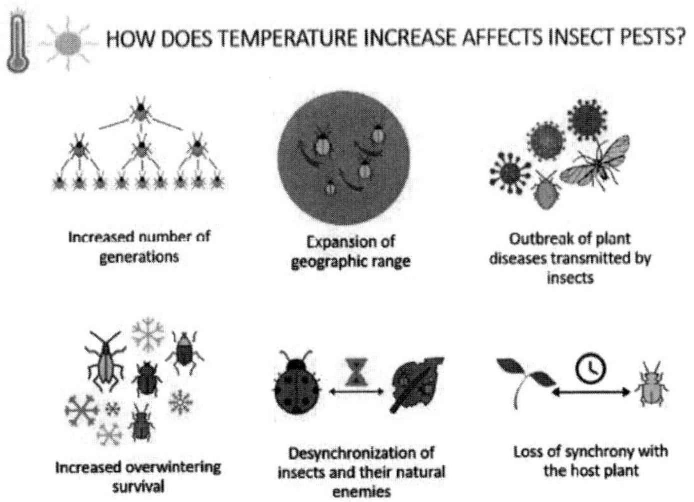


**Figure 1:** Effect of temperature on insect-pests

Previous studies have demonstrated that changes in insect species' development patterns and rising temperatures both promote herbivory (De Lucia *et al.*, 2008). Due to the current temperature being roughly ideal for pest formation and growth, it is predicted that insect populations in tropical zones will suffer a decline in growth rate, while insect populations in temperate regions will experience an increase in growth rate (Figure 1).

The same scientists' research indicates that higher temperatures stimulate the growth of pest populations in wheat, which is typically grown in temperate settings. In contrast, maize grown in temperate and tropical regions is anticipated to respond differently to the rise of insect pest populations (Deutsch *et al.*, 2018). Rice grown in tropical regions is predicted to reduce the expansion of pest populations. According to another study of over 1,000 insect species, between 15 to 37 percent of these species will go extinct by the year 2050 as a result of climate change brought on by global warming (Thomas *et al.*, 2004; Hance *et al.*, 2007). Expansion of geographic range, higher overwintering insect pest population survival rates, increased risk of introducing alien insect species to a new environment, increased incidence of plant diseases spread by insect pest vectors due to range expansion and rapid insect vector multiplication, decreased effectiveness of bio-agents like predators, parasitoids, etc. are some general effects of global warming on insect dynamics (Skendzic *et al.*, 2021).

## 2. Increased $CO_2$ concentration and insect pests

By enhancing photosynthetic activity, increased $CO_2$ may have an impact on plant physiology, resulting in greater plant growth and production.

Insects may then be indirectly impacted by changes in crop plant quality and quantity. Increased $CO_2$ levels can alter the chemical makeup of leaves, which may therefore have an impact on the nutritional value of foliage and the appeal of it to insects that consume plants as food (Lincoln, 2004). Additionally, these agricultural plants frequently store sugars and starches in their leaves, which alters the C:N ratio and decreases palatability (Cortrufo *et al.*, 1998). Increased $CO_2$ concentration may cause some insect pest species to consume more plants since nitrogen is essential for an insect's regular growth and development (Bezemer *et al.*, 1998).

Due to the fact that insect pests must consume more plant tissue to receive the same amount of food, this circumstance may result in an increase in plant damage. In reaction to a decrease in nitrogen content with compensatory feeding, foliage feeders like caterpillars and a few species of leaf miners frequently increase their consumption rates (DeLucia *et al.*, 2008) (Hamilton *et al.* 2005). According to a study, the potato leaf hopper, *Empoasca fabae*, Mexican bean beetle, *Epilachna varivestis*, etc. all exhibited compensatory insect eating in response to the rise in simple sugar content in soybean foliage (Hamilton *et al.*, 2005). Whiteflies and aphids, which feed on phloem, respond by speeding up population growth and reducing density (Figure 2) (Sutherst *et al.*, 2011).

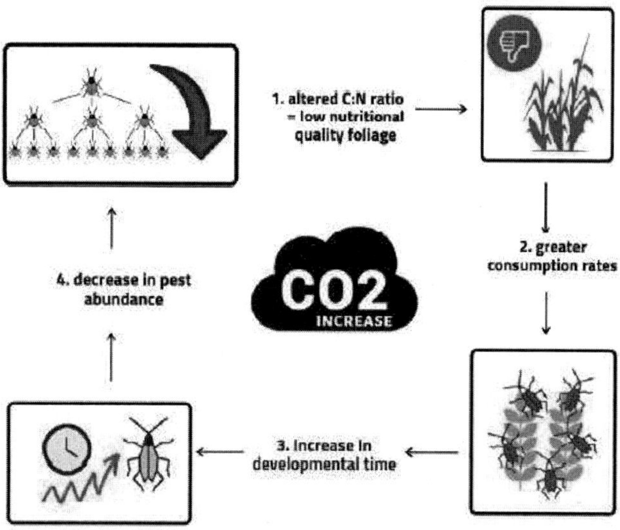


**Figure 2:** Effect of elevated $CO_2$ on insect pests

## 3. Insect pests in relation to changing patterns of precipitation

One of the most significant key climatic change elements is changes in precipitation (rainfall) in terms of quantity, severity, and frequency. The condition of overlapping precipitation has a direct impact on insect-pest species that overwinter in the soil. In a nutshell, severe precipitation can cause flooding and extended water stagnation in the area, which might endanger insect and pest survival and have an impact on their unpredictable conduct. The washing away of insect eggs and larvae is another potential impact of higher rainfall and flooding (Shrestha, 2019). Heavy rains can also wash away other soft-bodied insect pests as aphids, mites, jassids, whiteflies, etc. (Pathak *et al.*, 2012) (Figure 3). According to observations made in southern China, eastward summer migration of *Sogatella furcifera* is constrained by the pre-flood period's severe rainfall [Schneider *et al.*, 2022].

Through a number of methods, drought has a significant impact on insect pests that feed on foliage.

i) Dry climates may offer favourable environmental circumstances for the establishment and proliferation of insect pests that eat plants; and

ii) Plants under stress from drought tend to attract some insect pest species. For instance, when crop plants lose water through transpiration, water columns in the xylem separate or collapse, producing an ultrasonic acoustic emission that bark beetles from the Scolytidae family can easily detect;

iii) Plants under drought stress are more vulnerable to insect-pest attack because their production of protective secondary metabolites is reduced (Yihdego *et al.*, 2019).

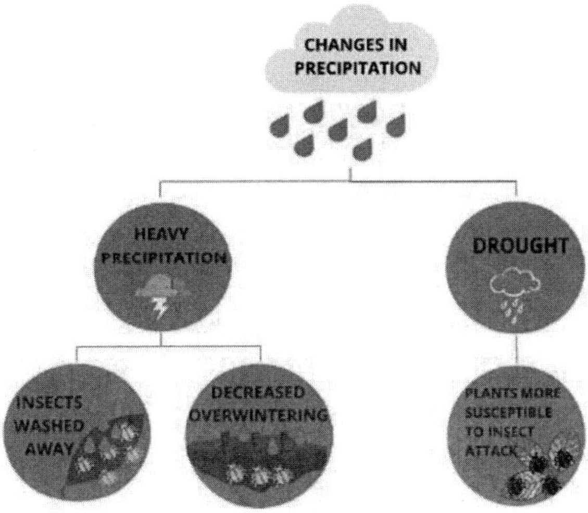


**Figure 3:** Effect of precipitation on insect pests

## 4. Risk of increased invasive insect-pest species

According to Shine *et al.* (2000), invasive alien species of insects are those taxa that are either deliberately imported (for example, as food, crops, ornamentals, pets, or livestock) or accidentally introduced by human activity outside of their natural environment. Invasive insects are typically pests of forestry, agriculture, stored goods, homes, and structures and are frequently parasite or disease carriers (Ward and Masters, 2007). Alien insects can find a larger range of acceptable habitats because invasive species typically have a wider tolerance range or bioclimatic range than native insects (Walther *et al.*, 2009). Most of the physiological processes of insect pest species are temperature-dependent, and they are known to be particularly vulnerable to climate change (Vermeij, 1996). According to the Convention on Biological Diversity (CBD, 2018), invasive alien species pose the largest danger to biodiversity worldwide and have a significant negative impact on forestry, agriculture, and aquatic ecosystems (Shrestha, 2019).

Sometimes, invasive insect introduction and transport might be directly impacted by climate change. Extreme weather phenomena, such as storms, strong winds, hurricanes, air currents, and wave action, may move pests to new geographic locations where they may find environmental conditions that are conducive to their establishment (FAO, 2020). For instance, during the 2005 hurricane season, the cactus moth (*Cactoblastis cactorum*) was transported from the Caribbean islands to Mexico, where it presented a serious ecological and financial danger to over 104 species of prickly pear cacti (*Opuntia* sp.) (Hill *et al.*, 2007) (Burgiel *et al.*, 2010). Phenotypic, behavioural, developmental, or physiological features can all be the manifestations of adaptation. The availability of food, pressure from predators or competitors, or variations in environmental factors (such as temperature, humidity, or photoperiod) can all influence physiological or behavioural plasticity (Abram *et al.*, 2017; Sgro *et al.*, 2016).

## 5. Decrease in the bio-efficacy of natural enemies

The success of bio-control initiatives would be hampered by potential severe impacts of climate change on the seasonal abundance and distribution pattern of both host insect pests and their natural foes (Thomson *et al.*, 2010). With the aid of host plant processes of availability and quality, herbivorous insect-pest species are naturally controlled by predators and parasitoids, which in turn interact to affect insect population dynamics, performance, and behaviour (Jamieson *et al.*, 2012). According to Evans *et al.* (2013), a rise in temperature interfered with the biological regulation of the cereal planthopper beetle (*Oulema melanopus*). *O. melanopus* development in this trophic level of the ecosystem was more impacted by global warming than that of its natural adversary, leading to a phenological shift between the adversary and the prey, and consequently a reduced biological control programme.

The ability of natural enemies to hunt their host insect pests is mostly influenced by their environmental tolerance in comparison to their phytophagous hosts and by how quickly they move (Thomson *et al.*, 2010). The growth and abundance of herbivore populations (host insects) are subsequently impacted by elevated $CO_2$ concentrations, altered precipitation patterns, and rising temperatures, which indirectly affect the supply of prey and hosts available for predation or parasitism (Lincoln, 1993) (Thomson *et al.*, 2010). According to Dyer *et al.* (2013), increased temperature and $CO_2$ had an impact on *Spodoptera exigua* ability to feed on lucerne crops. The host plants' host plant's lower nutritional quality caused the *S. exigua* larvae to develop more quickly. The local population of the ichneumon wasp (*Cotesia marginiventris*), a natural adversary, was wiped out at the same time because its larvae were unable to fully grow on their host body.

## 6. Insect-pest management strategies in climate change scenario

To lessen the negative effects of current pests and to lessen the possibility of the development of new illnesses and insect-pests, potential adaption solutions are explored. New improved IPM techniques, routine monitoring of insect pest population and distribution, and forecasting and model building for climate change are the most frequently mentioned strategies (Raza *et al.*, 2014) (Figure 4).

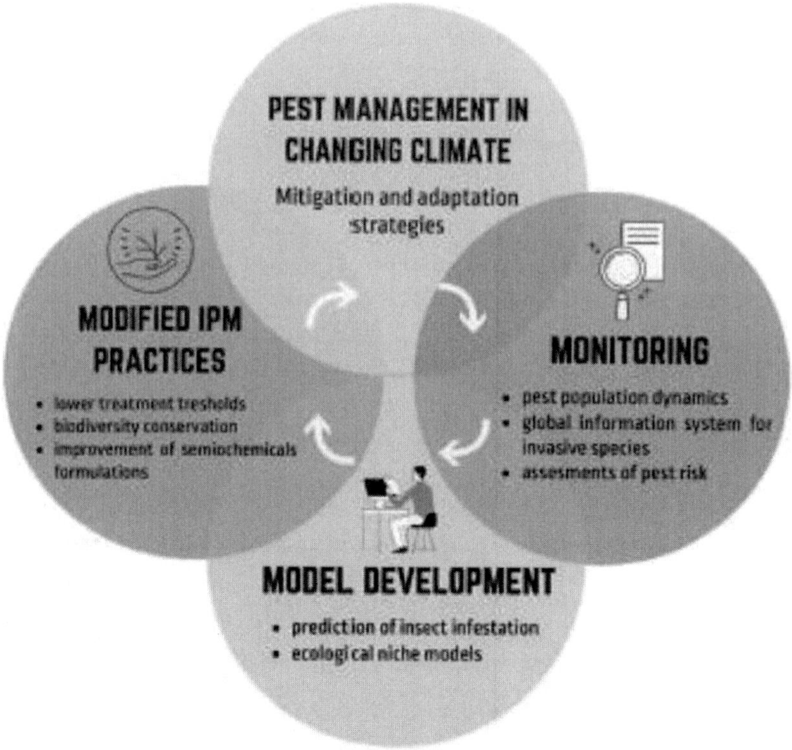


**Figure 4:** Strategic plan model for insect pest management in new environmental conditions

## 6.1 New modified IPM methods

The foundation of insect-pest management approaches in the framework of sustainable agriculture is upon the preventive measures, which must be completely exhausted before control or direct measures are employed. The necessity for control measures must be determined using cutting-edge technologies, such as forecasting techniques and thresholds that have been verified by science. When economically unaffordable losses cannot be avoided by preventive actions, chemical usage is the last, most effective option (Boller *et al.*, 2015).

The majority of IPM programmes focus on making choices based on in-depth understanding of the economic thresholds (ET), or the number of insect pests that can be tolerated before economic yield losses occur. IPM has historically evolved in the area of pest management, where the use of predetermined thresholds has produced positive outcomes. Economic thresholds are significant at IPM, but it has also been noted that they are occasionally not entirely relevant, sufficient, or even feasible. In order to lessen the effects of agricultural insect-pests on crops in a changing environment, modified farming practices and adaptive management measures are required. To reduce vulnerability to the possibility of an insect-pest outbreak, these measures may include growing different varieties of crops or doing so at different times of the year, or increasing biodiversity along field margins and borders to attract more natural enemies (Thomson *et al.*, 2010; Andrew *et al.*, 2017). Insects employ pheromones and allelochemicals to perceive their surroundings, which is a crucial process. According to Heuskin *et al.* (2011), they are crucial in the use of a variety of IPM techniques, including bio-control, push-pull tactics, surveillance, mating disruption, and trapping.

Furthermore, several bio-pesticides based on entomopathogenic bacteria, nematodes, fungi, and viruses are extremely sensitive to any form of environmental changes. Some of these control strategies may become less effective with rising temperatures and falling relative humidity, and synthetic insecticides are predicted to have a similar impact (Nihal *et al.*, 2020). The creation of novel approaches to pest management as well as potential revisions to insecticide formulations, attractants, and repellents should be prioritized in this context.

Wenda-Piesik *et al.* (2015), for instance, looked at the behaviour of *Tribolium confusum* in their study in terms of its attraction and repulsion to various concentrations of environmentally favourable volatile organic chemicals. Understanding how global warming affects the effectiveness of different chemical insecticides, their persistence in the environment (soil, water, and air), and the emergence of insecticide resistance in insect-pest populations is urgently needed (Vadez *et al.*, 2012). As a result, it becomes vital to take into account the use of effective bio-control chemicals or the introduction of new crop plant varieties that are insect-resistant through genetic engineering or breeding (Gomez *et al.*, 2020).

### 6.2 Regular monitoring of insect pest abundance and distribution

The availability of long-term data is one of the most crucial prerequisites for assessing the impact of climate change on the dynamics of insect-pest populations (Yamamura *et al.*, 2006). Preliminary evidence of biological reactions to climate change can be found by long-term monitoring of pest populations and behaviour, particularly in areas that are prone to it (Heeb *et al.*, 2019). Monitoring changes in the topographic distribution of host populations, disease vector dynamics, and insect vector dynamics at the local level is necessary. Climate change also makes it easier for invasive species to spread to new regions where they haven't previously been reported. To stop invasive species from becoming economic pests in new geographic areas, efficient monitoring and control methods are required (Heeb *et al.*, 2019) (Hellmann *et al.*, 2008). Therefore, both pest management and bio-security require the implementation of adaptive strategies.

In response to climate change, it is currently possible to strengthen existing insect-pest management techniques such detection, prediction, physical, chemical, and biological control (Heeb *et al.*, 2019). The U.S. Department of Agriculture (USDA) and the European and Mediterranean regions use monitoring at entry points and quick eradication as part of their early warning and fast response programmes. Additionally, farmers can prevent the onset and growth of anticipated pest problems by preemptively adopting certain pest preventive practices by monitoring climate and pests in conjunction with data on climate and pest risk forecasts (Heeb *et al.*, 2019).

## 6.3 Forecasting and model development for climate change

Climate models can be a useful tool for estimating the possible amount of change on a global scale when combined with the environmental requirements of a specific insect-pest species. Therefore, modelling pest risk and the effects of climate change on the host agricultural plant can help forecast how an insect infestation will turn out (Raza *et al.*, 2014). Correlative models and mechanistic models are two categories into which the models can be subdivided. Correlative models forecast prospective habitats for the species based on correlated values of environmental factors and records of occurrence. Some of the most popular correlative models are MaxEnt, Bioclim, Random Forest, etc. (Kumar *et al.*, 2014).

Foreseeing future changes in the spatial distribution of species, determining extinction rates, and establishing priorities for biodiversity conservation all need the application of correlative modeling (Porfirio *et al.*, 2014). These models determine statistical correlations between a given species' current geographic distribution and climate variables, which are then converted into climate change forecasts to suggest future habitats for that species (Pacifici *et al.*, 2015). In order to estimate the future risk of their introduction and establishment, the total area of these future climatically suitable regions can be compared to the current geographic ranges shown on maps that are frequently presented as the final output of correlative models (Evans *et al.*, 2015).

Mechanistic models are forecasting methods that make use of the environmental variable values of a specific location along with an understanding of the environmental tolerances of a specific species (Kumar *et al.*, 2014). In contrast to correlative models, mechanistic species distribution models look at how the environment affects physiologic performance in a particular area. Then, future species distributions are anticipated by an elimination process, where regions are left out of the final distribution if they significantly impair physiological performance to the point that it affects an organism's capacity to survive, grow, or reproduce (Kearney *et al.*, 2010). A semi-mechanistic modeling software programme called CLIMEX is one of the examples where it analyses physiological and behavioural data as well as values for climate variables to forecast which areas would be good habitats for a given species (Kriticos *et al.*, 2015).

## 7. Conclusion

The impact of climate change on insects is complicated since it affects the range, diversity, abundance, development, growth, and phenology of some insect species while hindering others. A wider variety of insect pest species is also anticipated to result in an increase in overall insect-pest outbreaks. Some pests will become more common and their range will widen as a result of better overwintering survival rates and the capacity to produce more generations. There will probably be more invasive insect pest species that can easily colonize new places, and there will probably be more insect-borne plant illnesses. The diminished effectiveness of biological control agents, or natural enemies, could be a significant issue in future insect-pest management programmes, which would be another adverse result of climate change. There is a substantial danger of severe financial losses to the producing group and a threat to global food security if climate change variables create an environment that is favorable for insect-pest infestations and subsequent crop destruction. Planning and developing solutions for adaptation and alleviation, such as adapted IPM techniques, monitoring of pests and the climate, and the use of predictive modelling tools, are therefore extremely important.

## References


1. Abram, P. K., Boivin, G., Moiroux, J., & Brodeur, J. (2017). Behavioural effects of temperature on ectothermic animals: unifying thermal physiology and behavioural plasticity. *Biological Reviews*, *92*(4), 1859-1876.

2. Andrew, N. R., & Hill, S. J. (2017). Effect of climate change on insect

pest management. *Environmental pest management: challenges for agronomists, ecologists, economists and policymakers*, 195-223.

3. Bale, J. S., Masters, G. J., Hodkinson, I. D., Awmack, C., Bezemer, T. M., Brown, V. K., & Whittaker, J. B. (2002). Herbivory in global climate change research: direct effects of rising temperature on insect herbivores. *Global change biology*, *8*(1), 1-16.

4. Bezemer, T. M., Jones, T. H., & Knight, K. J. (1998). Long-term effects of elevated CO 2 and temperature on populations of the peach potato aphid Myzus persicae and its parasitoid Aphidius matricariae. *Oecologia*, *116*, 128-135.

5. Boller, E. F., Avilla, J., Joerg, E., Malavolta, C., Wijnands, F. G., & Esbjerg, P. (2004). Integrated Production Principles and technical guidelines. *IOBC wprs Bulletin*, *27*(2).

6. Burgiel, S. W., & Muir, A. A. (2010). Invasive species, climate change and ecosystem-based adaptation: addressing multiple drivers of global change. Global Invasive Species Program (GISP): Washington, DC, USA, 2010; ISBN 9789290592877.

7. CBD. Convention on Biological Diversity. Invasive alien species: The application of classical biological control for the management of established invasive alien species causing environmental impacts. In Proceedings of the Conference of the Parties to the Convention on Biological Diversity, 14th Meeting, Sharm El-Sheikh, Egypt, 17–29 November 2018.

8. Cotrufo, M. F., Ineson, P., & Scott, A. (1998). Elevated $CO_2$ reduces the nitrogen concentration of plant tissues. *Global Change Biology*, *4*(1), 43-54.

9. DeLucia, E. H., Casteel, C. L., Nabity, P. D., & O'Neill, B. F. (2008). Insects take a bigger bite out of plants in a warmer, higher carbon dioxide world. *Proceedings of the National Academy of Sciences*, *105*(6), 1781-1782.

10. Deutsch, C. A., Tewksbury, J. J., Tigchelaar, M., Battisti, D. S., Merrill, S. C., Huey, R. B., & Naylor, R. L. (2018). Increase in crop losses to insect pests in a warming climate. *Science*, *361*(6405), 916-919.

11. Dukes, J. S., Pontius, J., Orwig, D., Garnas, J. R., Rodgers, V. L., Brazee, N. & Ayres, M. (2009). Responses of insect pests, pathogens, and invasive plant species to climate change in the forests of

northeastern North America: what can we predict? *Canadian journal of forest research*, *39*(2), 231-248.

12. Dyer, L. A., Richards, L. A., Short, S. A., & Dodson, C. D. (2013). Effects of $CO_2$ and temperature on tritrophic interactions. *PloS one*, *8*(4), e62528.

13. Evans, E. W., Carlile, N. R., Innes, M. B., & Pitigala, N. (2013). Warm springs reduce parasitism of the cereal leaf beetle through phenological mismatch. *Journal of Applied Entomology*, *137*(5), 383-391.

14. Evans, T. G., Diamond, S. E., & Kelly, M. W. (2015). Mechanistic species distribution modelling as a link between physiology and conservation. *Conservation physiology*, *3*(1).

15. FAO. Food and Agriculture Organization Plant Pests and Diseases in the Context of Climate Change and Climate Variability, Food Security and Biodiversity Risks. Available online: http://www.fao.org/3/nb088/nb088.pdf (accessed on 12 January 2020).

16. Gomez-Zavaglia, A., Mejuto, J. C., & Simal-Gandara, J. (2020). Mitigation of emerging implications of climate change on food production systems. *Food Research International*, *134*, 109256.

17. Hamilton, J. G., Dermody, O., Aldea, M., Zangerl, A. R., Rogers, A., Berenbaum, M. R., & Delucia, E. H. (2005). Anthropogenic changes in tropospheric composition increase susceptibility of soybean to insect herbivory. *Environmental entomology*, *34*(2), 479-485.

18. Hance, T., van Baaren, J., Vernon, P., & Boivin, G. (2007). Impact of extreme temperatures on parasitoids in a climate change perspective. *Annu. Rev. Entomol.*, *52*, 107-126.

19. Heeb, L., Jenner, E., & Cock, M. J. (2019). Climate-smart pest management: building resilience of farms and landscapes to changing pest threats. *Journal of Pest Science*, *92*(3), 951-969.

20. Hellmann, J. J., Byers, J. E., Bierwagen, B. G., & Dukes, J. S. (2008). Five potential consequences of climate change for invasive species. *Conservation biology*, *22*(3), 534-543.

21. Heuskin, S., Verheggen, F. J., Haubruge, E., Wathelet, J. P., & Lognay, G. (2011). The use of semiochemical slow-release devices in integrated pest management strategies. Biotechnol. Agron. Soc. Environ. 2011, 15, 459–470.

22. Jamieson, M. A., Trowbridge, A. M., Raffa, K. F., & Lindroth, R. L.

(2012). Consequences of climate warming and altered precipitation patterns for plant-insect and multitrophic interactions. *Plant physiology*, *160*(4), 1719-1727.

23. Karthik, S., Reddy, M. S., & Yashaswini, G. (2021). Climate change and its potential impacts on insect-plant interactions. *The Nature, Causes, Effects and Mitigation of Climate Change on the Environment*, *10*.

24. Kearney, M. R., Wintle, B. A., & Porter, W. P. (2010). Correlative and mechanistic models of species distribution provide congruent forecasts under climate change. *Conservation letters*, *3*(3), 203-213.

25. Kocmánková, E.; Trnka, M.; Juroch, J.; Dubrovský, M.; Semerádová, D.; Možný, M.; Žalud, Z.; Pokorný, R.; Lebeda, A. Impact of climate change on the occurrence and activity of harmful organisms. Plant Prot. Sci. 2010, 45, S48–S52.

26. Kriticos, D. J., Maywald, G. F., Yonow, T., Zurcher, E. J., Herrmann, N. I., & Sutherst, R. (2015). Exploring the effects of climate on plants, animals and diseases. *CLIMEX Version*, *4*, 184.

27. Kumar, S., Neven, L. G., & Yee, W. L. (2014). Evaluating correlative and mechanistic niche models for assessing the risk of pest establishment. *Ecosphere*, *5*(7), 1-23.

28. Lincoln, D. E. (1993). The influence of plant carbon dioxide and nutrient supply on susceptibility to insect herbivores. *Vegetatio*, *104*, 273-280.

29. Mafokoane, L. D., Zimmermann, H. G., & Hill, M. P. (2007). Development of Cactoblastis cactorum (Berg)(Lepidoptera: pyralidae) on six north American Opuntia species. *African Entomology*, *15*(2), 295-299.

30. Nihal, R. (2020). Global Climate change and its impact on integrated pest management. *Agro Econ. Int. J*, *7*, 133-137.

31. Pacifici, M., Foden, W. B., Visconti, P., Watson, J. E., Butchart, S. H., Kovacs, K. M., & Rondinini, C. (2015). Assessing species vulnerability to climate change. *Nature climate change*, *5*(3), 215-224.

32. Pathak, H., Aggarwal, P. K., & Singh, S. D. (2012). Climate change impact, adaptation and mitigation in agriculture: methodology for assessment and applications. *Indian Agricultural Research Institute, New Delhi*, *302*.

33. Porfirio, L. L., Harris, R. M., Lefroy, E. C., Hugh, S., Gould, S. F., Lee, G.,... & Mackey, B. (2014). Improving the use of species distribution models in conservation planning and management under climate change. *PLoS One*, *9*(11), e113749.

34. Raza, M. M., Khan, M. A., Arshad, M., Sagheer, M., Sattar, Z., Shafi, J.,... & Sattar, A. (2015). Impact of global warming on insects. *Archives of Phytopathology and Plant Protection*, *48*(1), 84-94.

35. Roth, S.K.& Lindroth, R.L. Elevated atmospheric $CO_2$: Effects on phytochemistry, insect performance and insect-parasitoid interactions. Glob. Chang. Biol. 1995, 1, 173–182

36. Schneider, L., Rebetez, M., & Rasmann, S. (2022). The effect of climate change on invasive crop pests across biomes. *Current Opinion in Insect Science*, 100895.

37. Sgrò, C. M., Terblanche, J. S., & Hoffmann, A. A. (2016). What can plasticity contribute to insect responses to climate change?. *Annual review of entomology*, *61*, 433-451.

38. Shine, C., Williams, N., & Gündling, L. (2000). *A guide to designing legal and institutional frameworks on alien invasive species* (No. 40). IUCN.

39. Shrestha, S. Effects of climate change in agricultural insect pest. Acta Sci. Agric. 2019, 3, 74–80.

40. Skendži´c, S.; Zovko, M.; Živkovi´c, I.P.; Leši´c, V.; Lemi´c, D. The Impact of Climate Change on Agricultural Insect Pests. Insects 2021, 12, 440. https://doi.org/10.3390/insects12050440

41. Sutherst, R. W., Constable, F., Finlay, K. J., Harrington, R., Luck, J., & Zalucki, M. P. (2011). Adapting to crop pest and pathogen risks under a changing climate. *Wiley Interdisciplinary Reviews: Climate Change*, *2*(2), 220-237.

42. Thomas, C. D., Cameron, A., Green, R. E., Bakkenes, M., Beaumont, L. J., Collingham, Y. C., & Williams, S. E. (2004). Extinction risk from climate change. *Nature*, *427*(6970), 145-148.

43. Thomson, L. J., Macfadyen, S., & Hoffmann, A. A. (2010). Predicting the effects of climate change on natural enemies of agricultural pests. *Biological control*, *52*(3), 296-306.

44. Vadez, V., Berger, J. D., Warkentin, T., Asseng, S., Ratnakumar, P., Rao, K. P. C., & Zaman, M. A. (2012). Adaptation of grain legumes to

climate change: a review. *Agronomy for Sustainable Development, 32*, 31-44.

45. Vermeij, G. J. (1996). An agenda for invasion biology. *Biological conservation*, *78*(1-2), 3-9.

46. Ward, N. L., & Masters, G. J. (2007). Linking climate change and species invasion: an illustration using insect herbivores. *Global Change Biology*, *13*(8), 1605-1615.

47. Wenda-Piesik, A., Piesik, D., Nowak, A., & Wawrzyniak, M. (2016). Tribolium confusum responses to blends of cereal kernels and plant volatiles. *Journal of Applied Entomology*, *140*(7), 558-563.

48. Yamamura, K., Yokozawa, M., Nishimori, M., Ueda, Y., & Yokosuka, T. (2006). How to analyze long-term insect population dynamics under climate change: 50-year data of three insect pests in paddy fields. *Population Ecology*, *48*, 31-48.

49. Yihdego, Y., Salem, H. S., & Muhammed, H. H. (2019). Agricultural pest management policies during drought: Case studies in Australia and the state of Palestine. *Natural Hazards Review*, *20*(1), 05018010.

# Chapter - 5

## Emerging Trends in Crop Genetics and Breeding

**Authors**

**Pattan Farooqkhan**

Ph.D. Scholar, Department of Genetics and Plant Breeding, Annamalai University, Chidambaram, Tamil Nadu, India

**Susmitha J**

Ph.D. Scholar, Department of Genetics and Plant Breeding, Annamalai University, Chidambaram, Tamil Nadu, India

# Chapter - 5

## Emerging Trends in Crop Genetics and Breeding

**Pattan Farooqkhan and Susmitha J**

**Abstract**

There are so many advancements in crop breeding which have allowed for creating highly resilient, adaptable and better performing crops. Some of the emerging trends in crop genetics that have revolutionized the crop improvement include gene editing, genomic selection and multi omics approach including genomics, proteomics, transcriptomics, and metabolomics. Omics approach is used in breeding for identifying favourable genetic variations, understanding gene expression patterns, and unravelling metabolic pathways and how omics data can be integrated with traditional breeding methods to accelerate the development of improved crop varieties. Gene editing techniques like CRISPR-Cas9, TALEN and ZFN, have gained tremendous attention in recent years due to their precision and efficiency in making targeted genetic modifications. Applications of these gene editing in crop breeding, such as creating novel genetic variations, enhancing crop resistance to diseases and pests, and improving nutritional content. There are challenges and ethical issues faced for using gene editing techniques in plants. Genomic selection is a predictive breeding approach that relies on the analysis of large-scale genomic data to estimate the genetic value of individuals in a breeding population. Genomic selection helps in precisely selecting the traits reducing the time and effort in crop improvement programmes. There are several advantages of genomic selection over traditional phenotypic selection, such as early prediction of breeding values and the ability to select for complex traits that are difficult to measure directly. Combined use of these gene editing techniques, genomic selection approach and omics technology helps in precisely incorporating genetic modifications and accelerating the crop improvement.

**Keywords:** Gene editing, genomic selection, proteomics, transcriptomics, metabolomics.

## 1. Introduction

Crop genetics and plant breeding play a vital role in addressing global food challenges that we face today. With a growing global population, diminishing arable land, and the increasing impact of climate change, it has become crucial to develop crops that are more resilient, productive, and sustainable. Recent advancements in crop genetics like gene editing where a gene can be inserted or deleted or modified, genomic selection which helps in selecting traits precisely using marker technology, Multi omics approach including genomics, proteomics, transcriptomics and metabolomics have accelerated the crop improvement programmes. The advent of omics technologies has revolutionized our understanding of the genetic makeup and functional dynamics of crops. Genomics, transcriptomics, proteomics, and metabolomics offer powerful tools for investigating the complex genetic networks underlying agronomically important traits (Maroli *et al.,* 2018). These high-throughput techniques allow the simultaneous analysis of thousands of genes, proteins, or metabolites, providing valuable insights into crop performance, adaptation, and response to stress (Singh *et al.,* 2021). Omics technologies also facilitate the identification of key genetic variations associated with desirable traits, leading to the development of improved crop varieties (Langridge and Fluery, 2011). Integrating omics data with traditional breeding methods enables breeders to make informed decisions and accelerate the breeding process (Rai 2022). In recent years, gene editing techniques have emerged as powerful tools for precise and targeted genetic modifications in crops. Among these techniques, CRISPR-Cas9 has gained significant attention due to its simplicity, efficiency, and versatility (Li *et al.,* 2017). CRISPR-Cas9 enables researchers to edit specific genes or regulatory elements, resulting in desired traits such as improved yield, enhanced resistance to diseases and pests, and enhanced nutritional content (Zhang *et al.,* 2021). Gene editing offers unprecedented opportunities to accelerate crop breeding by rapidly introducing beneficial genetic variations and overcoming the limitations of traditional breeding methods (Das *et al.,* 2022). However, the deployment of gene-edited crops also raises important regulatory and ethical considerations that need to be addressed for successful adoption. Genomic selection is another emerging trend in crop breeding that harnesses the power of large-scale genomic data to predict the breeding values of individuals within a population (He *et al.,* 2020). By analyzing genetic markers across the entire genome, genomic selection enables breeders to select individuals with the highest genetic potential for desired traits (Crain *et al.,* 2021). This approach has shown promising results in various crops, including maize, wheat, and rice, and has the potential to significantly accelerate the rate of genetic gain (Yadav *et al.,* 2020). Genomic selection offers advantages over traditional phenotypic selection, such as early prediction of breeding values, increased selection accuracy, and the ability to select for complex traits that are challenging to measure directly (Hickey *et al.,* 2017). These emerging trends hold immense promise for addressing global food challenges. By leveraging gene editing, genomic selection, and the multi-omics approach, we can accelerate the development of crops with improved yield, nutritional quality, and resilience to environmental stresses.

## 2. Gene editing techniques

Gene editing techniques have revolutionized the field of crop genetics and breeding by enabling precise modifications in the DNA of organisms. Three commonly used gene editing tools are CRISPR-Cas9, TALENs (Transcription Activator-Like Effector Nucleases), and ZFNs (Zinc Finger Nucleases). CRISPR-Cas9, in particular, has gained significant attention due to its simplicity, efficiency, and versatility in editing genes. It utilizes a guide RNA (gRNA) molecule that binds to a specific target sequence in the genome, guiding the Cas9 enzyme to create a double-strand break at the desired location. This break triggers the cell's natural DNA repair mechanisms, allowing for the introduction of precise genetic modifications. CRISPR-Cas9 has been successfully employed in crop species for various applications, including enhancing disease resistance, improving yield, and modifying plant traits (Haque *et al.,* 2018). TALENs, on the other hand, utilize engineered transcription activator-like effectors (TALEs) fused with a nuclease domain to induce targeted DNA cleavage. These TALENs recognize specific DNA sequences and create double-strand breaks, allowing for the introduction of desired genetic changes. TALENs have been used in crop species such as rice, wheat, and maize to achieve targeted gene modifications (Jaganathan *et al.,* 2018). ZFNs are another gene editing tool that uses engineered zinc finger proteins to recognize specific DNA sequences. The zinc finger proteins are linked to a nuclease domain, which induces double-strand breaks at the targeted genomic loci. This break triggers the cell's repair mechanisms, enabling precise modifications. ZFNs have been successfully applied in crop species like maize, soybean, and tobacco for targeted gene editing (Weeks *et al.,* 2016). These gene editing techniques have opened up new avenues for crop improvement by providing breeders with the ability to precisely modify specific genes and traits. They offer advantages over traditional breeding methods by allowing for targeted modifications without the need for lengthy backcrossing processes. However, it is important to consider ethical and regulatory aspects when deploying gene-edited crops. CRISPR-Cas9, TALENs, and ZFNs are powerful gene editing tools that have significantly advanced crop genetics and breeding. They provide precise and targeted modifications in the genome, offering tremendous potential for developing improved crop varieties.

**Applications of gene editing techniques**

- **TALEN: Disease resistance in tomato**

TALEN (Transcription Activator-Like Effector Nucleases) has been used to confer resistance to bacterial spot disease in tomato plants. By targeting specific genes associated with susceptibility to the bacterial pathogen, researchers successfully edited the genome of tomato plants, resulting in increased resistance to bacterial spot disease. This approach offers a promising solution for reducing crop losses caused by bacterial infections (Karmakar *et al.,* 2022).

- **ZFN: High oleic acid content in soybean**

Zinc Finger Nucleases (ZFNs) have been employed to develop soybean varieties with high oleic acid content. By targeting and modifying specific genes involved in fatty acid metabolism, researchers have successfully produced soybean plants with an increased proportion of oleic acid, which has desirable nutritional properties and improved stability for cooking and processing applications (Bonawitz *et al.,* 2019).

- **CRISPR: Herbicide resistance in corn**

CRISPR-Cas9 has been used to introduce herbicide resistance in corn crops. By targeting specific genes involved in herbicide sensitivity, researchers have modified corn plants to be resistant to specific herbicides. This trait allows farmers to effectively control weeds in cornfields, enhancing crop yield and reducing the reliance on broad-spectrum herbicides (Dong *et al.,* 2021).

- **CRISPR: Enhanced photosynthesis in rice**

Researchers have utilized CRISPR-Cas9 to enhance photosynthesis efficiency in rice plants. By targeting specific genes involved in the regulation of photosynthesis, they successfully edited the rice genome to improve the efficiency of carbon dioxide fixation and energy conversion during photosynthesis. This approach has the potential to increase rice yields and improve crop productivity (Caddell *et al.,* 2023).

These examples illustrate the diverse applications of gene editing technologies, including TALEN, ZFN, and CRISPR, in crop plants. These tools provide precise and efficient methods for targeted gene modifications, enabling the development of crops with desirable traits such as disease resistance, improved nutritional content, herbicide resistance, and enhanced photosynthesis. These advancements hold promise for sustainable agriculture and addressing global food challenges.

## 3. Genomic selection - Advancement in precision breeding

Genomic selection (GS) is a powerful approach in plant breeding that utilizes genomic information to predict the breeding values of individuals based on their DNA markers. It has revolutionized traditional breeding methods by enabling the selection of superior plants at an early stage, accelerating the breeding process, and enhancing genetic gain. Genomic selection leverages the vast amount of genetic information available in high-density marker datasets to estimate the breeding value of individuals for target traits. The process of genomic selection involves several key steps. First, a training population is established, consisting of individuals with phenotypic data for the target traits and genotypic data obtained through high-throughput genotyping platforms such as SNP arrays or sequencing. Next, a statistical model, such as the best linear unbiased prediction (BLUP) or Bayesian methods, is employed to estimate the relationship between the genetic markers and phenotypic values. The model is trained using the genotypic and phenotypic data from the training population. Once the model is established, it can be used to predict the breeding values of individuals in a validation or breeding population based solely on their marker data.

One notable example of the successful application of genomic selection in plant breeding is the development of high-yielding maize varieties. In a study by Zhao *et al.,* (2012), genomic selection was used to predict grain yield in maize using a training population with genotypic and phenotypic data. The researchers demonstrated that genomic selection significantly improved the accuracy of yield prediction compared to traditional selection methods. By incorporating genomic selection into their breeding program, breeders were able to efficiently select and advance individuals with high yield potential, resulting in the development of improved maize varieties. Another example is the implementation of genomic selection in rice breeding. In a study by Xu *et al.,* (2018), genomic selection was used to predict grain yield and other important agronomic traits in rice using a large training population. The researchers found that genomic selection greatly improved the accuracy of yield prediction compared to pedigree-based methods. By utilizing genomic selection, breeders can make more informed decisions regarding the selection of superior rice lines, leading to the development of varieties with improved yield and other desirable traits.

The integration of genomic selection into plant breeding programs has proven to be a valuable approach, enhancing the efficiency and precision of selection processes. It allows breeders to make early and accurate predictions of breeding values, enabling the selection of superior individuals at an early stage, and reducing the time and cost required for phenotypic evaluations. By leveraging genomic information, breeders can accelerate the breeding process, increase genetic gain, and develop improved crop varieties more efficiently.

**Applications of genomic selection in plant breeding**

1. **Early generation selection:** Genomic selection allows breeders to predict the performance of individuals at early stages, such as in the nursery or F1 generation, without relying solely on labour-intensive and time-consuming phenotypic evaluations. For instance, in maize breeding, genomic selection has been successfully used to predict grain yield in early generations, enabling the selection of superior lines before conducting field trials (Beyene *et al.,* 2019).
2. **Complex trait selection:** Genomic selection is particularly useful for selecting complex traits that are difficult or costly to measure directly. These traits often involve multiple genes and interactions with the environment. By leveraging large-scale genomic data, genomic selection can capture the genetic architecture underlying complex traits. For example, in wheat breeding, genomic selection has been employed to improve grain yield, quality, and disease resistance, which are complex traits influenced by numerous genes (Rutkoski *et al.,* 2016).
3. **Recurrent selection and hybrid breeding:** Genomic selection can significantly enhance the efficiency of recurrent selection and hybrid breeding programs. By accurately predicting the genetic value of individuals, breeders can strategically select parents to create more productive hybrids or recurrently select individuals for further breeding cycles. In maize hybrid breeding, genomic

selection has been applied to optimize the selection of parental lines, leading to improved hybrid performance (Technow *et al.*, 2014).

4. **Genomic-enabled pre-breeding:** Genomic selection can be employed in pre-breeding programs, which focus on the introgression of desirable traits from wild or unadapted germplasm into elite breeding lines. By using genomic information, breeders can efficiently select individuals with the desired trait introgression, reducing the time and resources required for phenotypic evaluations. Genomic selection has been successfully utilized in pre-breeding efforts for traits such as disease resistance in barley (Wang *et al.*, 2021).

5. **Genotype-by-environment interaction (GxE) modeling:** Genomic selection can account for genotype-by-environment interactions, enabling breeders to select for genotypes that exhibit stable performance across diverse environments. By incorporating environmental information into the genomic prediction models, breeders can identify genotypes that are adaptable and perform consistently across different conditions. This has been demonstrated in rice breeding, where genomic selection has been employed to improve yield stability and adaptation (Spindel *et al.*, 2015).

## 4. Omics approach

The multi-omics approach, which integrates various "omics" technologies such as genomics, transcriptomics, proteomics, and metabolomics, is revolutionizing our understanding of plant biology (Grobkinsky *et al.*, 2018). By analyzing multiple layers of biological information, researchers can unravel the intricate mechanisms underlying important traits and identify key genes and molecular pathways involved. The multi-omics approach provides valuable insights into the genotype-phenotype relationship and offers a holistic view of crop plants, facilitating the discovery of novel breeding targets and strategies.

### 1) Proteomics approach

Proteomics, the study of proteins and their functions. It allows researchers to understand the complex network of proteins involved in plant growth, development, and responses to various stresses. By identifying and characterizing these proteins, breeders can gain insights into important traits and design strategies to improve crop performance. Here are some key aspects of the proteomics approach in plant breeding, along with notable case studies:

**Protein PROFILING:** Proteomics enables the comprehensive profiling of plant proteins under different conditions, such as stress or developmental stages (Ghosh and Xu, 2014). This information helps breeders understand the dynamics of protein expression and identify key proteins associated with desired traits. For example, in a study by Rahman and Upadhyaya (2021), proteomic analysis of maize roots revealed differentially expressed proteins related to aluminium tolerance, providing valuable insights for breeding aluminium-resistant varieties.

**Marker-assisted selection:** Proteomics can be used to identify protein markers associated with specific traits of interest. These markers can serve as indicators for selecting superior plants in breeding programs. In a study by Du *et al.,* (2016), proteomic analysis of wheat grains identified unique protein markers linked to high grain weight, which could be used for marker-assisted selection to improve grain yield.

**Stress response mechanisms:** Proteomic analysis helps unravel the molecular mechanisms underlying plant responses to various stresses, such as drought, salinity, or pathogens. This knowledge can guide breeders in developing stress-tolerant crops. For instance, in a study by Vessal *et al.,* (2020), proteomic analysis of chickpea leaves exposed to drought stress revealed several drought-responsive proteins, providing valuable insights for breeding drought-tolerant varieties.

**Metabolic pathway analysis:** Proteomics combined with metabolomics provides a comprehensive view of metabolic pathways in plants. This integration enables breeders to identify key enzymes and regulatory proteins involved in important metabolic pathways, such as photosynthesis or secondary metabolite production. For instance, in a study by Xu *et al.,* (2021), proteomic and metabolomic analyses of soybean seeds identified proteins and metabolites related to seed oil accumulation, offering potential targets for breeding high-oil varieties.

### 2) Metabolomics

Metabolomics, the comprehensive study of small molecules or metabolites in a biological system. It allows researchers to analyse and quantify the metabolites present in plants, providing valuable insights into metabolic pathways and their association with desirable traits. By utilizing metabolomics, breeders can identify key metabolites and pathways associated with important agronomic traits, enabling targeted breeding strategies.

**Improving nutritional quality:** Metabolomics has been used to enhance the nutritional quality of crops. For instance, in a study by Wang *et* *al.,* (2022), metabolomic analysis of tomato fruits identified and quantified various metabolites, including antioxidants, flavour compounds, and vitamins. This information enabled breeders to select tomato varieties with improved nutritional profiles and desirable taste attributes.

Metabolomics has played a crucial role in the breeding of high-quality wheat. In a study by Ma *et al.,* (2022), metabolomic analysis of wheat grains identified and characterized the metabolic pathways associated with important grain quality traits, such as protein content, amino acid composition, and starch properties.

### 3) Transcriptomics

Transcriptomics, the study of all RNA molecules in a cell. It enables researchers to investigate gene expression patterns and regulatory networks in plants, providing valuable insights into the molecular mechanisms underlying important traits. By utilizing transcriptomics, breeders can identify key genes and pathways associated with desired traits, facilitating the development of improved crop varieties.

**Enhancing disease resistance:** Transcriptomics has been instrumental in developing disease-resistant crops. For example, in a recent study by Zhang *et al.,* (2019), transcriptomic analysis of banana plants infected with Fusarium wilt revealed key genes involved in disease resistance.

**Improving abiotic stress tolerance:** Transcriptomics has also been used to enhance abiotic stress tolerance in crops. A study by Le *et al.,* (2012) focused on transcriptomic analysis of soybean plants under drought stress. The researchers identified differentially expressed genes associated with drought tolerance and characterized the regulatory networks involved in stress responses.

**Nutritional enhancement:** Transcriptomics has played a significant role in improving the nutritional quality of crops. In a study by Xiao *et al.,* (2022), transcriptomic analysis of maize kernels identified genes related to the synthesis of essential amino acids, vitamins, and minerals. This helps in developing maize varieties with enhanced nutritional profiles, contributing to improved human health and nutrition.

### 4) Epigenomics approach

Epigenomics, the study of epigenetic modifications and their impact on gene expression and cellular function, has emerged as a valuable tool in plant breeding. Epigenetic modifications, such as DNA methylation and histone modifications, can influence gene activity without altering the underlying DNA sequence.

**Epigenetic regulation of stress responses:** Epigenomic studies have revealed the role of epigenetic modifications in plant responses to environmental stresses. For example, in a study by Zhao *et al.,* (2012), epigenomic analysis of rice plants exposed to drought stress identified differentially methylated regions associated with stress-responsive genes.

**Epigenetic variation and phenotypic diversity:** Epigenomic studies have highlighted the role of epigenetic variation in generating phenotypic diversity within plant populations. For instance, in a study by Bossdorf *et al.,* (2010), epigenomic analysis of Arabidopsis thaliana accessions revealed natural variation in DNA methylation patterns associated with diverse phenotypic traits.

## Multi omics

Multi-omics approaches involve the integration of multiple omics technologies, such as genomics, transcriptomics, proteomics, metabolomics, and epigenomics, to gain a comprehensive understanding of biological systems. These approaches provide a more holistic view of complex biological processes and interactions, to uncover valuable insights that would be difficult to obtain using a single omics analysis. Here are some common types of multi-omics approaches:

**Genomics-transcriptomics-proteomics (GTP):** This approach combines genomic, transcriptomic, and proteomic data to understand the relationship between genes, their expression patterns, and the resulting protein products. It provides insights into the regulation of gene expression, post-transcriptional modifications, and protein abundance. By integrating these layers of information, researchers can identify key genes and proteins associated with specific phenotypes or biological processes.

**Transcriptomics-proteomics-metabolomics (TPM):** This multi-omics approach focuses on the interactions between gene expression patterns, protein abundance, and metabolite profiles. It helps unravel the connections between gene regulation, protein function, and metabolic pathways. TPM analysis is particularly useful in understanding how changes in gene expression or protein abundance affect metabolic processes and contribute to phenotypic variations.

**Epigenomics-transcriptomics-proteomics (ETP):** ETP integrates epigenomic, transcriptomic, and proteomic data to elucidate the role of epigenetic modifications in gene expression and protein regulation. It enables the study of epigenetic marks, such as DNA methylation or histone modifications, and their impact on transcriptional and translational processes. ETP analysis provides insights into the epigenetic mechanisms underlying developmental processes, disease states, and environmental responses.

**Integrative multi-omics (IMO):** The integrative multi-omics approach combines multiple omics datasets, such as genomics, transcriptomics, proteomics, and metabolomics, into a single comprehensive analysis. It aims to integrate diverse layers of biological information to uncover complex relationships and regulatory networks. By merging data from different omics levels, researchers can identify key molecular players, pathways, and interactions that drive biological processes or influence phenotypic traits.

These multi-omics approaches offer a more comprehensive and systems-level understanding of biological systems by integrating data from multiple omics technologies.

## 5. Combined use of gene editing and genomic selection in plant breeding

The combined use of gene editing and genomic selection holds tremendous potential for accelerating breeding programs by precisely introducing desired genetic modifications into target organisms. Gene editing technologies, such as CRISPR-Cas9, enable targeted modifications in the genome, while genomic selection allows for the efficient identification and selection of individuals with desirable traits. By integrating these approaches, breeders can expedite the development of improved crop varieties and livestock breeds.

The combination of gene editing and genomic selection has been employed to enhance traits such as disease resistance, nutritional content, and stress tolerance. Researchers have used gene editing techniques to introduce specific genetic modifications associated with desired traits, such as knocking out disease susceptibility genes or enhancing the expression of genes involved in stress tolerance (Zaidi *et al.*, 2018). Genomic selection, on the other hand, assists in identifying individuals carrying these edited genes and predicts their potential performance for the target traits. This combined approach allows for the rapid development of new plant varieties with improved agronomic characteristics.

### Case studies regarding combined use of omics, genomic selection, and gene editing techniques in plant breeding

The combined approach of omics analysis, genomic selection, and gene editing has been employed in plant breeding to accelerate the development of improved crop varieties with desired traits. Several case studies highlight the successful integration of these techniques in plant breeding programs.

One notable example is the development of disease-resistant wheat varieties using a combined approach. In a study conducted by Dong *et al.,* (2018), genomic selection was used to identify wheat lines with high levels of resistance to the devastating fungal disease Fusarium head blight (FHB). Genomic selection models were developed using genotyping data from a training population and phenotypic data for FHB resistance. Subsequently, gene editing (CRISPR-Cas9) was employed to target and knockout a susceptibility gene associated with FHB. The edited lines identified through genomic selection exhibited significantly improved resistance to FHB, demonstrating the successful integration of genomic selection and gene editing for enhancing disease resistance in wheat.

Another case study involves the development of drought-tolerant maize varieties. In a study by Pan and Barrangou (2020), a combination of omics analysis, genomic selection, and gene editing was utilized. Initially, omics analysis, including transcriptomics and metabolomics, was performed to identify key genes and metabolic pathways associated with drought tolerance in maize. Genomic selection models were then developed using genotypic and phenotypic data from a diverse population. Finally, gene editing techniques were employed to introduce desired genetic modifications targeting the identified genes associated with drought tolerance. The integration of omics analysis, genomic selection, and gene editing resulted in the development of maize lines with improved drought tolerance traits.

## 6. Ethical considerations

Ethical considerations and regulatory frameworks play a crucial role in ensuring the responsible development and deployment of gene-edited crops. Gene editing technologies, such as CRISPR-Cas9, have the potential to revolutionize crop improvement by enhancing yield, nutritional value, and resilience. However, ethical concerns arise regarding potential unintended effects on ecosystems, biodiversity, and human health. It is essential to address these concerns through rigorous risk assessments, transparency, and stakeholder engagement. Regulatory frameworks need to be established to govern the research, development, and commercialization of gene-edited crops. These frameworks should encompass robust biosafety regulations, labelling requirements, and monitoring systems to ensure the safety and traceability of gene-edited crops. Additionally, ethical considerations, such as equitable access to improved crops and intellectual property rights, should be considered to promote fairness and social benefit. By addressing ethical considerations and implementing effective regulatory frameworks, we can foster responsible innovation in gene-edited crops while safeguarding public health, environmental sustainability, and societal interests.

## 7. Conclusion

The emerging trends in crop genetics and breeding such as gene editing, genomic selection, and omics techniques have revolutionized plant breeding through providing powerful tools for targeted genetic alterations, improved selection precision, and a comprehensive understanding of plant genomes. CRISPR-Cas9, TALENs, and ZFN gene editing techniques provide accurate and efficient means for introducing specific genetic alterations, expediting the development of new agricultural types. Genomic selection predicts breeding values using genomic information, allowing for early generation selection, complex trait selection, and optimization of recurrent selection and hybrid breeding programmes. Furthermore, omics methods such as genomics, transcriptomics, proteomics and epigenomics offer useful insights into the genetic basis of complex phenotypes and allow for the identification of candidate genes for breeding purposes helping in crop improvement.

## References


1. Beyene, Y., Semagn, K., Mugo, S., Tarekegne, A., Babu, R., Meisel, B.,... & Olsen, M. (2019). Genetic gains in grain yield through genomic selection in eight bi-parental maize populations under drought stress. Crop Science, 59(3), 1-13.

2. Bonawitz, N. D., Ainley, W. M., Itaya, A., Chennareddy, S. R., Cicak, T., Effinger, K.,... & Pareddy, D. R. (2019). Zinc finger nuclease-mediated targeting of multiple transgenes to an endogenous soybean genomic locus via non-homologous end joining. Plant biotechnology journal, 17(4), 750-761.

3. Bossdorf, O., Arcuri, D., Richards, C. L., & Pigliucci, M. (2010). Experimental alteration of DNA methylation affects the phenotypic plasticity of ecologically relevant traits in Arabidopsis thaliana. Evolutionary Ecology, 24, 541-553.

4. Caddell, D., Langenfeld, N. J., Eckels, M. J., Zhen, S., Klaras, R., Mishra, L.,... & Coleman-Derr, D. (2023). Photosynthesis in rice is increased by CRISPR/Cas9-mediated transformation of two truncated light-harvesting antenna. Frontiers in Plant Science, 14, 84.

5. Crain, J., DeHaan, L., & Poland, J. (2021). Genomic prediction enables

rapid selection of high-performing genets in an intermediate wheatgrass breeding program. The Plant Genome, 14(2), e20080.

6. Das, D., Singha, D. L., Paswan, R. R., Chowdhury, N., Sharma, M., Reddy, P. S., & Chikkaputtaiah, C. (2022). Recent advancements in CRISPR/Cas technology for accelerated crop improvement. Planta, 255(5), 109.

7. Dong, H., Huang, Y., & Wang, K. (2021). The development of herbicide resistance crop plants using CRISPR/Cas9-mediated gene editing. Genes, 12(6), 912.

8. Du, D., Gao, X., Geng, J., Li, Q., Li, L., Lv, Q., & Li, X. (2016). Identification of key proteins and networks related to grain development in wheat (Triticum aestivum L.) by comparative transcription and proteomic analysis of allelic variants in TaGW2-6A. Frontiers in Plant Science, 7, 922.

9. Ghosh, D., & Xu, J. (2014). Abiotic stress responses in plant roots: a proteomics perspective. Frontiers in plant science, 5, 6.

10. Großkinsky, D. K., Syaifullah, S. J., & Roitsch, T. (2018). Integration of multi-omics techniques and physiological phenotyping within a holistic phenomics approach to study senescence in model and crop plants. Journal of Experimental Botany, 69(4), 825-844.

11. Haque, E., Taniguchi, H., Hassan, M. M., Bhowmik, P., Karim, M. R., Śmiech, M.,... & Islam, T. (2018). Application of CRISPR/Cas9 genome editing technology for the improvement of crops cultivated in tropical climates: recent progress, prospects, and challenges. Frontiers in plant science, 9, 617.

12. He, T., & Li, C. (2020). Harness the power of genomic selection and the potential of germplasm in crop breeding for global food security in the era with rapid climate change. The Crop Journal, 8(5), 688-700.

13. Hickey, J. M., Chiurugwi, T., Mackay, I., & Powell, W. (2017). Genomic prediction unifies animal and plant breeding programs to form platforms for biological discovery. Nature genetics, 49(9), 1297-1303.

14. Jaganathan, D., Ramasamy, K., Sellamuthu, G., Jayabalan, S., & Venkataraman, G. (2018). CRISPR for crop improvement: an update review. Frontiers in plant science, 9, 985.

15. Karmakar, A., Taufiqa, S., Baig, M. J., & Molla, K. A. (2022). Increasing disease resistance in host plants through genome editing. Proceedings of the Indian National Science Academy, 88(3), 417-429.

16. Langridge, P., & Fleury, D. (2011). Making the most of omics for crop breeding. Trends in biotechnology, 29(1), 33-40.

17. Le, D. T., Nishiyama, R., Watanabe, Y., Tanaka, M., Seki, M., Ham, L. H.,... & Tran, L. S. P. (2012). Differential gene expression in soybean leaf tissues at late developmental stages under drought stress revealed by genome-wide transcriptome analysis. PloS one, 7(11), e49522.

18. Li, L., Song, L., Liu, X., Yang, X., Li, X., He, T.,... & Wei, Y. (2017). Artificial virus delivers CRISPR-Cas9 system for genome editing of cells in mice. ACS nano, 11(1), 95-111.

19. Ma, D., Xu, B., Feng, J., Hu, H., Tang, J., Yin, G.,... & Wang, C. (2022). Dynamic metabolomics and transcriptomics analyses for characterization of phenolic compounds and their biosynthetic characteristics in wheat grain. Frontiers in Nutrition, 9.

20. Maroli, A. S., Gaines, T. A., Foley, M. E., Duke, S. O., Doğramacı, M., Anderson, J. V.,... & Tharayil, N. (2018). Omics in weed science: a perspective from genomics, transcriptomics, and metabolomics approaches. Weed Science, 66(6), 681-695.

21. Pan, M., & Barrangou, R. (2020). Combining omics technologies with CRISPR-based genome editing to study food microbes. Current opinion in biotechnology, 61, 198-208.

22. Rahman, R., & Upadhyaya, H. (2021). Aluminium toxicity and its tolerance in plant: A review. Journal of Plant Biology, 64, 101-121.

23. Rai, K. K. (2022). Integrating speed breeding with artificial intelligence for developing climate-smart crops. Molecular Biology Reports, 49(12), 11385-11402.

24. Rutkoski, J., Poland, J., Mondal, S., Autrique, E., Pérez, L. G., Crossa, J.,... & Singh, R. P. (2016). Canopy temperature and vegetation indices from high-throughput phenotyping improve accuracy of pedigree and genomic selection for grain yield in wheat. G3: Genes, Genomes, Genetics, 6(9), 2799-2808.

25. Singh, R. K., Sood, P., Prasad, A., & Prasad, M. (2021). Advances in omics technology for improving crop yield and stress resilience. Plant Breeding, 140(5), 719-731.

26. Spindel, J., Begum, H., Akdemir, D., Virk, P., Collard, B., Redoña, E.,... & McCouch, S. (2015). Genomic selection and association mapping in rice (Oryza sativa): effect of trait genetic architecture, training

population composition, and significance thresholds in detecting quantitative trait loci (QTL) for grain yield. PLoS One, 10(10), e014113.

27. Technow, F., Riedelsheimer, C., & Schrag, T. A. (2014). Genomic prediction of hybrid performance in maize with models incorporating dominance and population specific marker effects. Theoretical and Applied Genetics, 127(7), 1-10.

28. Vessal, S., Arefian, M., & Siddique, K. H. (2020). Proteomic responses to progressive dehydration stress in leaves of chickpea seedlings. BMC genomics, 21(1), 1-15.

29. Wang, J., Forrest, K., & Pecota, K. (2021). Genomic selection enhances breeding efficiency for barley stripe rust resistance from wild barley. Crop Science, 61(1), 1-12.

30. Wang, X., Chen, F., Ma, L., Liao, X., & Hu, X. (2022). Non-volatile and volatile metabolic profiling of tomato juice processed by high-hydrostatic-pressure and high-temperature short-time. Food Chemistry, 371, 131161.

31. Weeks, D. P., Spalding, M. H., & Yang, B. (2016). Use of designer nucleases for targeted gene and genome editing in plants. Plant Biotechnology Journal, 14(2), 483-495.

32. Xiao, Y., Chen, M., Zheng, N., Xu, Z., Zhang, J., Hu, X.,... & Wang, J. (2022). Transcriptome Analysis Identifies Novel Genes Associated with Low-Temperature Seed Germination in Sweet Corn. Plants, 12(1), 159.

33. Xu, Y., Wang, X., Ding, X., Zheng, X., Yang, Z., Xu, C., & Hu, Z. (2018). Genomic selection of agronomic traits in hybrid rice using an NCII population. Rice, 11(1), 1-10.

34. Xu, Y., Yan, F., Liu, Y., Wang, Y., Gao, H., Zhao, S.,... & Li, J. (2021). Quantitative proteomic and lipidomics analyses of high oil content GmDGAT1-2 transgenic soybean illustrate the regulatory mechanism of lipoxygenase and oleosin. Plant Cell Reports, 40(12), 2303-2323.

35. Yadav, S., Jackson, P., Wei, X., Ross, E. M., Aitken, K., Deomano, E.,... & Voss-Fels, K. P. (2020). Accelerating genetic gain in sugarcane breeding using genomic selection. Agronomy, 10(4), 585.

36. Zaidi, S. S. E. A., Mukhtar, M. S., & Mansoor, S. (2018). Genome editing: targeting susceptibility genes for plant disease resistance. Trends in Biotechnology, 36(9), 898-906.

37. Zhang, D., Zhang, Z., Unver, T., & Zhang, B. (2021). CRISPR/Cas: A

powerful tool for gene function study and crop improvement. Journal of Advanced Research, 29, 207-221.

38. Zhang, L., Cenci, A., Rouard, M., Zhang, D., Wang, Y., Tang, W., & Zheng, S. J. (2019). Transcriptomic analysis of resistant and susceptible banana corms in response to infection by Fusarium oxysporum f. sp. cubense tropical race 4. Scientific reports, 9(1), 8199.

39. Zhao, Y., & Zhou, D. X. (2012). Epigenomic modification and epigenetic regulation in rice. Journal of Genetics and Genomics, 39(7), 307-315.

40. Zhao, Y., Gowda, M., Liu, W., Würschum, T., Maurer, H. P., Longin, F. H.,... & Reif, J. C. (2012). Accuracy of genomic selection in European maize elite breeding populations. Theoretical and Applied Genetics, 124, 769-776.

# Chapter - 6

## Innovations in Organic Farming System

**Authors**

**Pradosh Kumar Parida**

Ph.D. Scholar, Department of Agronomy, Tamil Nadu Agricultural University, Coimbatore, Tamil Nadu, India

**Chinmaya Sahoo**

[2]M.Sc. Scholar, Department of Agronomy, College of Agriculture Vellayani, Kerala Agricultural University, Thrissur, Kerala, India

**Jakkampudi Sowjanya**

M.Sc. Scholar, Department of Agronomy, College of Agriculture Vellayani, Kerala Agricultural University, Thrissur, Kerala, India

**Arindam Deb**

Ph.D. Scholar, Department of Agronomy, College of Agriculture Vellayani, Kerala Agricultural University, Thrissur, Kerala, India

# Chapter - 6

## Innovations in Organic Farming System

**Pradosh Kumar Parida, Chinmaya Sahoo, Jakkampudi Sowjanya and Arindam Deb**

### Abstract

Since time immemorial, India has been an agricultural economy, and till date, rural India thrives on agriculture. Indian agriculture has been mainly subsistence-based, which has trembled due to the population explosion. During the mid-20th century, to provide food for the huge Indian population, farmers finally adopted the technologies proposed by the Green Revolution. This had resulted in a spectacular yield increase; however, in the long run, many adverse effects of the green revolution on soil and the environment were recorded. As people became more aware of the health hazards caused by the ingestion of food loaded with chemicals, the organic revolution began. Organic farming sounded like a better alternative, but due to the reduction in yield over conventional farming, it has not been fully accepted. Several new innovations in this field have the potential to overcome this drawback. This chapter deals in detail with the concept of organic farming, innovations in this field, and types of certifications in India.

**Keywords:** Green revolution, health hazard, organic farming, certification.

### 1. Introduction

Organic farming is an agricultural approach that aims to produce food and other agricultural products using natural and sustainable methods. It is a holistic system that seeks to minimize the use of synthetic inputs, such as pesticides and fertilizers, while emphasizing the importance of soil health, biodiversity, and ecological balance (Giller *et al.*, 2021). The principles of organic farming revolve around creating a farming system that works in harmony with nature. Organic farmers focus on building and maintaining healthy soil through practices such as composting, crop rotation, and the use of organic matter. They strive to enhance soil fertility, structure, and biological activity, which in turn supports the growth of healthy plants (Yadav *et al.*, 2020). Certification plays a crucial role in organic farming. To label and market their products as organic, farmers must adhere to specific standards and regulations set by government bodies or certification organizations (Sapbamrer and Thammachai, 2021). These standards typically include requirements for soil and water management, pest and disease control, livestock welfare, and the avoidance of genetically modified organisms (GMOs).

The demand for organic products has been steadily increasing as consumers become more aware of the environmental and health impacts of conventional agriculture. Organic farming offers numerous benefits, including reduced chemical residues in food, enhanced soil fertility and long-term sustainability, and improved animal welfare (Diacono and Montemurro, 2011). While organic farming presents certain challenges, such as lower yields and higher labour requirements compared to conventional methods, ongoing research and technological advancements continue to support the development and optimization of organic farming practices.

## 2. Definition

**USDA:** organic farming is a system which avoids or largely excludes the use of synthetic inputs (such as fertilizers, pesticides, hormones, feed additives etc.) and to the maximum extent feasible rely upon crop rotations, crop residues, animal manures, off-farm organic waste, mineral grade rock additives and biological system of nutrient mobilization and plant protection.

**FAO:** Organic agriculture is a unique production management system which promotes and enhances agro-ecosystem health, including biodiversity, biological cycles and soil biological activity, and this is accomplished by using on-farm agronomic, biological and mechanical methods in exclusion of all synthetic off-farm inputs.

## 3. History

Until the 1920's, farmers produced food using natural means, controlling pests naturally and feeding the soil using traditional farming practices. After the Second World War with new research of certain chemical were capable of killing insects. Paul Miller developed DDT in 1939 and the new era of using a new class of insecticides began.

The beginning of organic agriculture to 1840 when the mineral plant nutrition theory was developed by Justus von Liebig.

Let's take a look at the significant dates in the organic farming history.

- 1924 - Rudolf Steiner publishes "Spiritual Foundations for Renewal of Agriculture.". This book explains the method that may be the first comprehensive organic farming system. The same year he gave eight talks on the spiritual foundation of agriculture, later called biodynamic agriculture.

- 1935 - Mokichi Okada established an agricultural system originally called, "no fertilizer farming" or "Nature Farming" in Japan.
- 1939 - Lord Northbourne first used the term "organic farming.". He derived the term from his concept of "the farm as organism" which he explains in details in his book *"Look to the Land"* (1940). Lady Eve Balfour launches the Haughley Experiment on farmland in England. This is the first scientific, side-by-side comparison of organic and conventional farming
- 1940 - Sir Albert Howard published "An Agricultural Testament," a book that focuses on composting methods. In this book he adopted Northbourne's terminology of "organic farming". Howard's work spread widely, and he became known as the "father of organic farming" for his work in applying scientific knowledge and principles to various traditional and natural methods
- 1942 - J.I. Rodale started publishing Organic Farming and Gardening magazine (founded Rodale Inc. in 1930). Today the magazine is known as Organic Gardening.
- 1947 - Sir Albert Howard published "The Soil and Health, A Study of Organic Agriculture.". This is the first book to include "organic" agriculture/farming in the title.
- 1962 - The scientist and naturalist Rachel Carson published "Silent Spring.". In the book, Carson writes about the effects of DDT and other pesticides on the environment.
- 1972 - The year marks the creation of IFOAM - the International Federation of Organic Agriculture Movements in Versailles, France.
- 1973 - Research Institute of Organic Agriculture (FiBL) was founded In Switzerland
- 1980 - IFOAM defines the basic standards and regulations relating to the certification of organic agriculture
- 1999 - Codex Alimentarius, a commission run by the UN's World Health Organization and the Food and Agriculture Organization, sanctions international guidelines to cultivate, process, market, and label organic foods (Anonymous- I, 2016).

## 4. Organic farming vs conventional farming

| Component | Organic farming | Conventional farming |
|---|---|---|
| Land system | Use of natural resources to improve soil health. Amount of water infiltration is > 15-20% than conventional farming, which helps to reduce erosion | Dependence on synthetic chemicals for primary nutrients. Amount of water infiltration is less, which increases erosion |
| Fertilizers | Only fertilizers obtained through plant based or biological sources are used | Synthetic chemicals used |
| Fuel material | Uses < 45% of energy than conventional farming | Consume more energy |
| Nutrient quality | Healthy and nutritive production | Significant loss of nutrient quality in produce. |
| Impact on soil | Healthy soil with sustainable production potential | Ignorant about soil health |
| Impact on the environment | Harmony with ecology | Toxic effect on ecology. |
| Genetic modified organisms | GMOs are not used | GMOs are used to increase the crop yield |
| Health safety | Production system is healthy for every component of ecology. | Detrimental to health, even for primary consumer. |
| Farming methods | Mix farming | Focused on crop production. |
| Lifestyle change for farmers | Towards sustainability, more positive direction | Short-sighted approach, threat to farmers livelihood |

(NPOP, 2021)

## 5. Status of organic farming

**5.1 World:** More than 76.4 million hectares of organic agricultural land are present worldwide as of 2021, including areas that are currently undergoing conversion. Australia (35.7 million hectares) is the countries with the largest areas of organic agricultural land. Organic farming occupied 1.6% of the world's agricultural land in 2021. Wild collection or al non-agricultural areas together covered 31.8 million hectares. So, the total organic area was 108.3 million hectares, of which 74.4 percent is agricultural land and crops, 24.3 percent is wild collections, 1% is non-agricultural land, and the remaining percent is forest.

**Table:** Organic agriculture statistics of the World in 2021

| Indicator | India (2022-23) | India (2020-21) | Top States (2020-21) |
|---|---|---|---|
| **States with organic Activities** | 2022-23: most of the states & UT in India | 2021: most of the states & UT in India | |
| **Organic agricultural land** | 2022-23: 10.17 million hectares (5.39 farm + 4.79 wild production) (2019-20: 3.67 million hectares) | 2020-21: 9.12 million hectares | MP (1.50 million hectares, 27%) Maharashtra (1.13 million ha) Gujarat (0.60 million ha) |
| **Increase of organic agricultural land 2020/2023** | 6.5 million hectares (ha); +63.91 % | 5.45 million hectares, +60% | |
| **Wild collection** | 2022-23: 4.79 million hectares | 2021-22: 4.39 million hectares | Chhattisgarh (2.99 million ha) Madhya Pradesh (0.87 million ha) Rajasthan (0.20 million ha) |
| **Organic farm production** | 2022-23: 2.97 million tonnes | 2020-21: 3.41 million tonnes | Madhya Pradesh (1.41 million t) Maharashtra (0.69 million t) Rajasthan (0.35 million t) |
| **Organic production from wild areas** | 2022-23: 0.02 million tonnes | 2020-21:0.02 million tonnes | Andaman & Nicobar (0.0036 million t) Goa (0.0035) Chhattisgarh (0.0031 million t) |
| **Total farmers** | 2022-23: 2.3 million | | |
| **Total operators** | 2022-23: 14523 no.s | | |
| **Organic market** | 2022-23: 0.32 million tonnes | 2021-22: 0.46 million tonnes | Madhya Pradesh (0.18 million t) Maharashtra (0.085 million t) Gujarat (0.06 million t) |
| | 2022-23: Rs. 5525.18 crore | 2021-22: Rs. 5249.32 crore | Madhya Pradesh (Rs. 1292.55 crore) Gujarat (Rs. 727.09 crore) Maharashtra (Rs. 696.71 crore) |
| | | | USA (Rs. 2217.84 crore) European Union (Rs. 2056.24 crore) Canada (Rs. 333.24 crore) |
| **Organic Produce in export** | Processed food including- Soya meal (61%), Oilseeds (12.85%), Cereals and millets (12.71%), Sugar (4.77%), Plantation crop products, Tea & Coffee (2.16 %), Spices and condiments (1.72%), Pulses (1.1%0) | | |
| **Number of states having sate organic certification agencies accredited by APEDA** | | 2020: 12 | MP, Gujarat, Telangana, Sikkim, Bihar, Karnataka, Odisha, Rajasthan, Uttarakhand, Chhattisgarh, Tamil Nadu and UP |

(FiBL survey, 2023)

**India:** In India, total organic area (cultivated + wild harvest) is 9.12

million hectares among them 4.73 million hectares is cultivated area (organic + in conversion) and 4.39 million hectares wild harvest collection area. Total production is 3.41 million t is from farm production and 0.02 million t is wild harvest production.

India ranked fifth globally in terms of organic agricultural area in 2021, fell to sixth in 2023, and now tops the list in terms of the number of producers overall (FiBl Survey, 2023)

**Table:** Organic agriculture statistics of India in 2020-21

<table>
<tr><th>Indicator</th><th>India</th><th>Top States</th></tr>
<tr><td>States with organic Activities</td><td>2021: 191 countries</td><td></td></tr>
<tr><td>Organic agricultural land</td><td>2020-21: 4.32 million hectares (2019-20: 3.67 million hectares)</td><td>MP (1.50 million hectares, 27%)<br>Maharashtra (1.13 million ha)<br>Gujarat (0.60 million ha)</td></tr>
<tr><td>Increase of organic agricultural land 2020/2021</td><td>0.18 million hectares (ha); +17.71%</td><td></td></tr>
<tr><td>Wild collection</td><td>2021-22: 4.39 million hectares (ha)</td><td>Chhattisgarh (2.99 million ha)<br>Madhya Pradesh (0.87 million ha)<br>Rajasthan (0.20 million ha)</td></tr>
<tr><td>Organic farm production</td><td>2020-21: 3.41 million tonnes</td><td>Madhya Pradesh (1.41 million t)<br>Maharashtra (0.69 million t)<br>Rajasthan (0.35 million t)</td></tr>
<tr><td>Organic production from wild areas</td><td>2021-22: 0.02 million t</td><td>Andaman & Nicobar (0.0036 million t)<br>Goa (0.0035)<br>Chhattisgarh (0.0031 million t)</td></tr>
<tr><td rowspan="3">Organic market</td><td rowspan="3">2021-22: 0.46 million t<br>2021-22: Rs. 5249.32 crore</td><td>Madhya Pradesh (0.18 million t)<br>Maharashtra (0.085 million t) Gujarat (0.06 million t)</td></tr>
<tr><td>Madhya Pradesh (Rs. 1292.55 crore)<br>Gujarat (Rs. 727.09 crore)<br>Maharashtra (Rs. 696.71 crore)</td></tr>
<tr><td>USA (Rs. 2217.84 crore)<br>European Union (Rs. 2056.24 crore)<br>Canada (Rs. 333.24 crore)</td></tr>
<tr><td>Organic Produce in export</td><td colspan="2">Processed food including- Soya meal (61%), Oilseeds (12.85%), Cereals and millets (12.71%), Sugar (4.77%), Plantation crop products, Tea & Coffee (2.16%), Spices and condiments (1.72%), Pulses (1.1%0)</td></tr>
</table>

| | | |
|---|---|---|
| **Number of states having sate organic certification agencies accredited by APEDA** | 2020: 12 | MP, Gujarat, Telangana, Sikkim, Bihar, Karnataka, Odisha, Rajasthan, Uttarakhand, Chhattisgarh, Tamil Nadu and UP |

(APEDA, 2020-21)

## 6. The concept of organic farming

Healthy crops are produced via organic farming, which is a sustainable agricultural method that is safe and kind to the environment. It relies on building a healthy, productive soil and raising a variety of crops, avoiding the use of synthetic chemical fertilisers and pesticides on the land. With a range of beneficial insects and other species acting as natural predators for crop pests and a rich soil of earthworms and microorganisms to keep the soil alive, the farm maintains a biologically balanced environment. Organic farmers avoid synthetic chemicals to decrease pollution and health issues.

## 7. Goals of organic farming

The two widely accepted goals are:

- Obtaining toxins-free food for long-term human health safety and
- Achieving a closed nutrient cycle for the correction of unfavourable soil nutrient balances.

## 8. Objective of organic farming

- To create nutrient-dense, nutritious, and superior food.
- To foster and strengthen biological cycles involving microbes, plants, and animals, soil flora and fauna, and to maintain and improve long-term soil fertility
- To support the conservation of soil and water
- To maintain and advance traditional and indigenous farming practices, varieties, etc.
- Utilise farm resources as much as possible
- Reduce any pollution that may be brought on by agricultural practices
- Take into account how farming practices affect society and the environment.

## 9. Principles of organic agriculture

i) **Principle of health:** Organic farming should maintain and improve the wellbeing of the soil, plants, animals, people, and planet as a whole.

ii) **Principle of ecology:** Organic Agriculture should be based on living ecological systems and cycle, should support, cooperate with them and help to sustain them.

iii) **Principle of fairness:** Organic Agriculture should build on relationships that ensure fairness with regard to the common environment and life opportunities.

iv) **Principle of care:** Organic Agriculture should be managed in a precautionary and responsible manner to protect the health and well-being of current and future generations and the environment. (IFOAM, 2023; NCOF, 2020-21)

## 10. Debated issues on organic agriculture

### 10.1 Can organic farming produce enough food for everyone?

Organic farming can even more productive than conventional farms, particularly in developing countries, and can be almost as productive as conventional farms in developed countries. In a high demand food production from a shrinking land resources scenario, we are unable to attempt organic farming in all ecosystems and regions. So, we should pick those regions that are still organic by default. In order to meet local and international demand for organic food as well as increase farm revenue, hill regions and the north-east region can be easily transformed into organic food production zones.

### 10.2 Is it possible to meet the nutrient requirement of crops entirely from organic sources?

It is impossible to supply all of the crop's nutritional requirements from organic sources in areas with intense cropping. To match the similar yield from the North Eastern Hill Zone, the Indo-Gangetic plane region required more conversion time (> 3 year) as well as more than twice the number of organic inputs.

### 10.3 Are there any significant environmental benefits from organic farming?

In addition to producing nutritious meals, organic farming places a strong emphasis on preserving and maintaining the ecology. Composting and recycling are used in the production of organic crops to reduce the risk for contamination from biomass, manure, and animal excrement.

### 10.4 Is the food produced by organic farming superior in quality?

Data from a 10-year European Union-funded experiment on organic farming showed that when compared to current chemical-based agriculture, there are 40% more beneficial chemicals in agricultural produce and 90% more in milk. Micronutrients, vitamins, and other quality components are known to be more abundant in organic produce.

### 10.5 Is organic farming labour intensive than conventional farming?

It is true that organic farming frequently requires more labour than conventional agriculture. There are actually a wide variety of labour-saving technologies and techniques that can be used. They consist of direct mulching with crop residues, reduced tillage, and the use of cover crops to prevent soil erosion and manage weeds.

### 10.6 Is organic farming economically feasible?

Organic farming is economically feasible in areas where resources are available within the farm and least dependent on external resources. Organic farming will be more cost-effective in areas like the hilly North Eastern region of India and other hilly regions where there is a lot of biomasses from the forest, weeds, crops, etc. Additionally, organic food should be cost-effective for the poor farmers because it is predicted to fetch a premium price (of at least 25%).

### 10.7 Is it possible to manage pest and disease in organic farming?

The most challenging task in organic farming is currently pest and disease management. However, these can be greatly controlled through experience, the selection of resistant cultivars, cultural practices, applications of biopesticides and ITK formulations, *etc*. (Babu *et al*., 2015)

## 11. Innovations in organic farming

This includes developing innovative technology and tactics to close the productivity gap between organic and conventional farming.

### 11.1 Crop production system

A crop production system encompasses the practices and techniques used to cultivate crops efficiently. This includes preparing the soil, choosing the right crops, using proper planting methods, managing irrigation and nutrients, controlling pests and diseases, and harvesting the crops effectively. In order to enhance organic farming, various innovations are being developed in crop production systems. These innovations focus on preserving or enhancing biodiversity through sustainable weed, pest, nutrient and water management approaches.

## 11.2 Nutrient management

In organic farming, several innovations have been developed to improve nutrient management and enhance soil fertility without relying on synthetic fertilizers.

**11.2.1 Compost and organic amendments:** Advanced composting techniques, such as aerated static pile composting or vermicomposting (using earthworms), enhance the microbial decomposition process and produce high-quality compost.

In a study, researchers found that compared to traditional composting, vermicompost had higher levels of available nutrients, such as nitrogen, phosphorus, and potassium, indicating its potential as a nutrient-rich organic amendment for organic farming (Parihar *et al.*, 2022)

**11.2.2 Nutrient cycling and cover crops:** Cover crops, such as legumes (e.g., clover, vetch) or grasses, are planted between cash crops or during fallow periods. These cover crops take up excess nutrients, fix atmospheric nitrogen, and scavenge nutrients from deeper soil layers. When terminated and incorporated into the soil, cover crops release nutrients, particularly nitrogen, through decomposition, enhancing nutrient cycling within the system.

The findings suggest that the organic farming system with winter cover crops (WCC) had considerably greater levels of dissolved nutrients C and N, pH, and moisture in the soil than the conventional farming system with WCC. (Murphy, 2019).

**11.2.3 Biofertilizers and microbial inoculants:** Biofertilizers contain beneficial microorganisms, including nitrogen-fixing bacteria (e.g., Rhizobium) and mycorrhizal fungi, which form symbiotic relationships with plant roots. These microorganisms facilitate nutrient uptake by plants, especially nitrogen and phosphorus, by enhancing root surface area, solubilizing nutrients, and improving nutrient cycling.

**11.2.4 Biofertilizers and microbial inoculants:** Biofertilizers contain beneficial microorganisms, including nitrogen-fixing bacteria (e.g., Rhizobium) and mycorrhizal fungi, which form symbiotic relationships with plant roots. These microorganisms facilitate nutrient uptake by plants, especially nitrogen and phosphorus, by enhancing root surface area, solubilizing nutrients, and improving nutrient cycling.

Recent research has highlighted the importance of arbuscular mycorrhizal fungus (AMF) in nitrogen uptake from the soil. However, new information suggests that AMF-inoculated plants respond effectively to a variety of environmental challenges such as salt, drought, nutrient deficiencies, alkaline conditions, cold stress, and severe temperatures. This results in higher yields of various crops and vegetables per hectare. (Begum *et al.*, 2019). The use of AMF-mediated growth and productivity increase in crop plants is especially significant in fulfilling the growing global population's food demands.

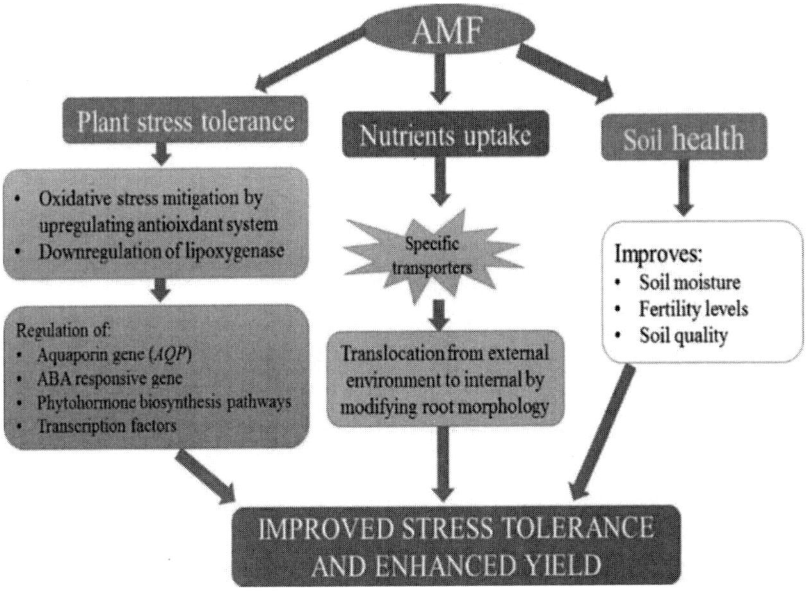


**Figure 1:** Functions of AMF in adverse abiotic stress conditions (Begum *et al.*, 2019)

**11.2.5 Nutrient-rich crop residues:** Conservation tillage and mulching techniques help retain crop residues on the soil surface. These residues, rich in nutrients, slowly decompose and release nitrogen, phosphorus, and other essential elements, providing a continuous nutrient supply to subsequent crops. Retaining crop residues also protects the soil from erosion, maintains soil moisture, and improves soil structure.

**11.2.6 Green manure intercropping:** This innovation involves growing leguminous cover crops, such as clover or vetch, alongside cash crops. Legumes have the unique ability to form symbiotic relationships with nitrogen-fixing bacteria, resulting in biological nitrogen fixation.

The legumes crimson clover or hairy vetch as intercrop increase soil nitrogen and its availability to the main crops, leading to enhanced growth and yield of the vegetables (Irrisarri *et al.*, 2021)

**11.2.7 Nutrient budgeting and planning:** This is a specialized software tool that integrates soil test results, crop nutrient requirements, and available organic nutrient sources. By utilizing this tool, farmers can generate nutrient management plans, optimize fertilizer application rates, and monitor nutrient inputs and outputs over time. This precision nutrient management approach has been shown to improve crop nutrition, reduce nutrient runoff, and enhance overall nutrient use efficiency on organic farms (Haney *et al.*, 2018).

**11.2.8 Phyto-accumulators:** A major issue in organic farming is the deficiency and imbalance of all essential elements, particularly for crops that require several nutrients in various proportions. It can be difficult to get the necessary nourishment from organic sources. Phyto-accumulators and Phytoextractors can fill this gap. These are the plant species that have the ability to take up and store particular nutrients from the soil (Adarsh *et al.*,2022).

## 11.3 Water management

Innovations in organic farming for water management aim to improve water conservation, reduce water pollution, and enhance overall water use efficiency. Some key innovations in this area include:

**11.3.1 Drip irrigation:** Drip irrigation is a micro-irrigation system that delivers water directly to the root zone of plants through emitters or drippers. This method reduces water loss through evaporation and deep percolation, leading to efficient water use.

**11.3.2 Rainwater harvesting:** Rainwater harvesting involves collecting and storing rainwater for agricultural use. It can be done through various methods such as rooftop rainwater harvesting or constructing ponds and reservoirs. By capturing and utilizing rainwater, organic farmers can reduce their dependence on groundwater and surface water sources.

A study conducted on organic olive orchards where rainwater harvesting systems were implemented, resulted in improved water availability during dry periods and increased crop productivity (Rossi *et al.*, 2022).

**11.3.3 Conservation tillage:** Conservation tillage practices minimize soil disturbance by reducing conventional ploughing. This helps maintain soil structure and organic matter content, which improves water infiltration and reduces surface runoff. Residue retention is an important aspect of conservation agriculture, which advocates at least 30 percent crop residue must be retained in the field. Difficulty in seeding in the presence of retained crop residue is a shortcoming, however the advent of new variants of zero-till seed-cum-fertilizer-drill such as, turbo seeder and happy seeder provides a promising solution (Gogoi *et al.,* 2023).

**11.3.4 Wetland and riparian buffer zones:** Wetland and riparian buffer zones are areas of vegetation established along water bodies to intercept and filter agricultural runoff. These buffer zones act as natural filters, reducing sediment, nutrient, and pesticide runoff, and improving water quality.

## 11.4 Organic weed management

Weeds in addition to increasing competition for water, sunshine, and nutrients and acting as host plants for pests and diseases, weeds have the tendency to reduce agricultural yields. For these reasons, they might be seen as a substantial concern. Farmers have been eradicating weeds from their fields with the help of herbicides ever since they were created. Herbicide use decreased weeding labour requirements while simultaneously increasing agricultural yields. Since the widespread use of agro-chemicals is alleged to have caused environmental and health issues, some farmers are now more interested in using organic weed management techniques. The use of herbicides has also been observed to occasionally result in the weeds being resistant to the herbicides, which can lead to some weed species taking over areas. Hence there is need of organic weed management.

**11.4.1 Biological weed management:** Biocontrol of weeds, refers to the use of living organisms, such as insects, pathogens, or other organisms, to suppress or manage the growth and spread of weeds in agricultural, horticultural, or natural ecosystems. It involves the deliberate introduction, augmentation, or manipulation of natural enemies or antagonists of weeds to reduce their population densities or mitigate their negative impacts.

**11.4.1.1 Allelopathy:** Allelopathy refers to the phenomenon where certain plants release chemical compounds into their environment that inhibit the growth or development of neighbouring plants. These chemicals, known as allelochemicals, can be released into the soil, air, or water, and can affect plants in proximity or even at a distance. Abbas *et al.* (2021), reported some allelopathic crops include barley, rye, annual ryegrass, buckwheat, oats, sorghum, sudan sorghum hybrids, alfalfa, wheat, red clover, and sunflower.

**11.4.1.2 Biocontrol agents** Biocontrol agents are living organisms or natural substances derived from living organisms that are used to control pests, diseases, or weeds in agriculture, forestry, and other industries. Biocontrol agents can include various organisms such as beneficial insects (predators and parasitoids), microorganisms (bacteria, fungi, viruses), nematodes, mites, and even certain plants.

**11.4.2 Cultural weed management:** Cultural weed management refers to the use of various agricultural practices and techniques to prevent, control, or manage weeds in a way that is consistent with cultural and environmental values.

**11.4.2.1 Crop rotation:** Crop rotation involves alternating different crops in a systematic sequence on the same land. In a meta-analysis of 54 studies conducted by Weisberger *et al.* (2019), it was found that diversification of crop rotations using the addition of more crops can significantly reduce weed density (49%) and keep Its high efficacy under varied environmental conditions and different crop production systems.

**11.4.2.2 Cover crop:** Cover crops can also modify the microclimate which affects pest populations and diseases. The cover crops are usually terminated before the planting of the main crops, but some management practices use cover crops simultaneously with the main crop Hooks *et al.*, (2010), reported that marigold (*Tagetes spp.*) can produce compounds such as terthienyl that are allelopathic to many species of plant parasitic nematodes.

**11.4.2.3 Stale seed bed:** Before sowing or planting crops, farmers employ the stale seed bed technique to manage weeds. It entails preparing the seedbed beforehand, permitting weed seeds to germinate, and then eliminating the weeds that have emerged before planting the intended crop.

11.4.2.4 **Mulching:** Mulching or covering the soil surface can prevent weed seed germination by blocking light transmission preventing seed germination. A scientific investigation carried out on vineyards practicing organic farming revealed that the implementation of cover crops and mulching resulted in decreased soil evaporation, increased water availability for grapevines, enhanced growth of vines, improved quality of berries, and also positively impacted soil organic carbon content, soil aggregate stability, and mitigation of soil erosion and greenhouse gas emissions (Abad *et al.*, 2021).

### 11.4.3 Thermal weed control

11.4.3.1 **Infrared weeders:** The use of infrared weeders is a development of flame weeding in which the target weeds are heated by the infrared radiation produced by the burners on ceramic or metal surfaces. To eliminate the weeds, some weeders combine infrared radiation with direct flames. Flame weeders are generally thought to be more efficient because they produce higher temperatures, but burner height and plant stage are also crucial. While covering a more precise area than a normal flame weeder, infrared weeders may take some time to heat up.

## 11.5 Precision agriculture

It involves the integration of various technologies, such as global positioning systems (GPS), geographic information systems (GIS), remote sensing, and data analytics, to gather precise information about soil conditions, crop growth, and other relevant factors.

Technologies make use of sensors, meteorological information, and remote sensing to precisely monitor crop health, soil nutrient levels, irrigation scheduling and water application. Real-time information on soil characteristics, nutrients availability, and plant growth parameters is made available by these devices. Farmers can customize their nutrient, irrigation, and weed management practices by using this knowledge, applying fertilisers or organic amendments, water, and herbicides only where and when necessary, minimising nutrient loss, and improving nutrient utilization.

11.5.1 **Variable rate technology (VRT):** VRT involves applying inputs, such as fertilizers and pesticides, at varying rates across a field, based on the specific needs of different areas. It considers soil characteristics, nutrient levels, and crop requirements, optimizing resource usage and minimizing environmental impacts.

**11.5.2 Data analytics and decision support systems:** Farmers can learn more about crop health, production forecasts, and pest/disease threats by analysing vast volumes of data gathered from diverse sources, including sensors, weather stations, and historical records. Based on these, decision support systems offer suggestions for the best course of action.

**11.5.3 Controlled environment agriculture (CEA):** Growing crops indoors or in protected spaces, like greenhouses or vertical farms, where the environment can be precisely regulated, is known as controlled environment agriculture (CEA). This enables year-round production, lower water use, and protection from pests and severe weather.

**11.5.4 Solar powered smart irrigation system:** The world's most plentiful energy source is solar energy. Solar energy is not only a sustainable source of energy but also a solution to the current energy issue. The water pump in this system is solar-powered, and a moisture sensor is used to automatically regulate the water flow. According to Harishankar *et al.* (2014), this technology saves water by reducing water losses and saves electricity by using less grid power.

**11.5.5 Agriculture drones:** These are used for crop surveillance, which can drastically minimize the usage of human power. Drones are commonly used for surveying and assessing the soil chemical composition, field status, crop spraying and irrigation, thereby reducing the cost of production.

## 12. Marketing

India has the highest number of organic producers in the world and accounted for 3rd highest increase in organic land during 2014-15, after Australia and USA (Karthikeyan *et al.,* 2019). The varied climatic conditions in India provide ample opportunities for the production of any kind of crop. This is a huge opportunity considering the demand and supply gap of organic produce. With the increase in standard of living and raising awareness about the harmful effects of chemicals in agriculture, the demand for organic food has skyrocketed, both in domestic and export market. India is already a big player in organic exports, best known for organic tea, apart from that Indian organic rice, vegetable, nuts, cereals, pulses, oilseeds and coffee are also revered in the world. In India, many farmers still are organic by default and/or depends on wild collect which is also organic nature. Efforts need to be made in development of proper supply chain in order to bring these products in the mainstream market and fetch them the deserved price. E-marketing of organic products is getting much popularity, especially after the pandemic. E-marketing facilitates timely sale, minimum loss and therefore good profit to the producer, and the consumers will get their desired and fresh product in their doorsteps (Deb and Parida, 2021).

## 13. Organic certification system in India

In India, two accreditation systems are there for authorizing certification and inspection agencies for organic certification. The core programme that governs the and defines the standards for organic certification is National Programme on Organic Production (NPOP). National Accreditation Body (NAB) is the apex body for certification and accreditation of inspection agencies. Regular inspections and a conversion period are 3 years or more

### 13.1 Participatory Guarantee System (PGS)

PGS is a type of free certification system for organic farmers suitable for domestic market. PGS – Participatory Guarantee System, as defined by IFOAM, are "locally focused quality assurance systems." PGS facilitates the direct participation of producers, consumers and other stakeholders. In India, PGS was set up in 2006 and was formally registered on April 2011 as PGS Organic Council. For a organic produce to be sold in our nation, it has to be either certified under NPOP or PGS-India.

### 13.2 Grower group certification

Previously, only individual farmers were taken into account when designing the Organic Standards and Certification process. Group Certification has recently been implemented, and has become widely popular among the Indian farming community.

Individual farmers can come together and form a group. The group must contain minimum of 25 and maximum of 500 members. They should have farmland in close proximity and follow similar production system. Such groups will develop the Internal Control System (ICS), which would be responsible for the primary inspection of the farming practices. The external inspection body verifies and evaluates the effectiveness of the ICS and certifies the group as a whole.

## 14. Conclusion

Without having a long-term negative impact on the environment or the health of the soil, organic farming may guarantee the availability of nutritious food of high quality. Inorganic farming is transitioning to organic farming as awareness of the damaging effects of chemicals on soil, health, environment, etc. grows. It is necessary to standardize the use of suitable organic sources for a certain crop and to look for lucrative markets for the produce. Growing support for organic farming as a potential key tool in the fight against poverty. All sustainable agriculture and rural development objectives are favourably impacted by organic farming, which also improves the socioeconomic standing of farmers and increases crop yield.

## References


1. Abad, J., Hermoso de Mendoza, I., Marín, D., Orcaray, L., & Santesteban, L. G. (2021). Cover crops in viticulture. A systematic review (1):

2. Abbas, T., Ahmad, A., Kamal, A., Nawaz, M.Y., Jamil, M.A., Saeed, T., Abid, M.A., Ali, H.H. and Ateeq, M., 2021. Ways to use allelopathic potential for weed management: A review. Int. J. Food Sci. Agric, 5, pp.492-498.

3. Adarsh, S., Mathew, S., Thomas, G., Roshni, A.S., Gopika, K.T., Nirosha, V. and Arya, P. (2022). Phyto-accumulators to mitigate micronutrient deficiency in organic farming. *International Journal of Agricultural Research*. 13, 1-14.

4. Anonymous I. (2016). The History of Organic Farming. https://www.organicnet.co/en/magazine/the-history-of-organic-farming#:~:text=The%20modern%20organic%20movement%20started%20in%20the%20early,Liebig%20%2812%20May%201803%20%E2%80%93%2018%20April%201873%29%29

5. APEDA (2020-23). Agricultural and Processed Food Products Export Development Authority. Statistics of the year 2020-23. https://www.apeda.gov.in/apedawebsite/organic/data.htm.

6. Babu, S., Singh, R., Avasthe, R. K., & Yadav, G. S. (2015). Organic farming: problems and prospects in North East India. Training manual on integrated farming system approaches for sustainable hill agriculture under changing climatic scenario. ICAR Research Complex for NEH Region, West Tripura.

7. Begum N, Qin C, Ahanger MA, Raza S, Khan MI, Ashraf M, Ahmed N and Zhang L (2019) Role of Arbuscular Mycorrhizal Fungi in Plant Growth Regulation: Implications in Abiotic Stress Tolerance. Front. Plant Sci. 10:1068. doi: 10.3389/fpls.2019.01068

8. ByjusSikkim: India's First Organic State: RSTV - In Depth. https://byjus.com/free-ias-prep/sikkim-indias-first-organic-state-rstv-in depth/#:~:text=Larger%20Background%201%20Back%20in%202003%2C%20Sikkim%20officially,goals%20have%20been%20rewarded%20by%20the%20United%20Nations. Accessed on June 10.

9. Deb, A., & Parida, P. K. (2021). Impact of COVID-19 on Agriculture and it Reconstruction with Special Reference to North-East India.

10. Dhyani, N. (2023). Fully Organic State in India: Sikkim. https://byjusexamprep.com/current-affairs/organic-state-in-india

11. Gogoi, M., Deb, A., and Kaman, N. 2023. Advanced waste management practices in agriculture and soil science. In: Kumar., C, Kiran., C., Bhanduprasad., D., and Jagadesh., M. (eds) *Advanced research trends in soil science*. Edition: First edition. Vital biotech publication 772, Basant Vihar, Kota, Rajasthan.

12. Haney, R.L., Haney, E.B., Smith, D.R., Harmel, R.D., & White, M.J. (2018). The soil health tool - Theory and initial broad-scale application. *Applied Soil Ecology, 125*, 162-168. Available: https://doi.org/10.1016/j.apsoil.2017.07.035.

13. Harishankar, S., Kumar, R.S., Sudharsan, K.P., Vignesh, U. And Viveknath, T., 2014. Solar powered smart irrigation system. Advance in electronic and electric engineering, 4(4), pp.341-346.

14. Hooks, C.R., Wang, K.H., Ploeg, A. And McSorley, R., 2010. Using marigold (Tagetes spp.) as a cover crop to protect crops from plant-parasitic nematodes. Applied Soil Ecology, 46(3), pp.307-320.

15. Implications on soil characteristics and biodiversity in vineyard. OENO One, 55(1), 295–312. https://doi.org/10.20870/oeno-one.2021.55.1.3599

16. Irisarri P, Imperial J, Lattanzi FA, Monza J, Palacios J, Sanjuan J and Grossman J (2021) Editorial: Maximizing Nitrogen Fixation in Legumes as a Tool for Sustainable Agriculture Intensification. Front. Agron. 3:796717. doi: 10.3389/fagro.2021.796717

17. Karthikeyan, M., Deyi, Z., & Ram, M. (2019). Analysis of trading opportunities and market trends of organic food products in South Asia

to the world: A case study of India. *Int J Multidiscip Res Dev*, *6*(3), 153-68.

18. NCOF (2020-21). National Centre for Organic and Natural Farming, Department of Agriculture & Farmers Welfare. Status of Organic farming. https://ncof.dacnet.nic.in/StatusOrganicFarming.
19. NPOP. (2021). National Centre for Organic and Natural Farming. Department of Agriculture & Farmers, Government of India. Welfare https://ncof.dacnet.nic.in/StatusOrganicFarming
20. Parihar, Parveen & Choudhary, Rakesh. (2022). Evaluation of the effect of Organic Waste on Nutrient Quality of Compost.
21. Raj Kishan Agrahari, Yuriko Kobayashi, Takashi Sonam Tashi Tanaka, Sanjib Kumar Panda & Hiroyuki Koyama (2021) Smart fertilizer management: the progress of imaging technologies and possible implementation of plant biomarkers in agriculture, Soil Science and Plant Nutrition, 67:3, 248-258, DOI: 10.1080/00380768.2021.1897479
22. Raj Kishan Agrahari, Yuriko Kobayashi, Takashi Sonam Tashi Tanaka, Sanjib Kumar Panda & Hiroyuki Koyama (2021) Smart fertilizer management: the progress of imaging technologies and possible implementation of plant biomarkers in agriculture, Soil Science and Plant Nutrition, 67:3, 248-258, DOI: 10.1080/00380768.2021.1897479
23. Rossi, Luca, Luca Regni, Sara Rinaldi, Paolo Sdringola, Roberto Calisti, Antonio Brunori, Francesca Dini, and Primo Proietti. 2020. "Long-Term Water Footprint Assessment in a Rainfed Olive Tree Grove in the Umbria Region, Italy" Agriculture 10, no. 1: 8. https://doi.org/10.3390/agriculture10010008
24. Uludag, A., Uremis, I. And Arslan, M., 2018. Biological weed control. In Non-Chemical Weed Control (pp. 115-132). Academic Press.
25. Weisberger, D., Nichols, V. And Liebman, M., 2019. Does diversifying crop rotations suppress weeds? A meta-analysis. PLoS One, 14(7), p.e0219847.

# Chapter - 7

## Climate-Smart Agriculture: Mitigating the Impacts of Climate Change


**Authors**

**Divya P**

Ph.D. Scholar, Department of Entomology, S.V. Agril College Tirupati, Acharya N.G. Ranga Agricultural University, Lam Guntur, Andhra Pradesh, India

**Chaitanya G**

Department of Entomology and Agril. Zoology, Institute of Agricultural Sciences, Banaras Hindu university Varanasi, Uttar Pradesh, India

**Abdul Mujeeb S**

Department of Entomology and Agricultural Extension Education, S.V. Agricultural College, Tiruapti, Acharya N.G. Ranga Agricultural University, Lam, Guntur, Andhra Pradesh, India

**Anuhya P**

Ph.D. Scholar, Department of Agricultural Extension Education, Acharya NG Ranga Agricultural University Lam Guntur. Adhra Pradesh, Andhra Pradesh, India

**Nikitha A**

Ph.D. Scholar, Department of Agricultural Extension Education, Acharya NG Ranga Agricultural University, Lam Guntur. Adhra Pradesh, India

# Chapter - 7

## Climate-Smart Agriculture: Mitigating the Impacts of Climate Change

**Divya P, Chaitanya G, Abdul Mujeeb S, Anuhya P and Nikitha A**

### Abstract

A growing global population and changing diets are driving up the demand for food. Nearly 690 million or 8.9 percent of the global population are hungry. The food security challenge will only become more difficult, as the world will need to produce about 70 percent more food by 2050 to feed an estimated 9 billion people. The environmental impacts of satisfying this increased food demand will be further aggravated by climate change. The problem also works in reverse. Agriculture is a major part of the climate problem. It currently generates 19–29% of total greenhouse gas (GHG) emissions. Around one-third of total production in the world is wasted, which in and of itself embodies a large carbon footprint. Climate-smart agriculture (CSA) refers to an approach that aims to transform and adapt agricultural systems to the challenges of climate change. It integrates sustainable agricultural practices, climate change adaptation strategies, and mitigation efforts to ensure food security, enhance resilience, and reduce greenhouse gas emissions. CSA recognizes the interdependence between agriculture, climate change, and sustainable development. Climate-smart agriculture is a holistic and multidisciplinary approach that recognizes the complex interactions between climate, agriculture, and sustainable development. By implementing CSA, countries can enhance agricultural productivity, improve resilience to climate change, reduce greenhouse gas emissions, and contribute to achieving the United Nations Sustainable Development Goals (SDGs).

**Keywords:** Food security, climate smart, greenhouse gases, SDG and sustainable development.

### Introduction

Climate-smart agriculture (CSA) refers to a set of practices and approaches aimed at sustainably increasing agricultural productivity, improving resilience to climate change, and reducing greenhouse gas emissions. It seeks to address the challenges posed by climate change on agricultural systems while ensuring food security and promoting rural development

Here are some key elements and practices associated with climate-smart agriculture:

1. **Sustainable intensification:** CSA promotes increasing agricultural productivity while minimizing environmental impacts. It involves the adoption of resource-efficient techniques such as precision farming, conservation agriculture, agroforestry, and organic farming. These practices optimize the use of water, fertilizers, and energy to reduce waste and increase efficiency.
2. **Climate change adaptation:** CSA emphasizes building the resilience of agricultural systems to climate change impacts. This involves using climate information and early warning systems to make informed decisions, adopting drought-tolerant and heat-resistant crop varieties, implementing water management strategies, and improving soil health and fertility. By enhancing adaptive capacity, farmers can better cope with changing climatic conditions and reduce vulnerability.
3. **Climate change mitigation:** CSA seeks to reduce greenhouse gas emissions associated with agricultural activities. It includes practices such as improving nutrient management, implementing agroforestry systems, adopting renewable energy sources, reducing deforestation, and minimizing post-harvest losses. These actions contribute to mitigating climate change by reducing emissions and enhancing carbon sequestration.
4. **Sustainable livelihoods:** CSA aims to improve the socio-economic well-being of farmers and rural communities. It focuses on enhancing market access, supporting value chain development, strengthening farmer organizations, providing access to finance and insurance, and promoting gender equality in agriculture. By creating opportunities for sustainable livelihoods, CSA contributes to poverty reduction and rural development.
5. **Knowledge and Innovation:** CSA emphasizes the importance of research, innovation, and knowledge sharing. It involves developing and disseminating climate-smart technologies, providing training and capacity building to farmers, fostering partnerships between

researchers, farmers, and policymakers, and integrating indigenous knowledge and local context. This facilitates the adoption of appropriate and context-specific practices.

6. **Policy and governance:** Effective policies and supportive governance frameworks play a crucial role in scaling up climate-smart agriculture. Governments, international organizations, and other stakeholders need to develop and implement policies that incentivize and support CSA practices. This includes providing financial incentives, establishing supportive regulatory frameworks, and integrating climate change considerations into agricultural policies and plans.

**Key principles of climate-smart agriculture**

1. **Sustainable intensification:** CSA focuses on increasing agricultural productivity while minimizing negative environmental impacts. It emphasizes efficient use of resources such as water, energy, and nutrients, and promotes the adoption of practices like conservation agriculture, precision farming, and integrated pest management.
2. **Adaptation:** CSA aims to enhance the resilience of agricultural systems to climate change impacts such as increased temperatures, altered rainfall patterns, and extreme weather events. It involves using climate information and resilient crop varieties, implementing water management strategies, diversifying crops, and promoting agro-forestry and other climate-resilient practices.
3. **Mitigation:** CSA seeks to reduce greenhouse gas emissions associated with agricultural activities. It involves adopting climate-smart technologies and practices that minimize emissions, such as improving fertilizer management, implementing agro forestry systems, using renewable energy sources, and reducing post-harvest losses.
4. **Food security:** CSA aims to ensure food security by improving agricultural productivity and supporting sustainable livelihoods for farmers. It focuses on increasing crop yields, improving access to markets and finance, promoting efficient value chains, and strengthening the capacity of farmers to cope with climate-related challenges.

To implement climate-smart agriculture, various stakeholders including farmers, researchers, policymakers, and development organizations collaborate to develop and disseminate climate-smart technologies, provide training and capacity building, and establish supportive policies and financial mechanisms. Additionally, integrating indigenous knowledge and local context is crucial for the successful implementation of CSA practices. Climate-smart agriculture plays a vital role in achieving the United Nations Sustainable Development Goals (SDGs), particularly goals related to zero hunger, climate action, and sustainable agriculture. It offers a pathway to build resilience, adapt to changing climate conditions, and reduce agriculture's contribution to climate change while ensuring sustainable food production for future generations.

**Impact of climate change on agriculture**

Economic losses from natural disasters are rising globally, and agriculture sector is highly vulnerable to these disasters. According to the United Nations Office for Disaster Risk Reduction (UNISDR) (2018), disaster-hit countries experienced direct economic losses to the tune of US$ 2908 billion during 1998–2017. Of the total losses, 77 percent were due to climate related disasters. Climate change impacts are more pronounced on agriculture sector in the recent past.

In India, the countrywide decline in major crop yields due to climate change effects between 2010 and 2039 could be as high as Nine percent, worsening further with time. The loss can be up to 35 percent for rice, 20 percent for wheat, 50 percent for sorghum, 13 percent for barley and 60 percent for maize depending on the location and future climatic scenario.

**Indian scenario**

Indian agriculture being a gamble of monsoons, crop yields continue to be climate sensitive and the fluctuations in temperature and rainfall pattern adversely affect crops productivity, thus threatening food security in India.

In India, climate change has triggered increase in temperatures by 0.6 °C to 25.1 °C between 1901 and 2018, caused shifts in monsoon pattern. Productivity of most crops is likely to decrease 10-40% by 2100 due to increases in temperature, rainfall variability, and decreases in irrigation water. The major impacts of climate change will be on rain fed or un-irrigated crops, which are cultivated in nearly 60% of cropland. A temperature rise by 0.5 degree C in winter temperature is projected to reduce rain fed wheat yield by 0.45 tons per hectare in India. Government of India's economic survey (2018) estimated that the annual loss of US$ 9-10 billion was due to the adverse effects of climate change.

**Climate smart agriculture: characteristics**

Contrary to conventional agricultural development, CSA systematically integrates climate change into the planning and development of sustainable agricultural systems.

CSA's three primary pillars are interrelated concerns-increased productivity, enhanced resilience and reduced emissions. However, the resultant trade-offs often cannot maximize the pillars simultaneously, only optimize them.

CSA maintains ecosystem services: Ecosystems provide the agricultural sector with a number of unpaid services –clean natural water, materials, food, sunlight etc. CSA attempts to ensure the sustainability of these services, preventing their degradation.

CSA is not a rigid set of particular practices, technologies or methodologies- it is only a concept amenable to adaptation. It has multiple entry points, ranging from the development of technologies and practices to the elaboration of climate change models and scenarios, information technology, insurance schemes, value chains and the strengthening of institutional and political enabling environments.

Climate-smart agriculture (CSA) also takes into account the management of insect pests in agricultural systems. Climate change can have significant impacts on insect populations, altering their geographic distribution, life cycles, and interactions with crops. Here are some key considerations regarding insect pests in the context of CSA:

1. **Integrated pest management (IPM):** CSA promotes the adoption of IPM practices to manage insect pests in a sustainable and environmentally friendly manner. IPM involves a combination of strategies that minimize the use of chemical pesticides while maximizing the use of biological control, cultural practices, and monitoring techniques. It emphasizes a holistic approach to pest management that considers the ecology of pests, their natural enemies, and the specific conditions of the farming system.
2. **Crop diversity:** CSA encourages the diversification of crops and planting systems as a means to reduce pest pressure. By growing a variety of crops or incorporating intercropping and agroforestry practices, farmers can disrupt pest lifecycles, enhance natural enemy populations, and create more resilient agro ecosystems.
3. **Resistant varieties:** Climate-smart agriculture promotes the use of

crop varieties that possess inherent resistance or tolerance to insect pests. Breeding programs can develop cultivars that are resistant to specific pests, reducing the reliance on chemical pesticides. Additionally, selecting local or indigenous varieties adapted to the prevailing climatic conditions can enhance pest resistance.

4. **Early warning systems:** Climate change can lead to shifts in the timing and intensity of pest outbreaks. Implementing early warning systems and pest surveillance networks can help farmers detect and respond to pest threats in a timely manner. By monitoring pest populations and weather patterns, farmers can take preventive actions, such as targeted insecticide applications or altering planting schedules, to mitigate pest damage.
5. **Climate information:** Access to climate information and forecasting models can aid in predicting and managing pest outbreaks. By understanding the relationship between climate variables and pest dynamics, farmers can make informed decisions regarding pest control measures, timing of planting, and crop selection.
6. **Farmer training and capacity building:** Providing farmers with training and technical assistance on pest identification, monitoring techniques, and IPM strategies is crucial. Farmer field schools, extension services, and community-based organizations can play a vital role in disseminating knowledge and building the capacity of farmers to effectively manage insect pests in a changing climate.

It's important to note that the specific pest management practices may vary depending on the regional context, crop type, and prevailing pests in a particular area. Therefore, it is recommended to consult local agricultural experts, extension services, and research institutions for region-specific guidance on implementing climate-smart pest management strategies.

Climate-smart agriculture (CSA) is increasingly important in the present era due to several key reasons:

1. **Climate change impact:** Climate change is already having significant impacts on agricultural systems worldwide. Rising temperatures, changing rainfall patterns, and increased frequency of extreme weather events pose challenges to food production and rural livelihoods. CSA provides a framework for adapting agricultural practices to these changing conditions, ensuring food security and sustainable development.

2. **Food security:** With a growing global population and increasing pressure on agricultural resources, ensuring food security is a paramount concern. CSA focuses on improving agricultural productivity, resource efficiency, and resilience to climate change. By adopting CSA practices, farmers can increase yields, reduce post-harvest losses, and enhance the availability and accessibility of nutritious food for local communities.

3. **Environmental sustainability:** Conventional agricultural practices often contribute to environmental degradation, including deforestation, soil erosion, water pollution, and greenhouse gas emissions. CSA promotes sustainable intensification, which emphasizes the efficient use of resources, soil conservation, biodiversity conservation, and reduced carbon emissions. By adopting CSA, agriculture can become more sustainable and contribute to climate change mitigation and environmental conservation.

4. **Resilience and adaptation:** CSA aims to enhance the resilience of agricultural systems to climate change impacts. By diversifying crops, improving water management, implementing agroforestry systems, and using climate-resilient varieties, farmers can reduce their vulnerability to climate-related risks such as droughts, floods, and pest outbreaks. CSA provides adaptive strategies to ensure that agricultural systems can withstand and recover from climate shocks.

5. **Economic benefits:** Climate-smart agriculture can bring economic benefits to farmers and rural communities. By improving productivity, reducing input costs, and enhancing market access, CSA practices can contribute to increased incomes and livelihood opportunities. Furthermore, the adoption of climate-smart practices can attract investment, promote sustainable value chains, and stimulate rural economic development.

6. **Sustainable development goals (SDGs):** Climate-smart agriculture aligns with several SDGs, including zero hunger (SDG 2), climate action (SDG 13), and sustainable agriculture (SDG 12). By implementing CSA, countries can make progress towards achieving these global goals, addressing interconnected challenges of poverty, hunger, climate change, and environmental degradation.

In summary, climate-smart agriculture is essential in the present era to address the challenges posed by climate change, ensure food security, promote sustainable development, and build resilience in agricultural systems. It offers a pathway to transform agriculture into a climate-resilient, environmentally sustainable, and economically viable sector.

**Table 1:** Climate-smart agricultural practices

| Crop management | Livestock Management | Soil and water management | Integrated food energy system |
|---|---|---|---|
| Intercropping to maximize space, pest control & cash crop<br>• Crop rotations should include legumes<br>• New crop varieties (e.g. drought, wind & flood tolerant)<br>• Improved storage and processing techniques<br>• Greater crop diversity<br>• Underground crops (e.g. yams, dasheen)<br>• Stake plants to reduce wind damage<br>• Composting and organic fertilizer<br>• Mulching crops<br>• Shade houSes | • Improve feeding strategies (e.g. cut 'n carry)<br>• Rotational grazing<br>• Grow suitable crops (with proper management) to feed animals e.g. Leucaena & gliricidia<br>• Manure treatment (well-rotted/ decomposed)<br>• Improved livestock health<br>• • Animal husbandry improvements | • Conservation agriculture (e.g. minimum tillage) • Contour planting<br>• Use mounds to plant on slopes<br>• Grass barriers (e.g. kush grass)<br>• Stone barriers<br>• Check dams<br>• Use bench/eyebrow terraces to plant on slopes<br>• Encase beds (pallets, bamboo)<br>• Water storage (e.g. rainwater harvesting)<br>• Improved irrigation (e.g. drips) | • Biogas<br>• Improved stoves<br>• Solar power<br>• Ram pumps for irrigation<br>• Gravity-fed irrigation system |

Key Government Initiatives on CSA National Innovation on Climate Resilient Agriculture

1. **(NICRA):** National Innovations in Climate Resilient Agriculture (NICRA) is a network project of the Indian Council of Agricultural Research (ICAR) launched in February, 2011 with an outlay of Rs.350 crores. The project aims to enhance the resilience of Indian agriculture, covering crops, livestock and fisheries to climatic variability and climate change through development and application of improved production and risk management technologies.
2. **National mission on sustainable agriculture (NMSA):** The Government is implementing the National Action Plan on Climate Change (NAPCC) which provides the overarching framework for climate actions, through national missions in specific areas. The National Mission for Sustainable Agriculture (NMSA), one of the Missions under NAPCC, includes programmatic interventions like Soil Health Card, Paramparagat Krishi Vikas Yojana, Mission

Organic Value Chain Development for Northeastern Region, Rainfed Area Development, National Bamboo Mission and Sub-mission on Agro-Forestry. The NMSA works through adoption of sustainable development pathway by progressively shifting to environment friendly technologies, adoption of energy efficient equipments, conservation of natural resources, integrated farming, etc. Besides, the NMSA aims at promoting location specific improved agronomic practices through soil health management, enhanced water use efficiency, judicious use of chemicals, crop diversification.

3. **The national adaptation fund for climate change (NAFCC):** The National Adaptation Fund for Climate Change (NAFCC) was established to meet the cost of adaptation to climate change for the State and Union Territories of India that are particularly vulnerable to the adverse effects of climate change. This Scheme was implemented during 2015-16 mainly for supporting concrete adaptation activities dealing with mitigating the adverse effects of global climate change in various sectors including agriculture. Under the NAFCC, various projects have been sanctioned in different states i.e., Punjab, Himachal Pradesh, Odisha, Manipur, Tamil Nadu, Kerala, Mizoram, Chhattisgarh, J&K, Meghalaya, Telangana, Andhra Pradesh etc.

4. **Climate smart village: Climate-smart village (CSV)** is an institutional approach to test, implement, modify and promote CSA at the local level and to enhance farmers" abilities to adapt to climate change. The CSV adopts a portfolio of interventions to tackle the climate challenges of the agriculture sector that cover the full spectrum of farm activities. Various Consultative Group on International Agricultural Research (CGIAR) centers and national research organisations are putting their concerted efforts into the CSVs in India. CSVs were piloted in two states of India: Karnal district of Haryana state and Vaishali district of Bihar state which later spread into districts of Punjab, Andhra Pradesh and Karnataka.

5. **Pradhan Mantri Krishi Sinchayee Yojna (PMSKY):** This Scheme was planned and formulated to give more priority on water conservation and its management in agriculture with the vision to extend the area under irrigation from 1 July 2015. The main motto of this Scheme is 'Har Khet Ko Paani' to improve water use efficiency. 'More crop per drop' scheme provide end-to-end

solutions in water on source creation, distribution channels and its management.

6. **Pradhan Mantri Fasal Bima Yojna (PMFBY):** The Pradhan Mantri Fasal Bima Yojana (PMFBY) introduced in the country from Kharif 2016 season is voluntary for States/UTs as well as for farmers. States/UTs can participate in the scheme keeping in view their risk perception and financial considerations etc. From inception of the scheme till 2020-21 cumulatively 2,938.7 lakh farmer applications for a sum insured of Rs. 10,49,342 crore have been enrolled under the scheme.

7. **Soil health card scheme:** This Scheme was launched in February, 2015 to issue soil health cards (SHC) to the farmers providing detailed information on test-based soil nutrient status of their own land along with recommended dose of fertilizers for improving productivity through judicious use of inputs. The Government of India targeted to issue 10.48 crores of SHCs since inception of the Scheme.

8. **National water mission (NWM):** A Mission was mounted to ensure Integrated Water Resource Management (IWRM) for conserving the water sources and minimizing its wastage, and also to optimize Water Use Efficiency (WUE) by 20 percent including agriculture sector.

9. **Paramparagat Krishi Vikas Yojna (PKVY):** It is an extended component of Soil Health Management (SHM) launched in 2015 under NMSA with the objective of supporting and promoting organic farming through adoption of organic village by cluster approach, which in turn result in improvement of soil health.

10. **Biotech-KISAN:** It is a scientist-farmer partnership scheme launched in 2017 for agriculture innovation with an objective to connect science laboratories with the farmers to find out innovative solutions and technologies to be applied at farm level. Under this scheme, so far 146 Biotech-KISAN Hubs have been established covering all 15 agroclimatic zones and 110 aspirational districts in the country. The scheme has benefitted over two lakhs farmers so far by increasing their agriculture output and income. Over 200 entrepreneurships have also been developed in rural areas.

11. **Neem coated urea:** It is a form of urea fertilizer coated with neem extracted material, which acts as a slow releaser of nitrogen

reducing the pest and disease infestation ultimately minimizing the usage of chemicals in farming by achieving the overall increase in crop yield.

12. **Sub-mission on agro-forestry:** This Mission was launched during 2016-17 with the objective of planting trees on farm bunds. Agro-forestry has the potential to bring sustainability in agriculture and also achieving the optimum productivity by mitigating the impact of climate change. National Livestock Mission: This Mission was initiated by the Ministry of Agriculture and Farmers' Welfare and got commenced from 2014-15 focusing mainly on livestock development through sustainable approach ultimately protecting the natural environment, ensuring bio-security, conserving animal bio-diversity and farmers' livelihood.

13. **Integrated pest management scheme (IPMS):** Implemented by the Ministry of Agriculture and Farmers Welfare, IPMS aims to promote the adoption of IPM practices among farmers. The scheme focuses on capacity building, training, and demonstration of IPM techniques to reduce pesticide usage and enhance the sustainability of agricultural systems.

These schemes, along with various state-level initiatives, are designed to address the challenges posed by pests in agriculture while ensuring the sustainability and resilience of farming systems in the face of climate change.

In India, several mobile applications have been developed to support climate-smart agriculture practices and provide information and guidance to farmers. Here are some notable apps related to climate-smart agriculture in India:

1. **Kisan Suvidha:** Developed by the Ministry of Agriculture and Farmers Welfare, this app provides farmers with information on weather updates, market prices, crop advisory, and plant protection. It helps farmers make informed decisions regarding pest management, irrigation, and other agricultural practices.

2. **AgriApp:** This app is designed to assist farmers in making informed decisions about crop management, including pest and disease control. It provides real-time weather updates, market prices, and information on agricultural practices, helping farmers adopt climate-smart techniques.

3. **Pusa Krishi:** Developed by the Indian Agricultural Research

Institute (IARI), this app offers farmers information on various crops, including pest management practices, crop production techniques, and weather-based advisories. It also provides solutions to specific crop-related problems.

4. **CropIn:** CropIn is a technology platform that offers a range of agriculture solutions, including climate-smart practices. It provides farmers with access to real-time weather information, pest and disease alerts, and personalized advisory services for improved crop management.
5. **mKisan:** Launched by the Ministry of Agriculture and Farmers Welfare, this app delivers agricultural information and advisories to farmers through SMS, voice messages, and multimedia content. It covers various aspects of agriculture, including crop protection measures, pest management, and climate resilience.
6. **Green pharmacy:** This app focuses on sustainable pest management and promotes the use of biopesticides and environmentally friendly solutions for pest control. It provides information on the usage, dosage, and efficacy of various biopesticides, helping farmers adopt safer and more sustainable pest management practices.

These apps serve as valuable tools for farmers to access real-time information, guidance, and advisory services related to climate-smart agriculture practices, including pest management, crop protection, and weather updates.

**Mitigation practices**

Climate-smart agriculture (CSA) encompasses a range of practices aimed at mitigating the impacts of climate change while ensuring sustainable agricultural production. Here are some key mitigation practices for climate-smart agriculture:

1. **Conservation agriculture:** Conservation agriculture involves minimum soil disturbance, permanent soil cover, and crop rotation or diversification. It helps improve soil health, increase water retention, and reduce greenhouse gas emissions from agricultural activities.
2. **Agroforestry:** Agroforestry involves integrating trees with crops or livestock systems. It sequesters carbon, provides shade, reduces soil erosion, and enhances biodiversity. Agroforestry systems contribute

to climate change mitigation by acting as carbon sinks and reducing the need for synthetic fertilizers.

3. **Precision farming:** Precision farming uses technology and data-driven approaches to optimize resource use, including water, fertilizers, and pesticides. It minimizes input wastage, enhances efficiency, and reduces greenhouse gas emissions associated with agricultural inputs.
4. **Efficient irrigation:** Improving irrigation efficiency is crucial for climate-smart agriculture. Practices such as drip irrigation, micro-irrigation, and scheduling irrigation based on crop needs help reduce water wastage, energy consumption, and associated greenhouse gas emissions.
5. **Nutrient management:** Proper nutrient management practices, such as balanced fertilization, use of organic fertilizers, and efficient nutrient application techniques, help minimize nitrogen losses, reduce greenhouse gas emissions (e.g., nitrous oxide), and enhance soil health.
6. **Integrated pest management (IPM):** IPM focuses on the prevention and control of pests through an ecological approach. It emphasizes the use of natural enemies, resistant crop varieties, crop rotation, and cultural practices, reducing the reliance on synthetic pesticides and associated environmental impacts.
7. **Livestock management:** Livestock production is a significant contributor to greenhouse gas emissions. Climate-smart livestock management includes improved feeding practices, better waste management, and the use of biogas digesters to capture and utilize methane emissions.
8. **Renewable energy adoption:** Using renewable energy sources, such as solar, wind, or biogas, for powering agricultural operations can significantly reduce greenhouse gas emissions associated with fossil fuel use.
9. **Carbon sequestration:** Implementing practices that enhance carbon sequestration in soils, such as cover cropping, agroforestry, and conservation tillage, helps remove carbon dioxide from the atmosphere and store it in agricultural ecosystems.
10. **Climate information services:** Access to timely and accurate climate information helps farmers make informed decisions regarding crop selection, planting schedules, and pest management.

Climate information services enable farmers to adapt their practices and mitigate climate risks effectively.

By adopting these mitigation practices, farmers can reduce greenhouse gas emissions, enhance resilience to climate change, conserve natural resources, and contribute to sustainable agricultural development.

## Conclusion

Climate-smart agriculture (CSA) is a vital approach that addresses the challenges posed by climate change while ensuring sustainable food production and livelihoods. Here are some key conclusions regarding climate-smart agriculture: Climate Change Adaptation, Mitigation of Greenhouse Gas Emissions, Sustainable Resource Management, Resilient Food Systems, Farmer Livelihoods, Environmental Benefits, Policy and Institutional Support, Knowledge and Information Sharing. In the end, climate-smart agriculture is a holistic approach that combines adaptation and mitigation strategies to address the challenges of climate change while ensuring sustainable agricultural development. By adopting CSA practices, farmers can enhance resilience, reduce emissions, promote resource efficiency, and contribute to a more sustainable and climate-resilient food system.

## References


1. Azadi, H., Moghaddam, S. M., Burkart, S., Mahmoudi, H., Van Passel, S., Kurban, A., & Lopez-Carr, D. (2021). Rethinking resilient agriculture: From climate-smart agriculture to vulnerable-smart agriculture. *Journal of Cleaner Production*, *319*, 128602.
2. Agricultural sustainability under emerging climatic variability: the role of climate-smart agriculture and relevant policies in India. (2020). *International Journal of Innovation and Sustainable Development*., 4-7.
3. Climate Change and Indian Agriculture: Impacts, Coping Strategies, Programmes and Policy. (2019). CAR, pp-16-21.
4. Heeb, L., Jenner, E., & Cock, M. J. (2019). Climate-smart pest management: building resilience of farms and landscapes to changing pest threats. *Journal of Pest Science*, *92*(3), 951-969.
5. Taylor, M. (2018). Climate-smart agriculture: what is it good for?. *The Journal of Peasant Studies*, *45*(1), 89-107.
6. https://www.worldbank.org/en/topic/climate-smart-agriculture

# Chapter - 8

## Integrated Agroforestry into Farming System

**Authors**

**Mamta Patange**

Assistant Professor, Vasantrao Naik Marathwada Krishi Vidyapeeth, Parbhani, Maharashtra, India

**Pundlik Waghmare**

Assistant Professor, Vasantrao Naik Marathwada Krishi Vidyapeeth, Parbhani, Maharashtra, India

**Balaji Choudhary**

M.Sc. Student, Vasantrao Naik Marathwada Krishi Vidyapeeth, Parbhani, Maharashtra, India

**Nilesh Maske**

Principal, MGM Nana Saheb Kadam College of Agriculture, Gandheli, Sambhajinagar, Maharashtra, India

# Chapter - 8

## Integrated Agroforestry into Farming System

**Mamta Patange, Pundlik Waghmare, Balaji Choudhary and Nilesh Maske**

**Abstract**

Agroforestry is the intentional integration of trees and shrubs with agricultural crops and/or livestock. It is a sustainable land use system that can provide a wide range of benefits, including increased crop yields, improved soil fertility, reduced soil erosion, enhanced biodiversity, and climate change mitigation.

Integrated agroforestry systems are those that combine multiple agroforestry practices within a single farming system. This can be done by planting different types of trees and shrubs in the same field, or by rotating crops and trees. Integrated agroforestry systems can be more productive and sustainable than traditional farming systems, and they can also provide a wider range of benefits.

This chapter reviews the benefits of integrated agroforestry systems, and discusses the concept and principles. Agroforestry sub-system, national agroforestry policy.

**Keywords:** integrated agroforestry, farming system, sustainable land use, foof security, biodiversity, soil health, rural livelihood.

**Introduction**

The global population will reach 11 billion by 2050 (Adam, 2021), hence achieving food security without hampering environmental quality is a global challenge for us (Babu *et al.*, 2022a). Climate and soil are the major deciding factor for agricultural productivity, profitability, and human wellbeing (Babu *et al.*, 2022b, Yadav *et al.*, 2021a). Climate adversities had negative impact on crop productivity, natural resources, and eventually food security, and environmental health (Đorđević *et al.,* 2018, Das *et al.*, 2022). Hence, there is a need to choose an appropriate crop and soil management practices that can potentially feed the world population without compromising environmental quality.

Appropriate tree-crop interaction can potentially improve the economic status, food, and nutritional security of small and marginal farmers through the combined production of food, feed, fiber, and firewood besides mitigating the climate change impact (Dollinger and Shibu, 2018, Omerkhil *et al.,* 2020). More specifically, agroforestry can mitigate the atmospheric greenhouse gases concentration while supporting local farmers to adapt to climate change (Yadav *et al.*, 2021b). Perennial fruit trees are known for their tolerance against environmental stresses with their ability to withstand weather aberrations and quick biomass gain under resource-limited situations (Carlsson *et al.*, 2017) over annual crops. Sustainably agroforestry system improves soil fertility by reducing the adverse effects of climatic change which always remains a threat under fragile agro-ecologies (Azembouh *et al.,* 2021). However, a suitable agroforestry system depends upon many tree-crop-environment interactions in the use of natural resources, complementarily or competition.

Many researchers have highlighted the promise of the adoption of agroforestry system as promising land-use systems because they conserve soil and water, reduce soil losses, and sustain higher production, productivity, and profitability (Awasthi *et al.*, 2010, Kuyah *et al.*, 2019, Yadav *et al.*, 2021b). The integrated agroforestry is a supportive from the economic side because it can generate additional employment, especially during the lean season, by providing resources for auxiliary industries (Yadav *et al.,* 2021b). AFSs comprising of both trees and field crops in alleys between two crops can lead to higher productivity under tropical conditions (Hall *et al.*, 2005, Kuyah *et al.,* 2019), and increased well-being (Thorlakson and Neufeldt 2012). The suitable amalgamation of appropriate tree species and annual field crops improves system resilience by minimizing negative impacts (Yadav *et al.*, 2021a). Further, AFSs can reduce the risk in agriculture, especially under marginal and fragile agro-ecologies.

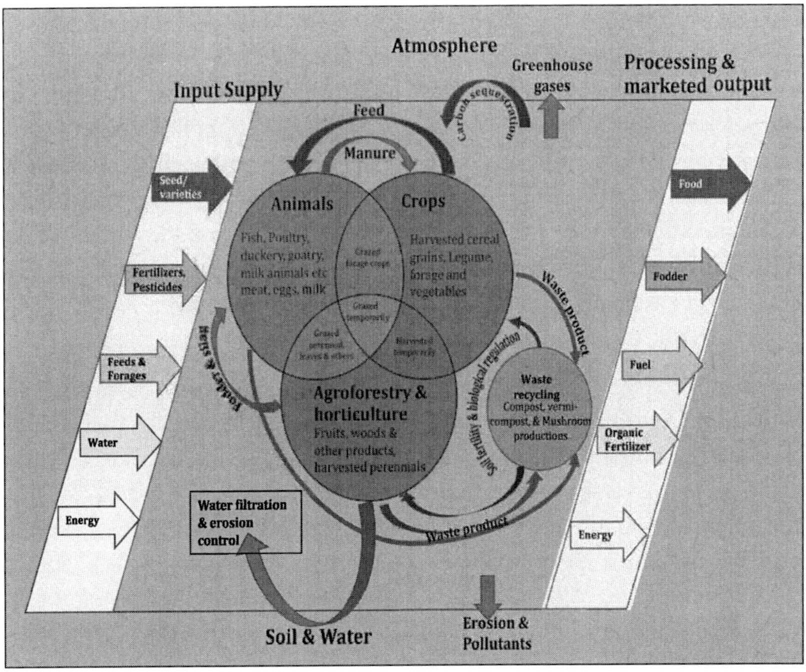


IFS involving agro-forestry based ecosystem process and its material flow

Agroforestry can be used to diversify and intensify farming systems through the integration of indigenous trees producing marketable timber and non-timber forest products, and is described in terms of an agroecological succession, in which climax agroforests are biodiverse, highly productive and profitable. The role of biodiversity in agroecosystem function is one of the keystones of sustainability. Complex agroforests that combine profitability with biodiversity are presented as a model worthy of expansion. However, little is known ecologically about how best to integrate agroforestry into the landscape or to what extent agroforestry can be used to link forest patches and expand bio-geographical islands.

## Concept of integrated agroforestry into farming system

Agroforestry refers to any of a broad range of land use practices where pasture or crops are integrated with trees and shrubs. This intentional combination of agriculture and forestry has multiple benefits, such as greatly enhanced yields from staple food crops, enhanced farmer livelihoods from income generation, increased biodiversity, improved soil structure and health, reduced erosion, and carbon sequestration. Trees in agroforestry systems can also produce wood, fruits, nuts, and other useful products with economic and practical value. Agroforestry practices are especially prevalent in the tropics, especially in subsistence small holdings areas with particular importance in sub-Saharan Africa. However, due to its multiple benefits, for instance in nutrient cycle benefits and potential for mitigating droughts, it has been adopted in the USA and Europe. Agroforestry shares principles with intercropping but can also involve much more complex multi-strata agroforests containing hundreds of species. Agroforestry can also use nitrogen-fixing plants such as legumes to restore soil nitrogen fertility. The nitrogen-fixing plants can be planted either sequentially or simultaneously.

Agroforestry systems can be advantageous over conventional agricultural and forest production methods through increased productivity, economic benefits, social outcomes and the ecological goods and services provided. Biodiversity in agroforestry systems is typically higher than in conventional agricultural systems. Agroforestry incorporates at least several plant species into a given land area and creates a more complex habitat that can support a wider variety of birds, insects, and animals. Agroforestry also has the potential to help reduce climate change since trees take up and store carbon at a faster rate than crop plants. Agroforestry integrates crops and/or livestock with trees and shrubs. The resulting biological interactions provide multiple benefits, including diversified income sources, increased biological production, better water quality, and improved habitat for both humans and wildlife. Farmers adopt agroforestry practices for two reasons. They want to increase their economic stability and they want to improve the management of natural resources under their care.

**Principles of integrated agroforestry into farming system**

- Farmer-centredness
- Aptitude to people, place, and purpose
- Synergy
- Efficiency
- Sustainability
- Diversity
- Adaptability
- Complementarity
- Symbiosis
- Management

- **Farmer-centeredness**
  - **Farmer-centeredness**: Agroforestry systems should be designed and implemented with the needs and priorities of the farmers in mind. This includes considering their goals, resources, and constraints.
- **Aptitude to people, place, and purpose**
  - **Aptitude to people, place, and purpose:** Agroforestry systems should be suited to the specific conditions of the people, place, and purpose for which they are intended. This includes factors such as climate, soil type, and market demand.
- **Synergy**
  - **Synergy:** Agroforestry systems should be designed to maximize the synergies between the different components of the system. This includes the interactions between trees, crops, and livestock, as well as the interactions between the system and the environment.
- **Efficiency**
  - **Efficiency:** Agroforestry systems should be efficient in their use of resources such as land, water, and labor. This can be achieved by carefully selecting the components of the system and by managing them effectively.
- **Sustainability**
  - **Sustainability:** Agroforestry systems should be sustainable in the long term. This means that they should be able to maintain their productivity and benefits over time, without degrading the environment.
- **Diversity**
  - **Diversity:** Agroforestry systems are diverse, both in terms of the species of plants and animals that are included, as well as the spatial arrangement of these species. This diversity helps to reduce the risk of pests and diseases, and it also provides a variety of products and services for farmers and other stakeholders
- **Adaptability**
  - **Adaptability:** Agroforestry systems are adaptable to a wide range of conditions, including different climate zones, soil types, and cropping systems. This makes them a valuable option for farmers in a variety of settings.

- **Complementarity**

    - **Complementarity:** The different components of an agroforestry system complement each other. For example, trees can provide shade for crops, which can help to reduce water loss and improve crop yields. Trees can also fix nitrogen, which can improve soil fertility.

- **Symbiosis**

    - **Symbiosis:** The different components of an agroforestry system interact in a mutually beneficial way. For example, the roots of trees can help to break down crop residues, which can provide nutrients for the crops. The leaves of trees can provide food for livestock, which can help to control pests and diseases.

- **Management**

    - **Management:** Integrated agroforestry systems require careful management. This includes choosing the right species of trees, crops, and livestock for the specific location; planting the trees and crops in the right way; and managing the system to ensure that it is productive and sustainable.

**Scope and Importance of Integrated Agroforestry into farming system**

The scope of integrated agroforestry in farming systems is vast. It can be used in a wide range of settings, from small-scale home gardens to large-scale commercial farms. It can be used to produce a variety of products, including food, fodder, fuelwood, timber, and medicinal plants.

**Opportunities**

- Growing demand for sustainable food production
- Increasing government support for agroforestry
- Development of new agroforestry technologies

**Challenges**

- Lack of awareness of the benefits of agroforestry
- Lack of financial incentives for farmers to adopt agroforestry
- Lack of technical knowledge about how to manage agroforestry systems

Some of the specific benefits of integrated agroforestry in farming systems include. Increased crop yields, Trees can provide shade for crops, which can help to reduce water loss and improve crop yields. Trees can also fix nitrogen, which can improve soil fertility. The scope of integrated agroforestry in farming systems is constantly expanding as new research is conducted and new technologies are developed. It is a promising approach to sustainable agriculture that has the potential to provide a wide range of benefits for farmers, the environment, and society as a whole.

**Benefits of integrated agroforestry into farming system**

**Improved soil health**

The roots of trees can help to break down crop residues, which can provide nutrients for the crops. The leaves of trees can help to improve soil structure and water infiltration, increment in soil nutrients through addition and decomposition of litter-fall.

**Economic benefits**

Agroforestry systems on croplands/farmlands bring significant economic benefits to the farmer, the community, the region or the nation. Increment in an maintenance of outputs of food, fuelwood, fodder, fertilizer and timber. Also, reduction in incidence of total crop failure, common to single-cropping or monoculture system. Increase in levels of farm incomes due to improved and sustained productivity.

**Reduced soil erosion**

Trees can help to prevent soil erosion by providing windbreaks and by stabilizing slopes in integrated agroforestry.

**Enhanced water infiltration**

The roots of trees can help to improve water infiltration, which can help to reduce the risk of flooding and drought.

**Increased biodiversity**

Agroforestry systems can help to increase biodiversity by providing habitat for a variety of plants and animals.

**Environment benefits**

Combining trees with food crops on cropland farms yield certain important environment benefits, both general ecological benefits and specific on-site benefits. Reduction of pressure on forest, more efficient recycling of nutrients by deep-rooted trees on the site, Reduction of surface run-off, nutrient leaching and soil erosion through impending effect of tree roots and stems of these processes.

**Reduced pest and disease pressure**

Trees can help to reduce pest and disease pressure by providing a physical barrier to pests and by attracting natural predators of pests.

**Improved microclimate**

Trees can help to improve the microclimate by providing shade, shelter, and windbreaks.

**Social benefits**

Besides the economics benefits, social benefits occur from increase in crop and tree product yields and in the sustainability of these products. Improvement in rural living standards from sustained employment and higher incomes, increased quality and diversity of food outputs, in nutrition and health.

**Increased income generation**

Agroforestry systems can help to increase income generation by providing a variety of products, such as food, fodder, fuelwood, timber, and medicinal plants.

**Improved food security**

Agroforestry systems can help to improve food security by providing a reliable source of food, even during times of drought or other environmental stresses.

**Agroforestry & sub-systems**

Agroforestry can be separated into three sub-system classifications - agrisilviculture, silvopastoral and agrosilvopastoral (WeiWei, 2014).

- Agrisilviculture combines annual and perennial crops with woody perennials (trees, shrubs, vines),
- Silvopastoral combines trees with pastures and animals, and
- Agrosilvopastoral combines crops, pastures, animals and trees.

**Agroforestry practices in different geographic regions**

- Temperate Region
  - ✓ Silvopasture
  - ✓ Allay cropping
  - ✓ Forest farming
  - ✓ Windbreaks, Shelterbelts, Living Fences
  - ✓ Riparian forest buffers

- Humid tropic region
    - ✓ Silvopasture
    - ✓ Home gardens
    - ✓ Shifting cultivation
    - ✓ Living fences and shade trees
- Trans- gangetic plane region

**Temperate region**

The temperate zone is home to much of the developed world and a significant portion of the world's agricultural land. The agroforestry techniques practiced in this region are well-suited to its climate, species, and biological structure. Common temperate zone agroforestry techniques include silvopasture, alley cropping, forest farming, windbreaks, shelterbelts and living fences, and riparian buffers.

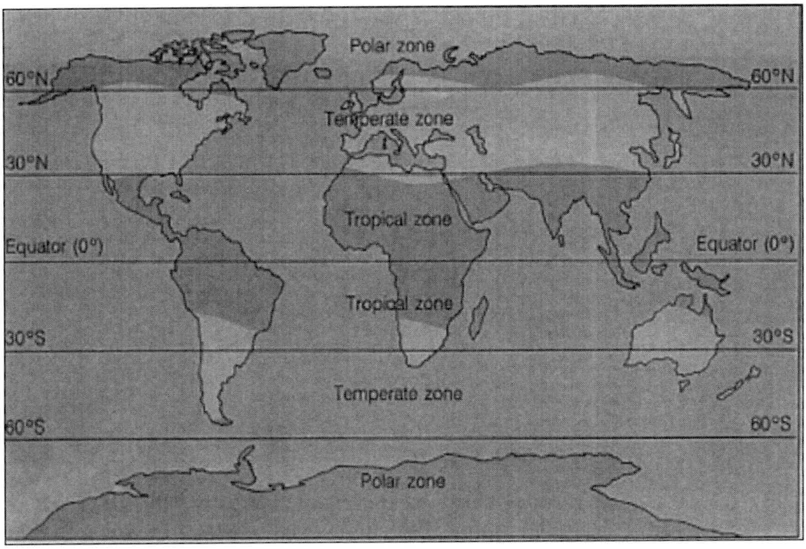


Map of World Climate Zones (Source: WebQuest Hawaii Education)

**Silvopasture**

Silvopasture is a system of land management that combines forestry and livestock production raised on the same plot of land (King, 1979). This system utilizes several agronomic principles such as fertilization, native pasture grasses, and rotational grazing systems with short grazing periods that maximize plant growth and harvest while avoiding damage to the tree crop (Brantly, 2013).

Silvopasture is a highly intensive agroforestry method that requires grazing and timber management that can involve tree pruning, grazing, haying, fertilization and more. There are several benefits of silvopasture, which have led to its increased use. Silvopasture systems reduce economic risk by producing multiple crops and products, create shorter timber rotations (due to forage fertilization), enhance tree growth (due to the ability of grazing animals to control competition for moisture, nutrients and sunlight), provide a cooler environment for livestock and allow for control of weeds and brush without herbicide applications.

It can also create high value timber products (resulting from pruning and tree management) that lead to higher diversification of income for farmers and increased income opportunities. While considerable scientific research depicting beneficial animal/tree interactions has been conducted, the issue of soil compaction and animal/soil interactions in silvopastures has not been scientifically evaluated to a great extent (Sharrow, 2007). This is an area where additional research could prove beneficial.

Integrated silvopasture system. (Reproduced by permission of the University of Minnesota Extension.)

**Allay cropping**

Alley cropping is an agroforestry practice wherein rows of trees or shrubs are planted at a spacing that provides an "alley" where perennial and annual agricultural crops are then grown. Agricultural crops within an alley cropping system are often referred to as intercrops. In this system, special care must be taken to ensure that crops and trees are compatible with one another as well as local climatic and geologic conditions. Alley cropping requires careful maintenance and pruning to limit the lateral spread of tree branches and to ensure that trees will provide the desired level of shade. The typical intercrops for this agroforestry system are row crops (corn, wheat, barley), forage crops (bluegrass, clover, alfalfa), specialty crops (dogwood, Christmas trees) and biomass crops (willows, birches) (Hodge, 1999).

The USDA states that alley cropping can be used to grow short-rotation woody crops or fast-growing woody plants, which can then be combined with forage to produce fuel wood and fodder. Alley cropping improves crop health through the creation of tree canopies that protect against wind damage and pests and also aid in pollination activities. It also improves soil health in areas prone to erosion, adds carbon to improve soil health, and enhances nutrient recycling. As with most other agroforestry techniques, alley cropping diversifies income for landowners who are able to harvest trees and other crops at different times throughout the year (Wotjkowski, 2006).

Alley cropping. (Reproduced by permission of the University of Minnesota Extension)

**Forest farming**

Forest farming, or multistory cropping, refers to the cultivation of specialty crops under the canopy of trees that provide a specific shade level. This agroforestry technique often involves thinning existing woodland to leave desired canopy trees. The remaining trees are used to create the right conditions for understory crops, while also providing the potential for timber production. The most common type of forest farming is done for specialty crops like mushrooms and maple syrup, but it can also be used for ginseng and ferns that are used for culinary, medicinal, or ornamental purposes.

Other examples of forest farming crops include fruits (blackberry, huckleberry), nuts (walnut, hazelnut, beechnut), medicinal herbs (mayapple, bloodroot), moss, and pinecones. Forest farming provides income from specialty crops while also creating revenue from high-quality trees that can be grown for wood products.

Forest farming. (Reproduced by permission of the National Agroforestry Center)

**Windbreaks, shelterbelts and living fences**

Windbreaks are linear plantings of trees or shrubs that are maintained to alter wind flow and regulate climate. They can protect people, property, and livestock, while also benefitting soil health and improving water conservation. There are three main types of windbreaks: field windbreaks, livestock windbreaks, and living fences.

- **Field windbreaks** protect crops from wind, reduce soil erosion, and increase pollination. They are typically planted in rows perpendicular to the prevailing wind.
- **Livestock windbreaks** are designed to help reduce animal mortality and manage feed consumption. They are typically planted in a circular or semi-circular pattern around livestock areas.

- **Living fences** reduce wind and snow on roads and can also spread snow uniformly across a field to ensure even soil moisture. They are typically planted in rows along property lines or roads.

In addition, windbreaks improve soil quality by preventing wind erosion, enhance air/water quality through the trapping of airborne pollutants, and reduce wind speed to decrease soil evaporation rates. The U.S. has a long history with windbreaks, including the planting of more than 100 million trees in response to the Dust Bowl conditions in the 1930s. Windbreaks can reduce energy costs around homes, maintain livestock health and wellbeing, produce specialty products, and improve crop yields.

Windbreak System (Source: USDA, National Agroforestry Center)

Thoughtful vegetation selection in a windbreak and its continued maintenance greatly influence associated benefits to people, livestock, wildlife, or pollinators. Windbreak systems are also useful for creating wildlife corridors and can provide carbon storage benefits. According to the National Agroforestry Center, planting 2.5 percent of tillable land in the American Great Plains with windbreaks could remove up to 80 million metric tons of carbon dioxide from the atmosphere over a period of twenty years (Wright, 2002).

**Riparian buffers**

Riparian buffers are strips of land along streams, rivers, and other waterways that are vegetated with trees, shrubs, and other plants. These buffers can be natural or planted, and they provide a variety of benefits, including water quality protection, crop production, flood control, wildlife habitat, and recreation opportunities. Riparian buffers help to filter out sediment, pesticides, and nutrients from agricultural runoff, which can protect aquatic resources and improve water quality. They can also help to stabilize stream banks and reduce erosion, which can protect crops and property from flooding. Additionally, riparian buffers provide important habitat for a variety of wildlife, including fish, birds, and mammals.

Forest stands planted along waterways can also regulate waterway temperature and decrease eutrophication rates by maintaining shade over the water (National Agroforestry Center, 2012). Woody plants and trees reduce shoreline erosion by absorbing energy from strong waves and keeping roots and soils in place. These buffers also foster habitat diversity and reduce the risk of floods while protecting water quality. The economic incentive of diversified income also makes riparian buffers a popular type of agroforestry technique within the temperate zone. This agroforestry technique requires intensive management to determine soil types, appropriate vegetation types and relative shoreline stability where riparian buffers will be placed.

**Humid tropic region**

The world's tropical zone is characterized by high amounts of rainfall and consists of a variety of environments from tropical rainforests to deserts. Due to the low nutrient content of some tropical soils, slash-and-burn techniques have traditionally characterized the agricultural approach in these regions (Vitousek, 1986).

A rising demand for wood products and specialty crops has led to increased interest in the implementation of sustainable agriculture and forestry management in recent years, including agroforestry systems. Commonly practiced humid tropical zone agroforestry techniques include home gardens, living fences and shade trees, silvopasture and shifting cultivation. In recent years, agroforestry techniques in the humid tropical zone have encompassed new intercropping and silvopasture systems that attempt to overcome issues with land and capital requirements for raising animals on cropland (Macdicken and Vergara, 1990).

**Home gardens**

Home gardens, also known as kitchen gardens, are a type of agroforestry system that is characterized by the intensive use of trees, shrubs, crops, and animals on a small plot of land. They are often found in tropical regions, where they are influenced by the natural strata of the rainforest. In the upper story of a home garden, trees that provide food, fodder, or fruit are typically grown.

The middle story is home to crops such as coffee, cacao, and fruit trees. The understory is typically used for growing root crops, legumes, and other vegetables. An important benefit of the home garden system is that it provides farmers with multipurpose crops and perennials. Planted trees reduce temperatures and wind speeds for understory crops, and provide shade and privacy for sensitive crops (MacDicken and Vergara, 1990).

Thai Home Garden Features diverse crops and rainwater storage for household use or watering (Source: Marten & Suutari, 2006)

**Living fences and shade trees**

Living fences are rows of trees or shrubs that are planted on agricultural land to provide a variety of benefits. They can help to protect crops from wind and sun damage, reduce erosion, and provide a source of wood, fruit, and medicine. Living fences in tropical regions are often made up of different species of trees and shrubs that are adapted to the local climate. Some of the most common perennials used in tropical living fences include gliricidia sepium, erythrina berteroana, hibiscus, and ipomoea. Effective shading practices in the tropical zone involve thinning natural forests and interplanting shade trees on crop plantations. The most common trees used for shading are albizia, cordia and ficus.

These practices are similar to the temperate zone forest farming technique. Multipurpose trees (MPTs) are frequently referenced in agroforestry resources and refer to the "trees and shrubs that are deliberately kept and managed for more than one preferred use" (Ramachandran, 1993). Leguminous trees are often planted in agricultural systems because they can fix nitrogen in the soil, which can improve crop yields. These trees can also be used as trellises for climbing plants, such as black pepper and vanilla. This dual-purpose use of leguminous trees can help to increase the productivity of agricultural land. Hedgerows may also be periodically pruned or pollarded to provide fodder or energy crops (Muschler, 1993).

**Trans- gangetic plane region**

This region is characterized by high land productivity and delicate water balance. Groundwater exploitation has already surpassed 100% of the utilizable balance. The yield level of rice and wheat crops has stagnated, and net profit has shown a diminishing trend, creating many complications in the agro-ecosystem.

In this scenario, farmers in this region need to diversify their crop rotation through appropriate and sustainable land use systems to increase their income. Fast-growing exotic tree species such as poplar (Populus deltoides) and eucalyptus (Eucalyptus tereticornis) have been introduced in farmers' fields to maximize income from a given piece of land in a short period of time. These tree species have proven to be an economically viable option for crop diversification in the trans- gang etic plane region.

A study of the comparative economic viability of poplar-based agroforestry systems with rice-wheat systems showed that agroforestry systems can more than double farmers' income in seven years, and triple their income if farmers are rewarded for the carbon sequestered by the agroforestry system. When the time value of money is considered with a discount rate of 12%, the estimated net incomes are ₹166,145, ₹364,142, and ₹518,131 per hectare, respectively, for rice-wheat systems, poplar-based agroforestry systems without carbon sequestration, and poplar-based agroforestry systems with carbon sequestration.

## Case studies

### Agroforestry systems provide firewood for livelihood improvement in guatemala country

Rural populations in Guatemala are facing increasing challenges due to climate change and poverty. They need to find ways to manage resources sustainably, produce enough food, and improve their quality of life. Agroforestry systems have been proposed as a potential solution.

A study in Camotán and Jocotán compared conventional cropping systems to enriched agroforestry systems. The results showed that the enriched agroforestry system produced 65% of the firewood, compared to only 7% from the conventional cropping system. Families using the conventional cropping system extracted 81% of their firewood from the forest, while families using the enriched agroforestry system extracted only 32%. This means that the enriched agroforestry system helped to reduce pressure on forests and provided an additional benefit to women, children, and young people who no longer had to travel long distances to collect firewood.

Families using the enriched agroforestry system also benefited from the knowledge they gained about tree management. This knowledge helped them to improve their cropping systems and increase their yields. Land tenure security was also found to be an important factor in whether or not families chose to invest in agroforestry systems. Agroforestry systems can play a key role in augmenting resources and improving quality of life in rural communities. They provide a sustainable way to produce food and firewood, and they can also help to reduce pressure on forests. (Siblet N, *et. al.,* 2019)

### Alder agroforestry farming system

Alder-based farming is a traditional system practiced in Nagaland. It involves planting alder trees alongside crops such as maize, Job's Tears, millet, potato, chili, pumpkin, and barley. Alder trees are non-leguminous trees that fix atmospheric nitrogen through nodules on their roots. This helps to improve soil fertility and productivity. Alder trees can grow well in altitudes ranging from 800 to 3,000 meters. Besides fixing atmospheric nitrogen, the litter of the trees add to the soil phosphorus, potassium, calcium, and other nutrients through the addition of biomass (Sharma and Singh, 1994).

Alder is a multipurpose tree that, besides improving soil fertility, provides timber for furniture and fuelwood. The value of the alder tree was recognized by the tribal farmers a long time ago, as trees that are more than 200 years old can be seen in the area. Agricultural crops, together with the alder trees, help form a very remunerative agroforestry system, and the ability of the tree to develop and retain soil fertility has been fully utilized by the tribal farmers of the Angami, Chakhasang, Chang, Yimchaunger, and Konyak tribes (Gokhle *et al.*, 1985). The Knononome village in the Kohima district of Nagaland is proud of its alder plantation and alder tree-based agriculture.

Alder trees are important, and a process called pollarding has been developed to multiply them. Alder trees are ready for pollarding after 6 to 8 years, when the bark becomes rough and develops cracks. Farmers cut the alder trees at a height of 2 to 2.5 meters, and the leaves and twigs are added to the soil. The trunk then sprouts new shoots called coppices.

One alder tree can produce approximately 100 to 200 coppices after pollarding. About 5 to 6 coppices are left on the main trunk to regenerate the tree. It has been estimated that if a village of 100 families set aside about 120 hectares of land to grow alder trees, all families would have enough cereals, vegetables, firewood, etc. Pollarding is done from November to January, and the fresh cut is covered with stones to protect it from frost damage. It is important to maintain alder trees to ensure their longevity. The trunk of the tree should be kept free of parasitic growth. In addition to pollarding, there are other practices that can help conserve crop diversity. For example, swidden farming is a practice that is supported by women and helps to preserve the diversity of crops. (Singh A., 2010)

Stumps of alder (Alnus nepalensis) under the alder agroforestry farming system

## Tropical agroforestry in amazonian ecuador

Traditional agroforestry systems in the humid tropics have been practiced by indigenous communities through the form of home gardens and shifting cultivation for a long time, but the influence of uncertain land tenure and lack of technological support has forced many indigenous groups to turn to other forms of agriculture. In an effort to encourage sustainable agroforestry practices among small producers, the Ecuadorian government implemented the Coca Agroforestry Project with funding from the U.S. Agency for International Development (Anderson, 1990). The government encouraged the use of three forms of agroforestry: shifting cultivation or chacra, modification of coffee plantations, and the addition of partial tree cover to cattle pastures. Chacra is similar to shifting agriculture in that both systems involve removing some trees while leaving the most valuable ones and using the understory to grow perennial crops. However, the difference is that a chacra is continuously managed rather than being periodical left fallow (Anderson, 1990).

The Coca Agroforestry Project was implemented in 1987 through on-farm demonstrations that provided direct training and assistance to Ecuadorian farmers with a goal of increasing productivity. A legume was designated as the choice cover crop and cattle feed. Lechero, or euphorbia cotinifolia was planted around trees to prevent cattle from eating the leaves. Living hedges were also installed to divide pastures into smaller units. Overall, the Coca Agroforestry Project proved that one of the biggest obstacles to successful agroforestry was the lack of available labor and expertise. When these factors were addressed through the hiring of local people and region-specific training, a successful agroforestry model for small producers was established.

Within the project, almost 200 on-farm demonstrations were completed and the Ecuadorian government oversaw the creation of a nursery that made planting stock and seeds available to the public. The farmer training was completed at local agronomy programs within universities and led by people who had been raised on farms and had practical knowledge of local crops and management strategies.

### Temperate agroforestry in Europe

In 2001, the European Union's Common Agricultural Policy (CAP) approved a Europe-wide intercropping system. Three years later, the European Commission launched Article 44 (also known as Measure 222) to support agroforestry. This measure was approved on a limited regional and national level by several countries. France was one of the countries that approved Article 44 in 2010. This gave farmers access to subsidies for agroforestry plots that included 30 to 200 trees per hectare.

Under Article 44, growing trees on cropland was considered a factor of production and was therefore eligible for subsidies. Since the adoption of Article 44 in France, the country has recognized the economic and ecological benefits of agroforestry. As a result, France has adopted tax and funding advantages for farmers who transition to this system.

### Agroforestry and the economy in China: Grain for green program

In 1999, the Chinese government launched the Grain for Green program to reduce soil degradation and erosion, and promote economic growth in rural areas. Under the program, the government provided grain, seedlings, or subsidies to farmers who agreed to convert their degraded cropland into forests and grassland.

The program has been very successful, and has helped to restore millions of hectares of degraded land. It has also helped to improve the livelihoods of millions of farmers, and has reduced the risk of flooding and landslides. The Grain for Green program attempted to isolate the economic factors that prevented farmers from adopting agroforestry techniques by subsidizing local governments to sponsor training programs, transportation and more. The response to the program by farmers led to the spread of fruit tree intercropping across the countryside (Liu, 2010). Eleven years after the program was initiated, more than 15 million hectares of cropland were converted into forest or grassland in almost twenty distinct Chinese provinces (Buttoud, 2013).

**National agroforestry policy**

**The major policy goals are**

- Setting up a National Agroforestry Mission or an Agroforestry Board to implement the National Policy by bringing coordination, convergence and synergy among various elements of agroforestry scattered in various existing, missions, programmes, schemes and agencies pertaining to agriculture, environment, forestry, and rural development sectors of the Government.
- Improving the productivity; employment, income and livelihood opportunities of rural households, especially of the smallholder farmers through agroforestry.
- Meeting the ever-increasing demand of timber, food, fuel, fodder, fertilizer, fibre, and other agroforestry products; conserving the natural resources and forest; protecting the environment & providing environmental security; and increasing the forest / tree cover, there is a need to increase the availability of these from outside the natural forests.

**Basic objectives**

- Encourage and expand tree plantation in complementarity and integrated manner with crops and livestock to improve productivity, employment, income and livelihoods of rural households, especially the small holder farmers.
- Protect and stabilize ecosystems, and promote resilient cropping

and farming systems to minimize the risk during extreme climatic events.

- Meet the raw material requirements of wood-based industries and reduce import of wood and wood products to save foreign exchange.
- Supplement the availability of agroforestry products (AFPs), such as the fuel-wood, fodder, non-timber forest produces and small timber of the rural and tribal populations, thereby reducing the pressure on existing forests.
- Complement achieving the target of increasing forest/tree cover to promote ecological stability, especially in the vulnerable regions.
- Develop capacity and strengthen research in agroforestry and create a massive people's movement for achieving these objectives and to minimize pressure on existing forests

**Strategy**

**1. Establishment of institutional setup at national level to promote agroforestry**

- An institutional mechanism, such as a Mission or Board is to be established for implementing the agroforestry policy. It will provide the platform for the multi-stakeholders to jointly plan and identify the priorities and strategies, for inter-ministerial coordination, programmatic convergence, financial resources mobilization and leveraging, capacity building facilitation, and technical and management support. Such an institutional arrangement will ensure that agroforestry gets equal treatment with other agriculture enterprises, because at present whether in the sphere of inputs, markets, institutional finance, or research and extension, agroforestry is at a sub-optimal level. A suitable mechanism for coordination and convergence with state agriculture, and forest departments as nodal agencies may be established. The Mission / Board may be provided a corpus in order to effectively leverage Rs. 4000-5000 crores annually from the on-going programmes.
- The Ministry of Agriculture has the mandate for agroforestry. Agroforestry Mission / Board will be located in the Department of Agriculture and Cooperation (DAC) in the Ministry of Agriculture (MoA). The Mission/ Board may comprise representatives of the Department of Agriculture Research & Education (DARE),

Ministry of Environment & Forest (MoEF), Ministry of Rural Development (MoRD), Ministry of New and Renewable Energy (MN&RE), International Centre for Research in Agroforestry (ICRAF, South Asia Office), Planning Commission, National Rainfed Area Authority (NRAA), representatives of Non-Governmental Organizations, Industry, NABARD, Agricultural Universities, State Governments, etc.

- The actual implementation may involve convergence and dovetailing with a number of programmes. They may include, the Mahatma Gandhi National Rural Employment Guarantee Programme (MGNREGA), Integrated Watershed Management Programme (IWMP), National Rural Livelihood Mission (NRLM), National Bamboo Mission, Kisan Mahila Sashatikaran Pariyojana, Rashtriya Krishi Vikas Yojana (RKVY), National Medicinal Plants Board (NMPB), Mission for Integrated Development of Horticulture (MIDH), National Green India Mission, Warehouse Development and Regulation Act 2007 (WDRA), CAMPA fund. State Government may identify a Nodal Department for implementing the Agroforestry Mission/Board. At the district level, nodal agency may be Agriculture Technology Management Agency (ATMA) or any other department to be identified by the State, with other partners such as the Krishi Vigyan Kendras (KVKs), Van Chetna Kendras, Farmer's Associations, NGOs, private sector, Self-Help-Groups, Farmers' Cooperatives, Famer Producer Organizations, and Panchayati Raj Institutions (PRIs) etc.
- Agroforestry research and development (R&D), including capacity development and pilot studies / testing and action research should be the responsibility of the ICAR. ICRAF should be involved as an important partner in agroforestry research & development. Strength of Indian Council for Forestry Research & Education (ICFRE) should also be capitalized in this endeavour.
- In the proposed institutional arrangement, the current stakes of the key ministries are to be respected and utilized. It is envisaged that existing structures at the state, district and sub-district levels will be utilized with necessary strengthening. The important dimension of the implementation strategy is that it will effectively leverage tree-plantation, afforestation, greening, nutrition strengthening, etc., which are components of major flagship programmes to achieve its objectives. That will be its uniqueness. That will be its strength.

## 2. Simple regulatory mechanism

- There is a need to create simple mechanisms / procedures to regulate the harvesting and transit of agroforestry produce within the State, as well as in various States forming an ecological region. There is also the need to simplify procedures, with permissions extended on automatic route as well as approval mode through a transparent system within a given time schedule. There are regulations imposed by multiple agencies of State governments (*viz.* Department of Forest, land revenue, other local bodies) on harvesting and transit which have negative implications on the growth of agroforestry. All these restricting regulations need to be identified and aligned with the proposed simplified mechanism.
- Considering that States may find practical difficulties to exempt all tree species planted under agroforestry in farmland from State regulations for harvesting and transit, a practical way forward could be to develop a bouquet of commonly grown tress across the country and develop a uniform harvest and transit rules. The IV Quinquennial Review Team (QRT, 2013) of NRCAF and AICRP on agroforestry has identified about twenty important multipurpose tree species at national level after thorough discussion with all the stakeholders, which can be the reference point while developing the bouquet.
- A Committee constituted by the Ministry of Environment & Forests under Shri Arun Kumar Bansal has submitted its report in2011on the regulatory regime to be followed. The relevant recommendations of the Committee may be considered for inclusion in the implementation Guidelines for the National Agroforestry Policy.
- Decentralized institutions of local governance, such as the PRI, Gram Sabhas, Joint Forest Management Committees (JFMCs), Eco Development Committees or other similar people's institutions, such as those under the Forest Rights Act (FRA), Panchayats (Extension to Scheduled Areas) Act 1996 (PESA) etc. may be considered for playing a role in the regulatory mechanisms. The National Agroforestry Policy should be consistent with the FRA, PESA and such Acts (*viz.* Chotanagpur Tenancy Act) under which such provisions of community rights for harvesting and transportation may have been already provided. A point of caution is that delegation of power to the decentralized institutions should

be accompanied with the capacity building measures to equip them (including the Gram sabhas) with the knowledge on agroforestry rules and regulations, quality control, etc. to avoid possible misuse of the power.

3. **Development of a sound database & information system**

- Security of land tenure is a critical issue. More than in other agricultural systems, due to the longer gestation period of trees relative to annual crops stability and security of tenure rights is a necessary condition for farmers to take up agroforestry. In the absence of such rights farmers would be reluctant to invest their labor and meagre capital resources in a crop that may yield benefits after several years. A clear guarantee of tenure rights can support a farmer's strategy to invest in trees. Only then can farmers – as investors – make plans with confidence that the parameters shaping their long-term vision will not change. Similar is the case with the rights of the tenant farmers on trees planted in the farm which they have been cultivating or share-cropping for long. Tenancy rights must be respected. Even the financial institutions usually provide credit based on demonstration of security of tenure. Trees on farmland may be considered as collateral security for the purpose of financing.

- There is need for a more efficient system of data recording as at present there is complete lack of data on agroforestry. Number of trees on farm, species grown, survival rate, number of trees planted /harvested, or quantity of timber/fuel wood/fodder produced is not available. The contribution of agroforestry to the national GDP is also not known. This poses a challenge for planning and implementation. However, in the land record system there is a provision of recording trees on farm. But it is not clear whether the data recorded is authentic and used for any purpose. Updating the old land record systems with the help of modern geo-spatial tools (remote sensing and geographic information system), for recording of basic data such as existing trees, planting of new trees, and harvesting, etc. needs to be undertaken.

4. **Investing in research, extension and capacity building and related services**

- There are over 30 research centers of the ICAR involved in agroforestry research, which is coordinated by the National

Research Centre for Agroforestry (NRCAF). There are claims of availability of technology and appropriate agroforestry models suitable for different agro-climatic regions of the country. Also, advance research knowledge and facilities are reported to be available in the private domain, especially in the pulp and timber industry. However, there appears to be very little replication of such knowledge and models on the ground. The non-existent extension system for agroforestry may be the key reason for non-adoption of technologies; however, one needs to reconfirm the robustness of the technologies which have been developed. Therefore, it is suggested to set aside some seed money to fund time-bound research projects with specific objectives delivered through location specific basic, strategic and applied research.

- Agroforestry cannot succeed without the willing support and cooperation of the people. It is essential, therefore, to inculcate in the people, a direct interest in agroforestry, its promotion and development. This can be achieved through the involvement of educational institutions, right from the primary stage. Farmers and interested people should be provided opportunities through institutions, such as the Krishi Vigyan Kendras, Trainers' Training Centres to learn agroforestry, agri-silvicultural and silvicultural techniques to ensure optimum use of their land and water resources. Short term extension courses and lectures should be organized in order to educate farmers. For this purpose, it is essential that suitable programmes are propagated through mass media, audio - visual aids and the extension machinery.
- There is need to develop yield and volume tables of agroforestry tree species. Likewise, the species with high carbon sequestration capacity need to be identified. Also, a common web-based platform may be setup to bring all research findings and available knowledge in the area of agroforestry for access of all stakeholders.

**5. Improving famers' access to quality planting material**

- Certification of nurseries, seeds and other planting materials for agroforestry is required to make available good quality planting material at the required scale. Institutional mechanism for registration of nurseries and their accreditation should be established. Private sector can play an important role in augmenting supplies of improved planting stock as demonstrated in case of poplar and clonal eucalyptus plantations. Hence, the private sector participation should be encouraged in production and development of supply chain of quality planting materials.

6. **Providing institutional credit and insurance cover for agroforestry**

   - There is need for setting up of special purpose vehicles /banking institutions to address specific needs of agroforestry sector. Agroforestry sector should also be benefitted with the provisions of interest subvention in the line of agricultural credit. Dedicated Farmers Producers Organizations (FPO) be promoted to organize the farmers and take up agroforestry at economies of scale.

7. **Facilitating increased participation of industries dealing with agroforestry produce**

   - The role of agroforestry/biomass-based industries in the promotion of agroforestry is crucial. Therefore, such industry sector needs to be encouraged and facilitated. The role of industries in the promotion of agroforestry can be tapped in multiple ways, especially in the areas of

     a) Production and supply chain development for high quality planting materials,

     b) Technology development and dissemination, especially for planting materials, processing, etc.

     c) Providing extension services to the farmers,

     d) Providing market information and future trends,

     e) Certification of nurseries, seeds and finished products for sustainable management practices,

     f) Developing agroforestry plantation on government lands on lease contract and in partnership with local people's institutions, etc.

   - Policies are therefore, required to recognize the role of industries and make enabling conditions for increasing the participation of the industries.

   - Agroforestry be treated as a priority area under Corporate Social Responsibility programmes.

8. **Strengthening farmer access to markets for tree products**

   - Barring few sporadic instances, the marketing system for agroforestry produce is unorganized. Marketing infrastructure similar to what is available for agricultural commodities, including market information be introduced with more private sector participation.

**9. Incentives to farmers for adopting agroforestry**

- Regional and thematic differentiation in the agroforestry policy needs to be minimized. Incentive and support structure, such as the input subsidy, interest moratorium, etc. during the gestation period for promotion of agroforestry be provisioned to encourage farmers. Clear directives for value chains development need to be incorporated for the agroforestry sector over a period of time. The role of Farmer Producer Organizations (FPOs) in the promotion and value chain development for agroforestry should be considered.

**10. Promoting sustainable agroforestry for renewable biomass-based energy**

- Emphasis needs to be on raising fast growing trees / bushes / grasses on marginal and degraded farmlands keeping in view their uses for meeting various energy requirements for making profitable agroforestry practices. This is of particular significance for meeting various energy needs of agroforestry itself, such as for irrigation, motive power, farm machines and processing industry. Therefore, it should be considered in conjunction with making of the provisions for financial incentives, especially for setting up of various renewable energy systems / devices.

## Conclusion

Enriched agroforestry systems in Guatemala provide greater benefits for families and forests than conventional cropping systems.

- They provide more firewood (65% vs. 7%)
- They provide more maize (26% more)
- Families using enriched agroforestry systems are less likely to sell their labor (15% vs. 48%)

This is important because Guatemala is a country that suffers from persistent poverty and is vulnerable to climate change. Enriched agroforestry systems can help to improve the livelihoods of families and protect forests in this vulnerable country.

## References


1. Adam D. How far will global population rise? Researchers can't agree. *Nature.* 2021; 597:462–465. doi: 10.1038/d41586-021-02522-6.
2. Anderson, Anthony B. 1990. Alternatives to Deforestation: Steps Toward Sustainable Use of the Amazon Rain Forest. New York: Columbia University Press
3. Awasthi, O.P., Singh, I.S., Verma, M.K., 2010. Fruit based diversified cropping system for arid region. Presented in national seminar on "precision farming in horticulture" during December 28–29, 2010, Horticulture College, Jhalawar, Rajasthan India; 12/2010.
4. Azembouh R.T., Nyong P.A., Kfuban Yerima B.P. Characterization of agroforestry systems and their effectiveness in soil fertility enhancement in the south-west region of Cameroon. *Curr. Res. Environ. Sustain.* 2021; 3:100024. doi: 10.1016/j.crsust.2020.100024.
5. Babu S., Mohapatra K.P., Yadav G.S., Lal R., Singh R., Avasthe R.K., Das A., Chandra P., Gudade B.A., Kumar A. Soil carbon dynamics in diverse organic land use systems in North Eastern Himalayan ecosystem of India. Catena. 2020; 194:104785. doi: 10.1016/j.catena.2020.104785.
6. Babu S., Rathore S.S., Singh R., Kumar S., Singh V.K., Yadav S.K., Yadav V., Raj R., Yadav D.D., Shekhawat K., Wani O.A. Exploring agricultural waste biomass for energy, food and feed production and pollution mitigation: A review. *Bioresour. Technol.* 2022; 360:127566. doi: 10.1016/j.biortech.2022.127566.
7. Babu S., Singh R., Yadav D., Rathore S.S., Raj R., Avasthe R., Yadav S.K., Das A., Yadav V., Yadav B., Shekhawat K., Upadhyay P.K., Yadav D.K., Singh V.K. Nanofertilizers for agricultural and environmental sustainability. *Chemosphere.* 2022 doi: 10.1016/j.chemosphere.2021.133451.
8. Buttoud, G. 2013. Advancing Agroforestry on the Policy Agenda: A Guide for Decision-Makers. Rome, Italy: Food and Agriculture Organization of the United Nations. (http://www.fao.org/3/a-i3182e.pdf)
9. Carlsson G., Martensson L.M., Prade T., Svensson S.E., Jensen E.S. Perennial species mixtures for multifunctional production of biomass on marginal land. *Bioenergy.* 2017; 9:191–201. doi: 10.1111/gcbb.12373.
10. Das A., Babu S., Datta M., Sanjeev K., Singh R., Avasthe R.K., Rathore S.S., Yadav S.K., Singh V.K. Restoring soil carbon in marginal land of

Indian Himalayas: Impact of crop intensification and conservation tillage. *J. Environ. Manage.* 2022 doi: 10.1016/j.jenvman.2022.115603.

11. Đorđević N.O., Todorović N., Novaković I.T., Pezo L.L., Pejin B., Maraš V., Tešević V.V., Pajović S.B. Antioxidant activity of selected polyphenolics in yeast cells: the case study of Montenegrin merlot wine. *Molecules.* 2018; 23:1971. doi: 10.3390/molecules23081971.
12. Gokhle A, M, Zeliang DK, Kevichusa R, Angami T, and Bendangnugsang S. 1985. The Use of Alder Tree. State Council of Education Research & Training, Government of Nagaland, Kohima, India. 20 pp
13. Hall N.M., Kaya B., Dick J. Effect of improved fallow on crop productivity, soil fertility and climate-forcing gas emissions in semi-arid conditions. *Biol. Ferti. Soils.* 2005:224–230. doi: 10.1007/s00374-005-0019-8.
14. Hodge, Sandra, H.E. Garrett, and Jerry Bratton. 1999. Alley Cropping: An Agroforestry Practice. Agroforestry Notes. National Agroforestry Center. (http://nac.unl.edu/documents/agroforestry notes/an12ac01.pdf)
15. King, K.F.S. July 21, 1979. Concepts of Agroforestry. Proceedings of an International Conference on International Cooperation in Agroforestry. Nairobi, Kenya.
16. Kuyah S., Cory W.W., Jonsson M., Sileshi G.W., Öborn I., Muthuri C.W., Luedeling E. Agroforestry delivers a win-win solution for ecosystem services in sub-Saharan Africa. A meta-analysis. *Agron. Sustain. Dev.* 2019; 39:47. doi: 10.1007/s13593-019-0589-8.
17. Liu, Can, and Wu, Bin. April 2010. Grain for Green Programme in China: Policymaking and Implementation? China Policy Institute, University of Nottingham. (https://www.nottingham.ac.uk/cpi/documents/briefings/briefing-60-reforestation.pdf)
18. MacDicken, Kenneth G., and Napoleon T. Vergara. 1990. Agroforestry: Classification and Management. New York: Wiley and Sons.
19. Marten, Gerry, and Amanda Suutari. 2006. Reversing Tropical Deforestation: Agroforestry and Community Forest Management (Nakhon Sawan Province, Thailand). EcoTipping Points Project. (http://www.ecotippingpoints.org/our- stories/indepth/thailand-watershed-forest-agroforestry-community-management.html)

20. Muschler, R.G., P.K.R. Nair, and L. Melendez. November 1993. Crown development and biomass production of pollarded Erythrina berteroana, E. fusca and Gliricidia sepium in the humid tropical lowlands of Costa Rica. Agroforestry Systems. Vol. 24, Issue 2, pp 123-143

21. National Agroforestry Center. (http://nac.unl.edu/documents/agroforestrynotes/an12ac01.pdf)

22. National Agroforestry Report 2014

23. Ramachandran Nair, P.K. 1993. An Introduction to Agroforestry. ICRAF and Kluwer Academic Publishers. (http://www.worldagroforestry.org/Units/Library/Books/PDFs/32_An_introduction_to_agroforestry.pdf?n=161)

24. Sharrow, Steven H. November 2007. Soil compaction by grazing livestock in silvopastures as evidenced by changes in soil physical properties. Agroforestry Systems. Volume 71, Issue 3, pp215-223.

25. Sibelet N., Posada K. E., Gutiérrez-Montes I. A., 2019. Agroforestry systems provide firewood for livelihood improvement in Guatemala. Bois et Forêts des Tropiques, 340: 91-102. Doi: https://doi.org/10.19182/bft2019.340. a31692

26. Singh Anurudh K 2010 Probable Agricultural Biodiversity Heritage Sites in India: VI. The Northeastern Hills of Nagaland, Manipur, Mizoram and Tripura. Asian Agri-History 14 (3):217-243.

27. Sharma UC and Singh KA. 1994. Agriculture with alder – a potential indigenous farming system of Nagaland. In: Proceedings of the National Seminar on Developments in Soil Science, 28 Nov to 1 Dec 1994, ICAR, New Delhi, India. ICAR, New Delhi, India. pp. 570–571.

28. Thorlakson T., Neufeldt H. Reducing subsistence farmers' vulnerability to climate change: evaluating the potential contributions of agroforestry in western Kenya. *Agric. Food Secur.* 2012; 1:15.. doi: 10.1186/2048-7010-1-15.

29. Vitousek, P M., and R L. Sanford. 1986. Nutrient Cycling in Moist Tropical Forest. Annual Review of Ecology and Systematics Vol. 17.1: 137-167.

30. Wojtkowski, Paul A. 2006. Introduction to Agroecology: Principles and Practices. New York: Food Products Press.

31. Wright, Bruce, Stuhr, Kimberly. March 2002. Windbreaks: An Agroforestry Practice. Agroforestry Notes. National Agroforestry Center. (http://nac.unl.Edu/documents/agrofore strynotes /an25w01.pdf)

32. Weiwei, Liu, *et al.* 2014. Traditional Agroforestry Systems: One Type of Globally Important Agricultural Heritage Systems. Journal of Resources and Ecology 5.4.

33. Yadav G.S., Das A., Kandpal B.K., Babu S., Lal R., Datta M., Das B., Singh R., Singh V.K., Mohapatra K.P., Chakraborty M. The food-energy-water-carbon nexus in a maize-maize-mustard cropping sequence of the Indian Himalayas: An impact of tillage-cum-live mulching. *Renew. Sustain. Energy Rev.* 2021 Doi: 10.1016/j.rser.2021.111602.

34. Yadav G.S., Kandpal B.K., Das A., Babu S., Mohapatra K.P., Devi A.G., Devi H.L., Chandra P., Singh R., Barman K.K. Impact of 28-year-old agroforestry systems on soil carbon dynamics in Eastern Himalayas. *J. Environ. Manage.* 2021; 283:111978. Doi: 10.1016/j. jenvman. 2021.111978.

# Chapter - 9

## Biotechnology Applications in Agriculture


**Authors**

**Dr. Khyati Savani**

Senior Research Fellow, Junagadh Agricultural University, J.A.U., Junagadh, Gujarat, India

**V.M. Parmar**

Ph.D. Student, Junagadh Agricultural University, J.A.U., Junagadh, Gujarat, India

# Chapter - 9

## Biotechnology Applications in Agriculture

**Dr. Khyati Savani and V.M. Parmar**

### Abstract

Biotechnology is the use of scientific procedures to alter and improve plants, animals, and microbes in order to increase their value. Agricultural biotechnology is the branch of biotechnology concerned with agricultural applications. Agricultural biotechnology has been used for a long time to improve agriculturally important organisms through selection and breeding. The production of disease-resistant varieties by cross-breeding multiple genotypes until the necessary disease resistance was present in the ensuing new variety is an example of conventional agricultural biotechnology. Biotechnology has played a significant role in agriculture by modifying genes, studying, and cloning diverse crops in order to create higher-quality food products, thereby enhancing human lives. Some of the potential applications include hybrid seeds, artificial seeds, photosynthesis improvers, stress resistant crops and plants, biofertilizers, and bio-pesticides. Biotechnology has the potential to provide benefits in a wide range of agricultural applications, including animal management, agricultural product storage, and maintaining present crop yields while lowering the usage of fertilizers, herbicides, and pesticides. Biotechnology provides a possible alternative to manufactured foods as well as an advance over traditional plant-breeding technologies. It, in conjunction with other sophisticated agricultural technology, provides an exciting and environmentally responsible solution to meet customer demand for sustainable agriculture.

**Keywords:** Agriculture, biotechnology, GM crops, tissue culture, biofertilizers, biopesticides.

### Introduction

Biotechnology is a technique that makes or modifies a product using living organism or compounds derived from these organisms. Biotechnology may be applied to all species of organism, from viruses and bacteria to plants and animals, and it is rapidly becoming an important component of modern medicine, agriculture, and business.

Agricultural biotechnology has been used for a long time to improve agriculturally important organisms through selection and breeding. Transgenic plants have been generated as a result of genetic engineering research in which genetic material is transported from one organism to another for the latter to exhibit a desired attribute. Modern agricultural biotechnology covers a variety of methods that scientists employ to understand and change the genetic make-up of organisms for application in agricultural product production or processing. Biotechnology is being employed to solve issues in all aspects of agriculture production and processing. Biotechnology is being used to develop low-cost disease-free planting materials for crops such as banana, as well as new diagnostic and treatment tools for plant and animal diseases.

Biotechnology is being utilized to accelerate plant, livestock, and fish breeding programs and to broaden the spectrum of qualities that can be addressed. Biotechnology is changing animal feeds and feeding practices to increase animal nutrition and reduce environmental waste. Biotechnology is utilized in disease diagnosis and vaccine manufacturing against animal diseases, but agricultural biotechnology has the potential to be the most powerful and beneficial to the poor.

**Applications of biotechnology in agriculture**

All living species have the ability to evolve naturally in order to adapt to changing environmental conditions. However, it takes hundreds of years to see any discernible progress. Then, through various ways, including biotechnology, man learnt how to domesticate and breed plants in order to grow food to his liking and needs.

Biotechnology is defined as a set of technologies that employs living organisms (or portions of organisms) to create or modify a product, improve plants, trees, or animals, or create microbes for specialized use. Agricultural biotechnology refers to crop and livestock enhancement using biotechnology technologies. Agricultural biotechnology includes a variety of tools and features from traditional breeding approaches, including bioinformatics, microbiology, molecular genetics, biochemistry, plant physiology, and molecular biology.

**The following biotechnology tools are important for agricultural biotechnology**

- The conventional plant breeding
- Plant tissue culture

- Molecular breeding and marker assisted selection
- Genetic engineering and genetically modified crops

### a) The conventional plant breeding

Farmers have been modifying the genetic makeup of the crops since the agriculture eight to ten thousand years ago. Early farmers saved the best-looking plants and seeds to grow the following year. Domesticated plant species have changed considerably in comparison to their wild cousins due to selection for traits such as faster growth, bigger yields, pest and disease resistance, larger seeds, or tastier fruits. Plant breeding began when man discovered that crop plants could be artificially mated or cross-pollinated to increase the plant's characteristics. Desirable traits from the parent plants may be blended in the progeny. Plant breeders learned better how to choose superior plants and breed them to generate new and improved kinds of various crops as the science of plant breeding advanced in the twentieth century. This has enhanced the production and quality of the plants we grow for food, feed, and fibre significantly.

For hundreds of years, the conventional plant breeding has been utilised to generate new crop varieties (Figure 1). However, with an expanding population, dwindling agricultural resources such as land and water, and an apparent plateauing of the yield curves of staple crops, conventional plant breeding can no longer meet global demand. As a result, new crop enhancement technologies should be developed and implemented.

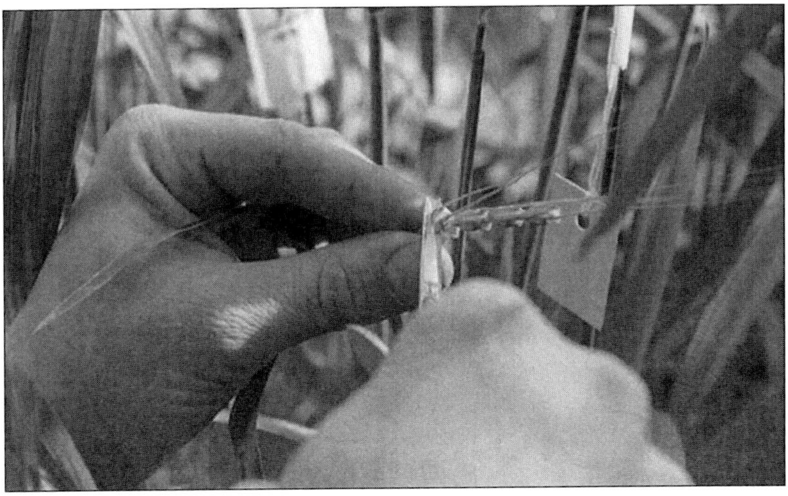

**Figure 1:** Conventional method of plant breeding (Source: Mac-Key & Qualset, 1986)

Since the beginning, the technology has had a huge impact on important crops such as rice, corn, wheat, cotton, and other crops such as numerous vegetables. With the demonstrated influence of hybrid seed technology, new techniques for hybrid breeding, such as cytoplasmic male sterility (CMS), were identified and used for self-pollinating crops. Cytoplasmic male sterility occurs when a plant is unable to produce functional pollen and must rely on another source of pollen to create seeds. This considerably simplifies large-scale hybrid seed production by eliminating the need for human pollination.

To produce hybrid seed, current hybrid seed technology employs three lines: (a) the A line, which contains a defective mitochondrial genome in the cytoplasm and a suppressed restorer gene, (b) the B line, which is genetically similar to the A line but contains normal cytoplasm and a suppressed restorer gene, and (c) the restorer line, a distinctly unrelated line that contains normal cytoplasm and an active restorer gene (dominant).Another hybrid seed method is the two line hybrid system, which relies on temperature and geographic location to impact the plant's nuclear genome, resulting in male sterility. Hybrid seed technology ensures hybrid vigour in progenies, but discovering and developing CMS lines takes a lot of effort and time.

The two-line hybrid system is another hybrid seed method that uses temperature and geographic location to influence the plant's nuclear genome, resulting in male sterility. Hybrid seed technology provides hybrid vigour in progenies, but identifying and developing cms lines is time-consuming and labour-intensive. This limits the number of new qualities that can be introduced to those already present in the species. Second, when plants are crossed, numerous traits, including those having negative effects on yield potential, are transferred together with the trait of interest. Breeders can overcome these issues through agricultural biotechnology.

**b) Plant tissue culture**

Plants typically reproduce sexually, with blooms and seeds to form the next generation. Pollen from the stamens (male part) of the same plant's flower (self-pollination) or another plant's flower (cross-pollination) fertilizes egg cells in the flowers. Each of these reproductive cells has DNA, which is genetic material. During sexual reproduction, DNA from both parents is mixed, resulting in offspring that are similar to the parents (in self-pollinated crops) or unique organisms (in cross-pollinated crops). Some plants and trees, require several years before flowering and producing seeds, making plant improvement challenging. To assist breeders in this effort, plant scientists have developed the science and art of tissue culture.

The growing of plant cells, tissues, or organs on specially formulated nutritional media is known as tissue culture. An entire plant can be regenerated from a single cell under the appropriate conditions. Plant tissue culture is a method that has been used for over 30 years. Tissue culture can be classified into different types based on the plant (explant) employed.

The anther culture is a tissue culture technology used to rapidly generate better cultivars (Figure 2). Pollen carries half of the genome (haploid), which spontaneously doubles (diploid) during cultivation. However, in some species, colchicine therapy is required to cause doubling. The doubling of the genome will allow the manifestation of recessive traits that were previously suppressed, hidden, or overlooked in plant breeding.

Immature pollen within the anther divides and generate a mass of dividing cells known as callus when placed in a specific media. Picking healthy calli (plural of callus) and placing them in another medium to create shoots and roots (regeneration). In the greenhouse, stable plantlets are allowed to grow and mature. Plant breeders can then choose the plants they want from the regenerated plants. Anther cultivation of F1 plants that are progenies in a certain breeding objective would allow for a much greater variety of regenerants. This is because the pollen's genetic makeup will be more varied than that of inbreeds, giving breeders a greater selection of features to choose from. This method has been used to successfully create twofold haploid rice, wheat, sorghum, barley and other field crops. Since 1995, the National Seed Industry Council of the Philippines has introduced rice varieties created by anther culture (AC). IRRI developed and commercialized the first AC-derived, salt tolerant cultivar PSBRc50 (Bicol) in 1995.Eight salt-tolerant types and two rainfed variants were produced by the Philippine Rice Research Institute.

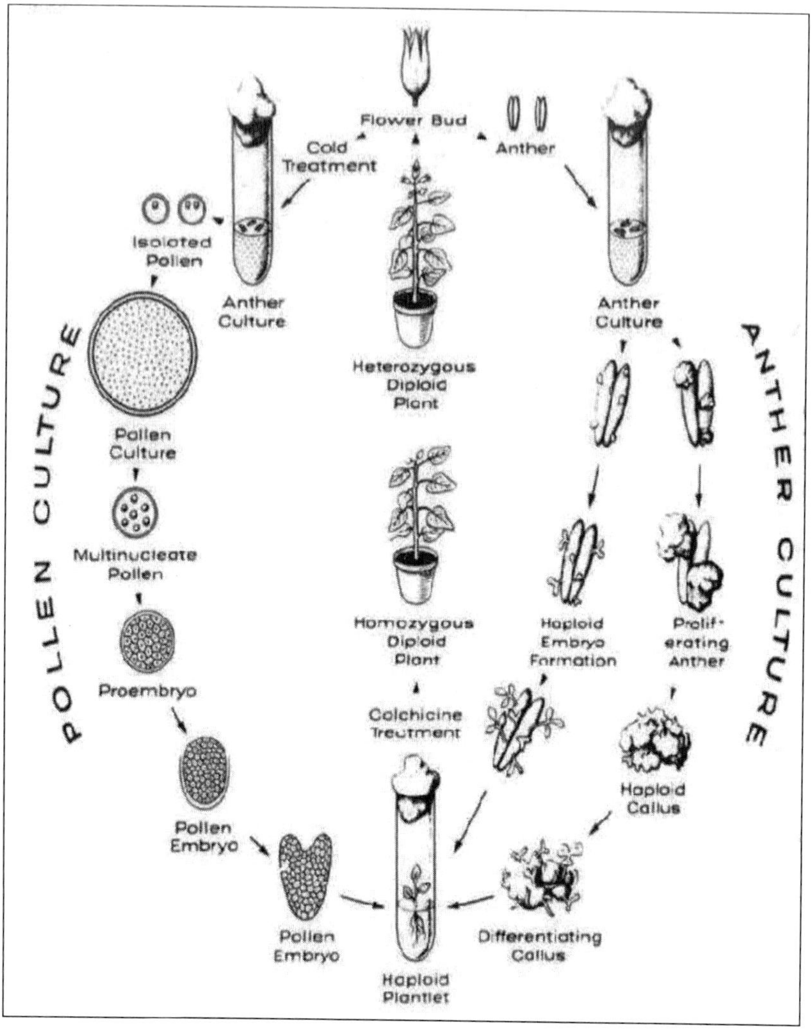


**Figure 2:** Procedure of anther culture (Source: Susanto *et al.,* 2013)

Micropropagation is a tissue culture process used to produce disease-free, high-quality planting material and to rapidly produce a large number of uniform plants. Young cells that are actively dividing (meristem) are implanted in a particular medium and treated with plant hormones to produce a large number of similar sister plantlets. Because the meristem divides faster than the virus, clean materials are propagated and hundreds of uniform plantlets are created in a short period of time.

Micropropagation allows for the production of clean and uniform planting materials in plantations such as oil palm, plantain, pine, banana, abaca, date and rubber tree; field crops such as eggplant, jojoba, pineapple and tomato; root crops such as cassava, yam and sweet potato; and many ornamental plants such as orchids and anthuriums. Micro propagated plants established faster, grew more aggressively and taller, had a shorter and more consistent production cycle, and produced larger yields than conventional propagules.

Embryo rescue is the growth of immature embryos of plants in a specific medium to prevent abortion and to promote germination (Figure 3). This is commonly utilised in breeding parental lines with divergent or incompatible genomes, such as when transferring crucial features of wild cousins into domesticated crops.

Wide crosses between the Asian *Oryza sativa* and the African rice *Oryza glaberrima* resulted in the production of a new rice plant type for West Africa (NERICA - New Rice for Africa). To stabilize the breeding lines, it uses embryo rescue in the initial breeding as well as successive back crossing work followed by anther culture. The new plants combined O. *sativa* parent yield features with *O. glaberrima* parent local adaption traits.

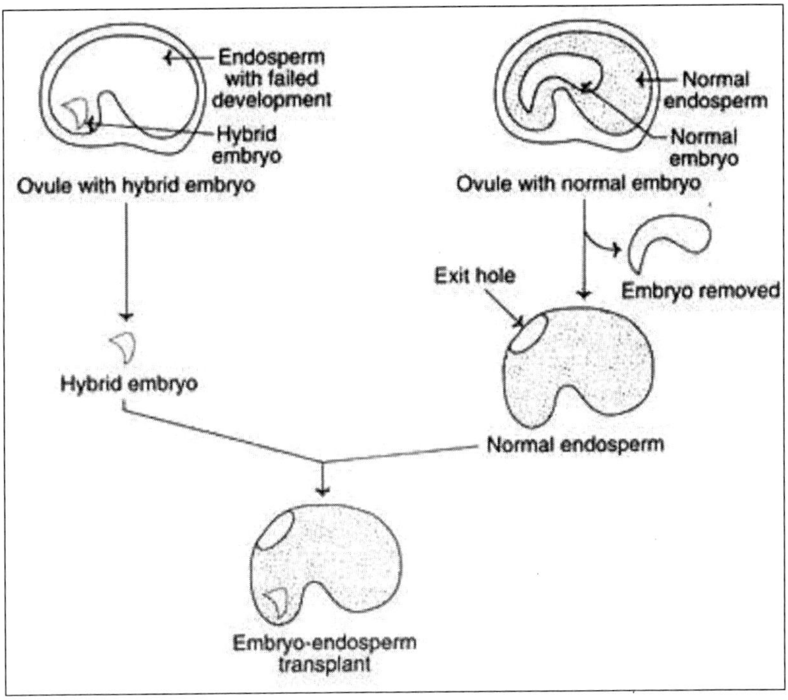


**Figure 3:** Embryo rescue (*Source*: Cisneros *et al.*, 2013)

Wild rice are a rich source of pest and abiotic stress resistance characteristics. Embryo rescue was used at the International Rice Research Institute to promote the transfer of bacterial blight resistance genes from wild rice *Oryza longistaminata* to variety IR24, resulting in a bacterial blight resistant line (IRBB21). *Oryza rufipogon* provides tungro resistance to a variety of rice cultivars. The embryo produced in the cross between very salt tolerant wild rice *Oryza coarctata* and cultivated rice variety IR56 was saved at IRRI and used to build a new super salt tolerant rice.

Plant tissue culture is at the bottom of the agricultural biotechnology food chain. However, the ability of the plant to regenerate a new plant is a crucial requirement in the production of enhanced crops via agricultural biotechnology. Plant tissue culture is a simple process that many developing countries have already mastered. Its use requires simply sterilized workplace, nursery, green house, and qualified personnel. Unfortunately, tissue culture is time consuming, labour intensive, and can be costly.

### c) Molecular breeding and marker assisted selection

Developing new crop types is a lengthy process that might take anywhere from 10 to 25 years depending on the crop. However, agricultural biotechnology applications have significantly cut the time it takes to bring them to market. It currently takes 7-10 years to develop new crop varieties. Marker assisted selection (MAS) is one of the strategies that scientists use to make it easier and faster to pick plant features.

Plant qualities and physical characteristics are encoded in the plant's genetic material, deoxyribonucleic acid (DNA). DNA is found in pairs of chromosomes (strands of genetic material), one from each parent. Genes are distinct portions of each chromosome that govern the plant's traits. The genome of a plant is made up of all of its genes. Some characteristics, such as flower colour, may be determined by a single gene. Other, more complex traits, such as crop yield or starch content, may be controlled by a number of genes. Plant breeders have traditionally chosen plants based on observable or measurable features known as phenotypes. However, this process can be tough, sluggish, environmentally impacted, and costly - not only in terms of development, but also for the economy, as farmers suffer crop losses (Figure 4).

Plant breeders currently use molecular marker-assisted selection as a shortcut. Scientists utilise molecular markers, which are small strings or sequences of nucleic acid that make up a segment of DNA, to assist identify certain genes. The markers are found near the desired gene's DNA sequence.

Because the markers and genes are close together on the same chromosome, they tend to stay together with each generation of plants. This is known as genetic linkage. Scientists can use this relationship to forecast whether a plant will have the desired gene. If researchers discover the gene's marker, it implies that the gene itself exists. Scientists can develop a map of the markers and genes on certain chromosomes when they understand where each marker occurs on a chromosome and how close it is to a specific gene. The position of markers and genes, as well as their distance from other known genes, are depicted on this genetic linkage map. In only one generation of plant breeding, scientists can create precise maps.

Previously, scientists used traditional approaches to create relatively rudimentary genomic maps. Long ago, it was discovered that when plant generations were crossed, some features regularly emerged together in the new generations (genetic linkage). However, because researchers could only focus on a few features in each cross-breeding effort, it took several crosses to obtain even a relatively simple genetic map. Researchers can analyze a tiny amount of tissue from a newly germinated seedling using very detailed genetic mapping and a greater understanding of the molecular structure of a plant's DNA. They don't have to wait for the seedling to mature before testing for the existence of the given characteristic. Scientists can determine whether a seedling carries the relevant gene after analysing the tissue using molecular techniques. If not, they can swiftly move on and focus investigation on another seedling, eventually working only with plants that have the desired characteristic.

Molecular markers are also utilised to determine a line's or variety's genetic profile. Molecular approaches are utilised to scan the genomic makeup of the plant using random primers. The data is entered into a computer programme, which analyses the relationship of one line to another. The genetic variety of the lines' information is used to select for extremely unrelated parents useful for hybrid seed technology. The data will also include information on the line's parentage, potential qualities, and the plant's unique identification, which will be important for germplasm collection databases.

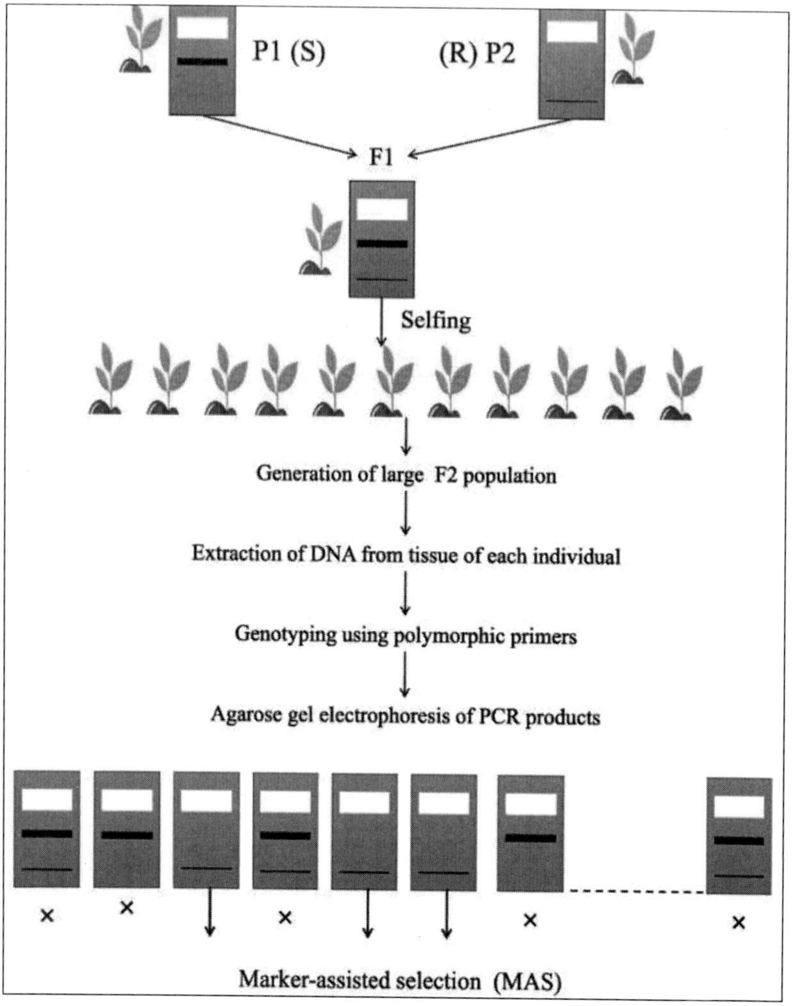


**Figure 4:** Procedure of Marker Assisted Selection (Source: Hasan *et al.*, 2021)

However, it should be noted that molecular breeding through marker assisted selection is somewhat limited in scope when compared to genetic engineering or modification because:

1) It only works for traits already present in a crop;
2) It cannot be used effectively to breed crops with long generation time, like citrus; and
3) It cannot be used effectively with crops that are clonally propagated because they are sterile or their offspring do not resemble the paternal parent.

Many mainstays are included, such as yams, bananas, plantains, sweet potatoes, and cassava.

### d) Genetic engineering and genetically modified crops

One of the recombinant DNA-based agricultural biotechnology methods used today is genetic engineering. The term "genetic engineering," which is sometimes interchanged with terms like "gene technology," "genetic modification," or "gene manipulation," refers to the procedure of changing an organism's genetic makeup utilizing "recombinant DNA technology." This entails the removal, insertion, and modification of DNA fragments that contain one or more desirable genes using laboratory equipment and specialized enzymes. Genetic engineering differs from conventional plant breeding in that it allows for the manipulation of specific genes as well as the transfer of genes between species that would not typically interbreed.

With normal plant breeding, there is little to no assurance that any specific gene combination will be obtained among the millions of crosses produced. Because the genes of both parents are combined and rearranged more or less randomly in the offspring, undesirable genes can be passed alongside beneficial genes, or while one desirable gene is obtained, another is lost. These issues restrict the advancements that plant breeders can make, costing time and money in the process (Figure 5).

In contrast, genetic engineering enables the direct transfer of one or a small number of genes, between species that are either closely related or unrelated. Not all methods of genetic engineering include introducing DNA from different organisms. Another way to alter plants is by deleting or turning off specific genes and genetic regulators (promoters).

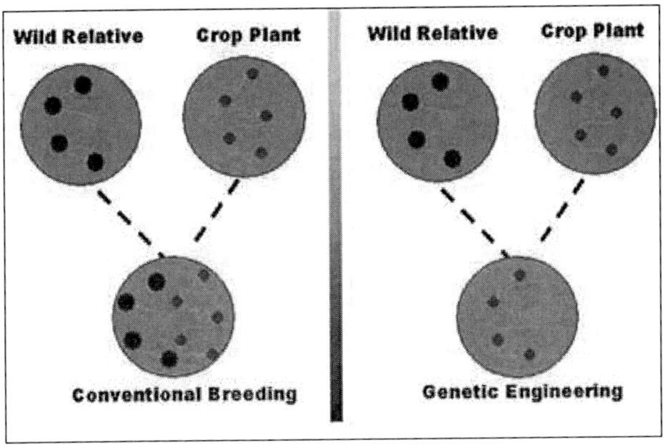


**Figure 5:** Conventional breeding vs genetic engineering (Source: Zarafi & Dauda, 2021)

Genetic engineering techniques are used only after all other techniques have been exhausted and when:

1) The trait to be introduced is not present in the crop's germplasm;
2) The trait is extremely difficult to improve using conventional breeding methods; and
3) It will take an extremely long time to introduce and/or improve such trait in the crop using conventional breeding methods.

Modern plant breeding is a multidisciplinary and coordinated process that employs and integrates a wide range of tools and elements from traditional breeding techniques, bioinformatics, biochemistry, molecular genetics, molecular biology, and genetic engineering.

**Transgenic crop development**

Although genetic engineering involves many diverse and complex procedures, its underlying concepts are rather straightforward. It is, nevertheless, critical to understand the biochemical and physiological processes of action, gene expression regulation, and the safety of the gene and gene product to be used. A set of five steps must be completed successfully to complete the genetic engineering process:

1. Nucleic acid (DNA/RNA) Extraction
2. Gene cloning
3. Gene Design and Packaging
4. Transformation by direct or indirect method
5. Detection of Inserted Genes

Alfalfa, *Argentine canola*, bean, carnation, chicory, cotton, creeping ben tgrass, eggplant, flax, maize, melon, papaya, petunia, plum, *Polish canola*, poplar, potato, rice, rose, soybean, squash, sugar beet, sugarcane, sweet pepper, tobacco, tomato and wheat are among the 27 commercially planted transgenic crops. More than one attribute can be integrated into a plant using genetic engineering, which is known as stacked traits. These crops are now maize, cotton and soybeans with herbicide and insect tolerance. Commercially accessible transgenic crops with combination characteristics include herbicide-tolerant and insect-resistant maize and cotton. Stacking different genes for one feature makes the crop more resilient to pest/disease attack and tolerating more herbicides. The use of a refuge is another technique for improving the technology's sustainability. Effective refuge systems for various transition events have been researched by technology developers. These are thoroughly addressed with farmers in order to ensure proper application, and they are routinely monitored for any resistant insects or weeds.

Commercial GM crops have improved crop output to this point, but a variety of goods that will have a more immediate impact on food quality, a clean environment, pharmaceutical manufacturing, and livestock feeds are also in the works. The following are some examples of these products: rice with higher levels of iron and beta carotene; long-lasting bananas that ripen faster on the tree and can therefore be harvested earlier; maize with improved feed value; delayed-ripening papaya; papaya resistant to the papaya ringspot virus; tomatoes with high levels of flavonols, which are potent antioxidants; drought-tolerant maize and wheat; maize with improved phosphorus availability; arsenic-tolerant plants.

## Conclusion

Agriculture biotechnology applications aid in long-term food production. Many fields of conventional agricultural research benefit from biotechnology rather than being replaced by it. It provides a number of tools to help us better understand and manage genetic resources for food and agriculture. As an example, by minimising the requirement for pesticide spray and fertilizer, new cultivars with high yield and tolerance to abiotic stress can be developed. These transgenic crops can grow in a variety of environmental conditions, have higher nutritional value, and can also be used to produce vaccines and healthcare products for the human population. Tissue culture, micropropagation, and DNA marker assisted techniques are fundamental tools for transgenic variety. These instruments are already being used in breeding and conservation programmes, as well as to aid in the diagnosis, treatment, and prevention of plant and animal illnesses. Biotechnology also addresses environmental issues that have an impact on agriculture, either directly or indirectly.

## References


1. Cisneros, A., Garcia, R. B., & Tel-Zur, N. (2013). Creation of novel interspecific-interploid Hylocereus hybrids (Cactaceae) via embryo rescue. *Euphytica*, *189*, 433-443.
2. Coghlan, A. (2003). 'Protato' to feed India's poor. *New Scientist*, *177*(2376), 7-7.
3. Desamero N.V, Chico M.V, Fernando I.C., Mariano, M. V., Pastor, H.M., Cabusora C.C.,... & Padolina, T. F. (2007). Genetic enhancement

of *in vitro*-culture derived tungro resistance rice breeding lines. *Philippine Journal of Crop Science (Philippines).*

4. DeVries, J., & Toenniessen, G. H. (2001). *Securing the harvest: biotechnology, breeding, and seed systems for African crops*. Cabi.

5. Eckardt, N.A. (2006). Cytoplasmic male sterility and fertility restoration.

6. Gelvin, S. B. (2003). Agrobacterium-mediated plant transformation: the biology behind the "gene-jockeying" tool. *Microbiology and molecular biology reviews*, *67*(1), 16-37.

7. George, E. F., Hall, M. A., & De Klerk, G. J. (2008). Plant propagation by tissue culture 3rd Edition. *The Netherland, The Back Ground Springer.*

8. Germana, M. A. (2011). Anther culture for haploid and doubled haploid production. *Plant Cell, Tissue and Organ Culture (PCTOC)*, *104*, 283-300.

9. Goto, F., Yoshihara, T., Shigemoto, N., Toki, S., & Takaiwa, F. (1999). Iron fortification of rice seed by the soybean ferritin gene. *Nature biotechnology*, *17*(3), 282-286.

10. Hasan, N., Choudhary, S., Naaz, N., Sharma, N., & Laskar, R. A. (2021). Recent advancements in molecular marker-assisted selection and applications in plant breeding programmes. Journal of Genetic Engineering and Biotechnology, 19(1), 1-26.

11. James, C. (2002). *Global review of commercialized transgenic crops: 2001 feature: Bt cotton* (Vol. 26). Ithaca, NY: ISAAA.

12. Khan, S., & Khan, J. (2010). Drought tolerant wheat cultivar (Raj) for rainfed areas of KPK, Pakistan. *Pak. J. Agri. Sci*, *47*(4), 355-359.

13. Kitch, L., Koch, M., & Sithole-Niang, I. (2002). Crop biotechnology: A working paper for administrators and policy makers in sub-Saharan Africa. *Plant Production and Protection. FAO. Harare, Zimbabwe.*

14. Kole, C. (Ed.). (2011). *Wild crop relatives: genomic and breeding resources: tropical and subtropical fruits*. Springer Science & Business Media.

15. Kunz, K. (2002). Vegetable breeding for market development: East-West Seeds 1982-2002. *(No Title).*

16. Lemaux, P. G. (2008). Genetically engineered plants and foods: a

scientist's analysis of the issues (Part I). *Annu. Rev. Plant Biol.*, *59*, 771-812.

17. Lopez-Bucio, J., De la Vega, O. M., Guevara-Garcia, A., & Herrera-Estrella, L. (2000). Enhanced phosphorus uptake in transgenic tobacco plants that overproduce citrate. *Nature biotechnology*, *18*(4), 450-453.

18. Mac-Key, J., & Qualset, C. O. (1986). Conventional methods of wheat breeding. *Genetic Improvement in Yield of Wheat*, *13*, 7-24.

19. Milu, M., 2008. Smallholder adoption and economic impacts of tissue culture banana in kenya. Department of Agricultural, Food and Resource Economics: Michigan State University.

20. Odland, W., Baumgarten, A., & Phillips, R. (2006). Ancestral rice blocks define multiple related regions in the maize genome. *Crop science*, *46*, S-41.

21. Phillips, R. L. (2006). Genetic tools from nature and the nature of genetic tools. *Crop science*, *46*(5), 2245-2252.

22. Phillips, R. L., Odland, W. E., & Kahler, A. L. (2007). Rice as a reference genome and more. In *Rice Genetics V* (pp. 3-15).

23. Rines, H. W., Molnar, S. J., Tinker, N. A., & Phillips, R. L. (2006). Oat. *Cereals and millets*, 211-242.

24. Robinson, C. (2001). *Genetic modification technology and food: consumer health and safety*. Ilsi Europe.

25. Schnable, P. S., & Wise, R. P. (1998). The molecular basis of cytoplasmic male sterility and fertility restoration. *Trends in plant science*, *3*(5), 175-180.

26. Senadhira, D., Zapata-Arias, F. J., Gregorio, G. B., Alejar, M. S., De La Cruz, H. C., Padolina, T. F., & Galvez, A. M. (2002). Development of the first salt-tolerant rice cultivar through indica/indica anther culture. *Field Crops Research*, *76*(2-3), 103-110.

27. Susanto, D., Ibrahim, A. M., & Hussin, Z. E. S. M. (2013). Pollen and anther cultures as potential means in production of haploid kenaf (Hibiscus cannabinus L.). Int J Adv Sci Eng Informat Technol, 3(1), 38-40

28. Tanaka, Y., Brugliera, F., Kalc, G., Senior, M., Dyson, B., Nakamura, N.,... & Chandler, S. (2010). Flower color modification by engineering of the flavonoid biosynthetic pathway: practical perspectives. *Bioscience, biotechnology, and biochemistry*, *74*(9), 1760-1769.

29. Thitisaksakul, M., Jiménez, R. C., Arias, M. C., & Beckles, D. M. (2012). Effects of environmental factors on cereal starch biosynthesis and composition. *Journal of Cereal Science*, *56*(1), 67-80.

30. Van Tran, D., & Duffy, R. (2003). Proceedings of the 20th session of the International Rice Commission. Bangkok, Thailand, 23-26 July 2002. Sustainable rice production for food security. In *International Rice Commission, Sess. 20, Bangkok (Thailand), 23-26 Jul 2002*. FAO.

31. Ye, X., Al-Babili, S., Kloti, A., Zhang, J., Lucca, P., Beyer, P., & Potrykus, I. (2000). Engineering the provitamin A (β-carotene) biosynthetic pathway into (carotenoid-free) rice endosperm. *Science*, *287*(5451), 303-305.

32. Zarafi, A. B., & Dauda, W. P. (2019). Exploring the importance of fungi in agricultural biotechnology. Int. J. Agric. Sci. Vet. Med, 7, 1-12.

# Chapter - 10

## Remote Sensing and GIS for Agricultural Monitoring and Management


**Authors**

**Amol Rajkumar Pawar**
PhD Scholar, Department of Agronomy, Lovely Professional University Phagwara, Punjab, India

**Dipak Gite**
PhD Scholar, Department of Agronomy, Lovely Professional University Phagwara, Punjab, India

**Sushant Sukumar Patil**
Assistant Professor, Department of Agronomy, Lovely Professional University Phagwara, Punjab, India

**Mayur Patil**
PhD Scholar, Department of Agronomy, Lovely Professional University Phagwara, Punjab, India

# Chapter - 10

# Remote Sensing and GIS for Agricultural Monitoring and Management

**Amol Rajkumar Pawar, Dipak Gite, Sushant Sukumar Patil and Mayur Patil**


## Abstract

Remote sensing and GIS are powerful agriculture management tools. Remote sensing uses satellites, aircraft, or drones to collect Earth surface data. GIS collects, analyses, and displays geographical data. These technologies help farmers, academics, and policymakers optimize resource management and agricultural practices by providing vital data into crop health, soil conditions, and environmental factors. This article introduces remote sensing and GIS for agriculture. Satellite, aerial, drone, hyperspectral, multispectral, thermal, and radar remote sensing are covered. Each method captures data on vegetation health, soil moisture, and land cover changes, which are essential for crop evaluation and agricultural practice improvement. GIS helps remote sensing organise, analyze, and visualize spatial data. It combines attribute data like crop types, soil conditions, and weather with geographic data like maps, satellite photos, and aerial photographs. Stakeholders may assess land use changes, monitor crop growth and yield, and make educated precision agricultural decisions using GIS. The paper also covers remote sensing and GIS applications in agricultural monitoring and management. It encompasses crop health evaluation utilizing vegetation indices, disease and pest detection and mapping, irrigation monitoring, water resource management, and crop yield estimation and forecasting. Remote sensing and GIS are used in land use planning, soil mapping, land degradation assessment, watershed management, and environmental impact assessment. Real-world case studies and examples show how remote sensing and GIS work in agriculture. It also addresses data availability and quality, technical capacity and infrastructure, integration, and interoperability issues. These initiatives emphasize collaboration, customization, and continual monitoring and evaluation to inform future remote sensing and GIS applications for agricultural monitoring and management. Remote sensing and GIS improve decision-making, resource management, and sustainable agriculture. These technologies can boost crop output, conserve resources, and enhance agricultural systems.

**Keywords:** Remote sensing, GIS, agriculture management, satellites, aircraft.

## Introduction to remote sensing and GIS

Remote sensing and Geographic Information Systems (GIS) are two powerful technologies that have revolutionized agricultural monitoring and management. Remote sensing involves the acquisition of information about the Earth's surface without direct physical contact. It utilizes sensors mounted on satellites, aircraft, or drones to capture data in the form of images or other measurements. GIS, on the other hand, refers to the collection, management, analysis, and visualization of spatial data. Remote sensing plays a crucial role in agriculture by providing valuable information about crops, soils, and the environment. It utilizes various types of sensors, such as optical, thermal, and radar sensors, to capture data related to vegetation health, soil moisture, and land cover changes. These data can be analyzed to assess crop conditions, monitor changes in land use, and identify potential risks or opportunities for improved agricultural practices. Sishodia *et al.,* (2020)

GIS complements remote sensing by providing a framework for organizing and analyzing spatial data. It integrates geographic data, such as maps, satellite imagery, and aerial photographs, with attribute data, including crop types, soil properties, and weather information. GIS allows for the creation of informative layers of information, enabling farmers, researchers, and policymakers to visualize, analyze, and interpret data in a spatial context. The integration of remote sensing and GIS in agriculture offers numerous applications and benefits. It enables the mapping and monitoring of crop growth, health, and yield, supporting precision agriculture practices. By analyzing remote sensing data in conjunction with other spatial datasets, farmers can identify areas of crop stress, optimize resource allocation, and make informed decisions about irrigation, fertilization, and pest management.

Moreover, remote sensing and GIS facilitate the assessment of land use and land cover changes in agriculture. By comparing historical and current imagery, farmers and land managers can monitor changes in land productivity, identify areas of deforestation or urbanization, and develop strategies for sustainable land use. The integration of remote sensing and

GIS in agriculture is of paramount importance because it enhances decision-making processes and enables a more holistic understanding of agricultural systems. By combining the spatial information provided by remote sensing with the analytical capabilities of GIS, stakeholders can make informed decisions for efficient resource management, risk assessment, and sustainable agricultural practices. Sishodia *et al.,* (2020)

**a) Remote sensing techniques for agricultural monitoring**

Remote sensing techniques play a crucial role in agricultural monitoring, providing valuable information about crop health, soil conditions, and environmental factors. Here are some key remote sensing techniques used in agriculture:

**1. Satellite-based remote sensing systems and sensors**

- Satellites equipped with various sensors capture data about the Earth's surface from space. They provide wide coverage, repetitive observations, and data archives for historical analysis.
- Optical sensors, such as MODIS (Moderate Resolution Imaging Spectroradiometer), provide data in different spectral bands, enabling vegetation health and land cover analysis.
- Synthetic Aperture Radar (SAR) sensors use microwave frequencies to penetrate clouds and provide information on soil moisture, crop structure, and surface roughness. Mani. (2018)

**2. Aerial and drone-based remote sensing technologies**

- Aerial platforms, including manned aircraft or unmanned aerial vehicles (UAVs), offer flexibility in data acquisition at higher spatial resolutions than satellites.
- Drones equipped with cameras, multispectral sensors, or LiDAR (Light Detection and Ranging) scanners can capture high-resolution imagery for detailed crop monitoring and analysis.
- Aerial and drone-based remote sensing allow for localized assessments, quick response to field conditions, and the creation of detailed crop maps. Mani. (2018)

**3. Hyperspectral and multispectral imaging for crop analysis**

- Hyperspectral sensors capture data across numerous narrow and contiguous spectral bands, providing detailed spectral information for crop analysis.

- Multispectral sensors capture data in a few broad spectral bands, often including the visible, near-infrared, and thermal ranges.
- Hyperspectral and multispectral data can be used for vegetation indices, such as NDVI (Normalized Difference Vegetation Index), to assess crop health, monitor vegetation stress, and detect nutrient deficiencies or disease outbreaks.

**4. Thermal and radar remote sensing for agricultural applications**

- Thermal sensors measure the emitted or reflected thermal radiation from the Earth's surface, providing information on crop water stress, evapotranspiration rates, and soil moisture content.
- Radar sensors, such as Synthetic Aperture Radar (SAR), emit microwave signals and measure their interaction with the Earth's surface. Radar can penetrate cloud cover, providing data on soil moisture, crop height, and surface roughness.

These remote sensing techniques, whether satellite-based, aerial, or drone-based, offer valuable information for agricultural monitoring and management. By utilizing hyperspectral, multispectral, thermal, and radar sensors, farmers and researchers can assess crop health, monitor soil conditions, and make informed decisions for optimized resource management and improved agricultural practices. Mani. (2018)

**b) GIS applications in agricultural management**

Geographic Information Systems (GIS) play a vital role in agricultural management, providing a powerful framework for spatial data analysis, visualization, and decision-making. Here are some key applications of GIS in agriculture:

**1. Introduction to GIS and its relevance in agriculture**

- GIS combines spatial data, such as maps and satellite imagery, with attribute data, including crop types, soil characteristics, and weather information.
- GIS enables the integration, analysis, and visualization of this data, allowing farmers, researchers, and policymakers to make informed decisions based on spatial relationships and patterns. Shelestov *et al.,* (2013)

**2. Use of GIS for mapping and monitoring agricultural land**

- GIS is instrumental in mapping and monitoring agricultural land. It facilitates the creation of accurate digital maps that depict land parcels, land use patterns, and infrastructure.

- By integrating remote sensing data with GIS, land cover changes, crop rotation patterns, and encroachments can be monitored over time, providing valuable insights for land management and policy planning. Shelestov *et al.,* (2013)

**3. Spatial analysis and modeling in agricultural decision-making**

- GIS enables spatial analysis, which involves examining relationships between different spatial variables, such as crop yields, soil types, and weather patterns.
- Through spatial modeling, GIS can simulate various scenarios, helping farmers and decision-makers evaluate the potential impacts of different agricultural practices, land use changes, or climate conditions. This aids in making informed decisions for resource allocation and risk assessment. Shelestov *et al.,* (2013)

**4. Precision agriculture and variable rate technology using GIS**

- GIS is a cornerstone of precision agriculture, which involves optimizing resource use for maximum efficiency and productivity.
- By integrating GIS with sensor technologies, such as yield monitors, soil moisture sensors, and remote sensing data, farmers can implement variable rate technology. This technology enables the precise application of inputs, such as fertilizers and irrigation, tailored to the specific needs of different areas within a field. Shelestov *et al.,* (2013)

**5. Crop yield estimation and prediction with GIS**

- GIS facilitates crop yield estimation and prediction by incorporating multiple data sources, including historical yield data, weather data, soil characteristics, and remote sensing information.
- By analyzing these data layers spatially, GIS can generate predictive models that estimate crop yields, enabling farmers to anticipate production levels, plan harvest schedules, and make informed marketing and distribution decisions.

In summary, GIS plays a crucial role in agricultural management by enabling mapping and monitoring of agricultural land, conducting spatial analysis and modeling, supporting precision agriculture practices, and facilitating crop yield estimation and prediction. By leveraging the power of GIS, farmers and stakeholders can make data-driven decisions, optimize resource utilization, and improve agricultural productivity and sustainability.

### c) Crop monitoring and management

Crop monitoring and management are critical aspects of agricultural practices, and remote sensing and GIS-based approaches have proven invaluable in these areas. Here are the key applications of remote sensing and GIS in crop monitoring and management:

#### 1. Remote sensing and GIS-based approaches for crop monitoring

- Remote sensing, in combination with GIS, provides a comprehensive approach to monitor crops. Satellite imagery, aerial photography, and drone-based data collection techniques are employed to capture detailed information about crop conditions.
- GIS facilitates the integration and analysis of these data, allowing farmers and researchers to monitor crop growth, detect anomalies, and assess spatial variations in crop health and productivity. Seelan *et al.,* (2003)

#### 2. Crop health assessment using vegetation indices

- Vegetation indices, derived from remote sensing data, are widely used to assess crop health and vigor. The most common index is the Normalized Difference Vegetation Index (NDVI), which measures the density and vigor of vegetation based on the reflectance of near-infrared and red light.
- By analyzing NDVI and other vegetation indices, farmers can identify areas of stress, nutrient deficiencies, or disease outbreaks, enabling targeted interventions and resource allocation. Seelan *et al.,* (2003)

#### 3. Detection and mapping of crop diseases and pests

- Remote sensing and GIS enable the detection and mapping of crop diseases and pests. Changes in spectral signatures captured by remote sensing data can indicate the presence of diseases or pests in crops.
- By analyzing these spectral changes using GIS, experts can map the spatial distribution of diseases or pests, allowing for targeted monitoring and management strategies, such as localized treatment or pest control measures. Seelan *et al.,* (2003)

4. **Water resource management and irrigation monitoring**

- Remote sensing and GIS play a crucial role in water resource management and irrigation monitoring. Remote sensing data, including thermal imagery and soil moisture measurements, can provide insights into crop water requirements and identify areas of water stress.
- By integrating these data with GIS, farmers can optimize irrigation practices, monitor water usage, and ensure efficient water resource management, leading to improved water conservation and crop productivity.

5. **Crop yield estimation and forecasting using remote sensing data**

- Remote sensing data, combined with GIS and other ancillary data, can be utilized to estimate and forecast crop yields. By analyzing vegetation indices, weather data, soil conditions, and historical yield records, predictive models can be developed.
- These models, integrated within GIS, allow for spatially explicit crop yield estimation and forecasting. Farmers can use this information to make informed decisions about harvest schedules, resource allocation, and marketing strategies.

In summary, remote sensing and GIS-based approaches are instrumental in crop monitoring and management. They enable crop health assessment, disease and pest detection, water resource management, and crop yield estimation. By leveraging these technologies, farmers can enhance their decision-making processes, optimize resource allocation, and achieve improved crop productivity and sustainable agricultural practices. Seelan *et al.,* (2003)

**d) Land use planning and natural resource management**

Land use planning and natural resource management are crucial components of sustainable development. Remote sensing and GIS have emerged as powerful tools in these areas, offering valuable insights and facilitating informed decision-making. Here are the key applications of remote sensing and GIS in land use planning and natural resource management:

1. **Application of remote sensing and GIS in land use planning**

- Remote sensing data, such as satellite imagery and aerial photography, combined with GIS, provide valuable information for land use planning. It enables the identification, classification, and mapping of different land cover and land use types.

- By analyzing spatial patterns, land suitability, and environmental factors, GIS facilitates the assessment of land suitability for various purposes, such as agriculture, urban development, or conservation.

2. **Soil mapping and soil fertility assessment with remote sensing**

- Remote sensing, in conjunction with GIS, is utilized for soil mapping and fertility assessment. By analyzing spectral reflectance patterns and topographic characteristics, remote sensing data can provide information about soil composition and fertility.

- GIS enables the integration of soil data with other spatial data layers, such as land cover, slope, and drainage, supporting soil fertility assessment and guiding appropriate soil management practices. Seelan *et al.,* (2003)

3. **Assessment of land degradation and erosion**

- Remote sensing and GIS help in the assessment and monitoring of land degradation and erosion processes. Remote sensing data can detect changes in land cover, vegetation density, and topographic features associated with erosion.

- By integrating remote sensing data with GIS, land degradation hotspots can be identified, erosion vulnerability can be assessed, and appropriate conservation measures can be planned and implemented.

5. **Watershed management and conservation using GIS**

- GIS plays a vital role in watershed management and conservation. It enables the delineation and mapping of watershed boundaries, identification of land uses within the watershed, and assessment of water quality and quantity.

- By analyzing various spatial layers, such as slope, land cover, and soil types, GIS supports the identification of critical areas for conservation, planning of erosion control measures, and optimizing water resource management practices.

6. **Environmental impact assessment and monitoring**

- Remote sensing and GIS are utilized in environmental impact assessment (EIA) and monitoring. They provide spatial data on land cover changes, habitat fragmentation, and ecological indicators.

- GIS facilitates the integration of environmental data with socio-economic data, supporting the analysis of potential impacts and aiding in decision-making processes related to infrastructure development, natural resource extraction, and conservation initiatives.

In summary, remote sensing and GIS play a significant role in land use planning and natural resource management. They support soil mapping, land degradation assessment, watershed management, and environmental impact assessment. By harnessing the power of remote sensing and GIS, stakeholders can make informed decisions, implement sustainable land use practices, and ensure the conservation and efficient management of natural resources.

**e) Case studies and practical examples**

Remote sensing and GIS applications in agriculture have been successfully implemented in numerous real-world projects, showcasing their potential and impact. Here are a few examples:

**1. Crop disease monitoring and management**

- In a project in India, remote sensing and GIS were used to monitor and manage crop diseases. Satellite imagery and field data were integrated into a GIS platform to detect and map disease outbreaks.
- By analyzing the spatial patterns of disease occurrence and integrating weather data, farmers received timely alerts and guidance on disease management strategies, leading to reduced crop losses and improved yields.

**2. Precision agriculture and variable rate technology**

- In the United States, precision agriculture techniques were implemented using remote sensing and GIS technologies. A combination of satellite imagery and ground sensor data were integrated into a GIS platform.
- By analyzing variations in crop vigor, soil moisture, and nutrient levels, farmers could implement variable rate technology for precise application of inputs, resulting in optimized resource use, increased crop yields, and cost savings.

**3. Land use planning and conservation**

- In Brazil, remote sensing and GIS were utilized for land use planning and conservation in the Amazon rainforest. Satellite imagery and GIS were used to map land cover changes, identify deforestation hotspots, and prioritize areas for conservation.

- The project provided valuable information for policy-makers and land managers, leading to improved land use planning, targeted enforcement efforts, and the preservation of critical ecosystems.

**Challenges faced and lessons learned**

While remote sensing and GIS applications in agriculture have shown great potential, several challenges have been encountered:

1. **Data availability and quality**
   - Obtaining reliable and up-to-date data, including satellite imagery and ground-based information, can be a challenge. Access to high-resolution imagery and accurate ground truth data is essential for accurate analysis.
2. **Technical capacity and infrastructure**
   - Implementing remote sensing and GIS technologies requires technical expertise and suitable infrastructure. Adequate training and capacity building initiatives are necessary to enable effective utilization of these tools.
3. **Integration and interoperability**
   - Integrating data from multiple sources and ensuring interoperability between different software platforms can be challenging. Standardization and data sharing protocols are crucial for seamless integration.

**Lessons learned from these projects include**

1. **Collaboration and stakeholder engagement**
   - Successful implementation requires collaboration among researchers, farmers, policymakers, and technology providers. Engaging stakeholders from the beginning helps ensure the relevance and effectiveness of the applications.
2. **Customization and scalability**
   - Tailoring remote sensing and GIS solutions to local conditions and specific crop requirements enhances their effectiveness. Scalability of the technologies should also be considered to enable wider adoption and impact.

3. **Continuous monitoring and evaluation**

- Regular monitoring and evaluation of the remote sensing and GIS applications are crucial to assess their effectiveness, identify areas for improvement, and make necessary adjustments for better outcomes.

In conclusion, real-world examples demonstrate the successful application of remote sensing and GIS in agriculture, ranging from disease monitoring to precision agriculture and land use planning. However, challenges related to data availability, technical capacity, and integration exist. By addressing these challenges and incorporating lessons learned, the potential of remote sensing and GIS in agriculture can be fully realized, leading to improved decision-making, enhanced productivity, and sustainable agricultural practices.

## f) Future perspectives and emerging technologies

Recent advancements in remote sensing and GIS technologies, along with the integration of artificial intelligence (AI) and machine learning (ML), hold tremendous promise for the future of agricultural monitoring and management. In this section, we will discuss the latest developments, potential future applications, and emerging trends in the field.

### 1. Advancements in remote sensing and GIS technologies

a) High-resolution Imaging: Remote sensing technologies now offer higher spatial and spectral resolution, allowing for more detailed and accurate monitoring of agricultural landscapes. This enables precise identification of crop health, disease outbreaks, and other relevant parameters.

b) Sensor Fusion: Integration of data from multiple sensors, such as optical, thermal, and radar, provides a more comprehensive understanding of agricultural systems. Sensor fusion enhances the ability to detect subtle changes in crop conditions, soil moisture, and other critical factors.

c) Unmanned Systems: Unmanned aerial vehicles (UAVs) and autonomous ground-based systems are being increasingly used for data collection. Equipped with various sensors, they capture detailed information for monitoring and analysis in a cost-effective and efficient manner.

### 2. Potential future applications and trends in agricultural monitoring and management

a) Crop Phenotyping and Genomics: Remote sensing technologies combined with genomic analysis can accelerate crop breeding programs by providing rapid and non-destructive assessments of crop traits, such as yield potential, disease resistance, and nutrient uptake efficiency.

b) Real-time Monitoring and Decision Support Systems: Integration of remote sensing, GIS, and real-time data streams enables the development of decision support systems. These systems can provide farmers with instant feedback on crop health, weather patterns, market trends, and optimal management practices.

c) Predictive Modeling and Risk Assessment: Advances in AI and ML techniques allow for the development of predictive models for crop yield estimation, disease outbreak prediction, and risk assessment. These models can help farmers and policymakers make proactive decisions and implement preventive measures.

d) Climate Change Adaptation: Remote sensing and GIS can contribute to climate change adaptation strategies in agriculture. By monitoring temperature, precipitation, and soil moisture patterns, these technologies can aid in the identification of climate-resilient crops, efficient water management, and adaptive land-use planning.

## 3. Integration of artificial intelligence and machine learning in remote sensing and GIS

AI and ML techniques are being increasingly integrated into remote sensing and GIS workflows, enhancing their capabilities. AI and ML algorithms can automate image analysis, detect patterns, and extract valuable information from large datasets. They can assist in land cover classification, disease detection, yield prediction, and anomaly detection. The integration of AI and ML with remote sensing and GIS facilitates data-driven decision-making, accelerates data processing, and improves the accuracy and efficiency of agricultural monitoring and management practices.

In conclusion, the future of agricultural monitoring and management holds great potential with the advancements in remote sensing and GIS technologies. Higher-resolution sensors, sensor fusion, and unmanned systems are improving data quality and accessibility. Potential future applications include crop phenotyping, real-time monitoring systems, predictive modeling, and climate change adaptation. Additionally, the integration of AI and ML enhances the capabilities of remote sensing and GIS, enabling faster and more accurate analysis. By embracing these advancements, stakeholders in the agricultural sector can make informed decisions, optimize resource allocation, and achieve sustainable and productive agricultural systems.

## g) Conclusion

This chapter has explored the applications of remote sensing and GIS in agricultural monitoring and management. The key points discussed can be summarized as follows:

1. Remote sensing and GIS technologies provide valuable tools for capturing, analyzing, and visualizing spatial data in agriculture.
2. Remote sensing techniques, including satellite imagery, drones, and sensors, enable the collection of data on crop health, land cover, and environmental parameters.
3. GIS facilitates the integration, analysis, and modeling of spatial data, enabling informed decision-making in agricultural management.
4. The integration of remote sensing and GIS allows for comprehensive analysis, database development, and decision support systems in agriculture.
5. Real-world case studies have demonstrated successful applications in crop monitoring, disease detection, land use planning, and natural resource management.
6. Challenges include data availability, technical capacity, and integration, which need to be addressed for effective implementation.
7. Lessons learned emphasize the importance of collaboration, customization, scalability, and continuous monitoring and evaluation.

The importance of remote sensing and GIS in agricultural monitoring and management cannot be overstated. These technologies provide valuable insights into crop health, land use patterns, and environmental factors, enabling farmers, researchers, and policymakers to make informed decisions. They support precision agriculture, efficient resource allocation, and sustainable practices, leading to increased productivity and reduced environmental impact.

To further enhance the application of remote sensing and GIS in agriculture, the following recommendations can be made:

1. Improve data availability and quality by enhancing satellite imaging

capabilities, promoting open data initiatives, and establishing partnerships with data providers.

2. Invest in training and capacity building programs to enhance technical skills and promote widespread adoption of remote sensing and GIS technologies.
3. Foster collaboration among stakeholders, including farmers, researchers, policymakers, and technology providers, to ensure relevance, accessibility, and usability of the applications.
4. Develop standardized protocols for data sharing, interoperability, and integration to facilitate seamless analysis and decision-making.
5. Conduct further research to explore advanced techniques, such as machine learning and artificial intelligence, for automated analysis and decision support in agricultural monitoring and management.
6. Promote awareness and knowledge sharing through workshops, conferences, and publications to disseminate best practices and successful case studies.

By implementing these recommendations, remote sensing and GIS can continue to revolutionize agricultural monitoring and management, contributing to sustainable agricultural practices, increased productivity, and food security.

**References**


1. Mani. 2018 Remote S. A, Remote sensing and GIS in agriculture and forest resource monitoring, *Jugal Kishore Mani and A. O. Varghese*, year (2018).

2. Seelan, S., Laguette, S., Casady, G.M., & Seielstad, G.A. (2003). Remote sensing applications for precision agriculture: A learning community approach. *Remote Sensing of Environment, 88*, 157-169.

3. Shelestov, A.Y., Kravchenko, A.N., Skakun, S., Voloshin, S.V., & Kussul, N.N. (2013). Geospatial information system for agricultural monitoring. *Cybernetics and Systems Analysis, 49*, 124-132.

4. Sishodia, R. P., Ray, R. L., & Singh, S. K. (2020). Applications of remote sensing in precision agriculture: A review. *Remote Sensing, 12*(19), 3136. doi:10.3390/rs12193136

# Chapter - 11

## Water Management Strategies for Efficient Irrigation


**Authors**

**Anu**

Ph.D. Scholar, Department of Agronomy, CCS Haryana Agricultural University, Hisar, Haryana, India

**Arun**

M.Sc. Scholar, Department of Agronomy, CCS Haryana Agricultural University, Hisar, Haryana, India

**Anil Kumar**

Ph.D. Scholar, Department of Soil Science, CCS Haryana Agricultural University, Hisar, Haryana, India

# Chapter - 11

## Water Management Strategies for Efficient Irrigation

**Anu, Arun and Anil Kumar**


### Abstract

The effective and sustainable utilization of water resources for agricultural purposes is made possible by water management systems. The first step is planning irrigation, which entails precisely timing and calculating the quantity of water required for crops at various growth phases. Adopting water-saving irrigation methods is another crucial tactic. These methods, which minimize evaporation and runoff by delivering water directly to the root zone of plants, include drip, sprinkler, and precision irrigation. Effective irrigation management techniques also help to maximize water use. Harvesting rainwater can help supplement irrigation water needs. Reusing and recycling water are equally important tactics. Reducing water demand and relieving pressure on freshwater supplies are two benefits of recycling and treating wastewater from multiple sources, such as treated urban wastewater or agricultural runoff. Water management is improved by encouraging soil health through techniques like cover crops and conservation tillage. These methods improve soil structure, increase water infiltration, and decrease erosion, which results in effective water usage and conservation.


**Keywords:** Water, irrigation, harvesting, management.

### 1. Introduction

It is impossible to exaggerate the value of water in agriculture because it is a resource that is necessary to maintain life. Water is essential for plant growth, development, and productivity in agriculture, which is the main source of food production. Ample water supplies are immediately impacted by agricultural practices and are essential for guaranteeing food supply and livelihoods. It acts as a medium for numerous biological and physiological processes, water is a crucial requirement for plant existence. Irrigation becomes essential for agriculture in regions with irregular or insufficient natural rainfall. By augmenting precipitation and providing that plants receive the right amount of water at each stage of growth, drip irrigation systems provide controlled water delivery to crops. Ample irrigation aids in preserving the ideal soil moisture levels, avoiding water stress, and encouraging strong plant growth. Crop productivity and output are directly impacted by water availability. Reduced agricultural yields, inferior produce, as well as crop failure can result from a lack of water or drought conditions. On the other hand, effective irrigation and water management techniques can maximize crop output and raise agricultural productivity.

A shortage of water in agriculture causes a number of challenges and issues that endanger agricultural systems all over the world. One of the key issues is the disparity between the water supply and the expanding agricultural water demands. Due to changing dietary choices and an increasing global population, there is a growing need for production of food, which calls for more water resources. However, the quantity of water that can be used in agriculture is severely constrained by a paucity of freshwater supplies and competing demands from other businesses. Another issue is the inequitable distribution of water resources spanning both space and time. Water availability can differ greatly amongst places, with some experiencing chronic scarcity of water and others suffering with seasonal shortages.

Furthermore, because of altered rainfall patterns, an increase in the number of droughts, and heightened competition for water resources, climate change exacerbates the problems associated with water shortage. Agriculture's ineffective use of water management practices exacerbates the problem of water scarcity. Water can be lost by evaporation and deep percolation in substantial amounts using outdated irrigation techniques like flood irrigation. Furthermore, inefficient irrigation planning and over-irrigation can waste water, trigger waterlogged soil and soil salinity issues, and lower agricultural output. Water scarcity in agriculture is further complicated by insufficient water infrastructure and restricted access to irrigation technologies. The majority of agricultural workers are smallholder farmers, who frequently lack access to contemporary irrigation systems and water-saving technologies. This reduces their capacity to manage water resources effectively and makes them more susceptible to the effects of water scarcity.

Comprehensive water management techniques that encourage sustainable water usage in agriculture are required to address these issues. The key to reducing the problems of shortages of water in agriculture is to implement effective irrigation systems, embrace water-saving technology, encourage water recycling and reuse, and improve water governance and policy. Agricultural systems may become more resilient, efficient, and environmentally friendly in the face of issues related to water scarcity by prioritizing sustainable water management practices.

Strategies for managing water are essential for tackling issues with sustainability, agricultural productivity, and water availability. The increasing demand on water sources and the need for ensuring their sustainable usage are what drive the necessity of water management methods in agriculture. Water shortage has emerged as a major global issue as a result of expanding urbanization, rising populations, and conflicting needs from different industries. The stress on water supplies is increased by ineffective agricultural practices such obsolete irrigation techniques and excessive water use. Additionally, the issues of water supply in agriculture are made more difficult by the effects of climate change, such as altered patterns of rainfall and a heightened occurrence of extreme weather events. Water management solutions that maximize water use, boost water efficiency, and encourage appropriate irrigation practices are crucial for addressing these problems. Agricultural systems can adapt to shifting circumstances, conserve water supplies, and ensure long-term profitability and sustainability by putting into practice efficient water-saving methods, such as precision irrigation, water-saving techniques, and modern technologies for tracking and regulating water use. Water management techniques are crucial for the welfare of both the agricultural sector and the larger community since they also support sustainability, financial stability, and food safety.

**Techniques to improve water management strategies**

**1. Water use efficiency**

The amount of output from agriculture or yield from crops that is obtained for per unit of water used is referred to as water use efficiency. It determines how well irrigation water is used in agricultural productivity. High efficiency in water use means that a large percentage of the water given to crops is effectively used and helps to promote the best possible plant development and output.

**Techniques to improve water use efficiency**

To increase water use efficiency in agriculture, a number of methods and procedures can be used. These methods seek to maximize the efficiency of each unit of water that is utilized, optimize water application, and reduce losses. Among the often-employed methods are:

1) **Irrigation scheduling:** Water could be delivered when plants want

it most with the help of effective irrigation planning that takes into account crop water needs and soil moisture conditions. This minimizes water losses from deep percolation and run-off and prevents over-irrigation.

2) **Drip irrigation:** Drip irrigation minimizes evaporation and reduces water waste by delivering water straight to the root zone of plants. With this approach, water is applied precisely, exactly where it is required, and weed growth is minimized.

3) **Sprinkler irrigation:** Sprinkler irrigation uses overhead sprinklers to distribute water over the crop canopy. It can be adjusted to provide uniform coverage, minimizing water wastage due to evaporation or wind drift.

4) **Precision irrigation:** By giving real-time information on soil moisture levels and weather conditions, sensor-based irrigation systems like soil moisture sensors and weather-based controllers aid in irrigation optimization. Due to this, farmers can precisely apply water where it is needed, preventing both under- and over-irrigation.

5) **Mulching:** Mulching is the practice of adding organic or synthetic resources to the soil surface surrounding plants. Mulch aids in decreasing evaporation, inhibiting the growth of weeds, and improving soil moisture retention, all of which serve to decrease the quantity of water needed for irrigation.

6) **Crop selection and breeding:** Choosing crop varieties that are adapted to local conditions and have better drought tolerance can help optimize water use efficiency. Additionally, investing in crop breeding programs that focus on developing water-efficient varieties can lead to significant improvements in water use efficiency.

7) **Conservation tillage:** Conservation tillage practices, such as minimum tillage or no-till farming, reduce soil disturbance and help retain soil moisture. This improves water infiltration and reduces water loss through evaporation, enhancing water use efficiency.

8) **Water recycling and reuse:** Implementing wastewater treatment and reuse systems or capturing rainfall can assist minimize dependency on source of freshwater and improve water usage efficiency. Irrigation using treated wastewater reduces the need for freshwater resources.

## 2. Water-saving irrigation techniques

Water-saving irrigation systems strive to optimize agricultural water usage by minimizing water losses and increasing water application efficiency. These strategies aid in the conservation of water resources, the reduction of water waste, and the increase of agricultural output. Here are some examples of typical water-saving irrigation techniques:

1) **Drip irrigation:** Micro-sprinkler irrigation involves the use of small sprinklers that deliver water in a fine spray close to the ground. It provides uniform coverage and reduces water loss due to wind drift and evaporation. Micro-sprinklers are particularly effective for irrigating orchards, vineyards, and other tree crops.
2) **Micro-sprinkler irrigation:** Micro-sprinkler irrigation makes use of miniature sprinklers that spray water in a thin mist near to the ground. It offers consistent covering and decreases the loss of water due to evaporation and wind drift. Micro-sprinklers are very useful for watering fruit trees, vines, and different crops.
3) **Subsurface drip irrigation (SDI):** Subsurface drip irrigation is burying drip lines beneath the soil surface to supply water straight to the root zone. This approach also lowers weed growth and water evaporation. SDI is especially beneficial in locations with a high rate of evaporation and salty soils.
4) **Sprinkler irrigation management:** Water saving requires proper control of sprinkler irrigation systems. Techniques such as changing the sprinkler pattern and minimising spraying or runoff can assist optimize water application and decrease water losses.
5) **Variable rate irrigation (VRI):** VRI entails altering the duration and rate of irrigation to meet the demands of different sections within a field. VRI guarantees that water is administered exactly when and where it is needed, maximizing water usage efficiency by taking into account changes in types of soil, topography, and crop requirements.
6) **Surge irrigation:** Surge irrigation is a method of applying water in alternate periods of irrigation and non-irrigation. This permits the soil to dry out partially between watering cycles, minimising the loss of water through deep percolation and enhancing water infiltration efficiency.

### Irrigation management

Irrigation management is the design, execution, and monitoring of irrigation practices to guarantee effective agricultural production and optimal water utilization.

**Types of irrigation systems (Surface irrigation, overhead irrigation, localized irrigation)**

There are many types of systems for irrigation used in the agricultural sector, each having its own set of benefits and applications. Irrigation systems are classified into three types:

**1. Surface irrigation**

Surface irrigation is one of the oldest and most widely used irrigation methods. It involves applying water to the soil surface and allowing it to infiltrate and move across the field by gravity. Common surface irrigation techniques include:

- **Furrow Irrigation:** Water is supplied between crop rows in tiny channels or furrows. It is appropriate for crop rows and orchards.
- **Basin Irrigation:** Water is supplied to basins that are level or gradually sloping around the plants. It is often utilised for trees and paddy field floods.
- **Border Irrigation:** Long and rectangular beds with high edges are watered. It is ideal for row crops, especially those grown on flat terrain.

Surface irrigation is easy and affordable, however it is inefficient owing to water losses due to evaporation and runoff. Surface irrigation, with correct management, can still be beneficial, particularly under specific crop and soil situations.

**2. Overhead irrigation**

Overhead irrigation involves spraying water from above onto the crop, generally via sprinklers. It includes techniques such as:

- **Sprinkler irrigation:** Water is supplied using sprinkler heads, which spray water on the crop in a way similar to rainfall. It offers enough coverage and is appropriate for a broad spectrum of cultivars and field conditions.
- **Center pivot irrigation:** It consists of a lengthy spinning arm with sprinklers pivoting around a central point. It is widely utilised in large-scale agriculture, especially for field crops.
- **Lateral move irrigation:** Similar to center pivot irrigation, lateral move systems consist of a series of sprinklers that move laterally

across the field. It is suitable for rectangular or irregularly shaped fields.

### 3. Localized irrigation

Water is applied straight to the root zone of the plant using localized irrigation, reducing water losses and increasing efficiency. The following are examples of common localized irrigation techniques:

- **Drip irrigation:** Water is delivered directly to the roots of plants via tiny emitters. It applies water precisely, lowers evaporation, and inhibits weed development. Drip irrigation is appropriate for a broad variety of crops, especially in places with low availability of water or high salt.
- **Micro-sprinkler irrigation:** Micro-sprinkler systems, like sprinkler irrigation, distribute water in a thin spray straight to the crop. It is great for fruit orchards, grapevines, and other fruit trees because of its superior water distribution.
- **Subsurface irrigation:** SDI is a method of supplying water straight to the root zone by laying drip lines beneath the soil surface. It decreases evaporation and minimizes water loss.

Localized irrigation saves water, lowers waste, and enables accurate water and nutrient control. However, careful layout and upkeep are required to assure the system's optimal operation.

**Best management practices for irrigation**

1. Irrigation planning based on agricultural water needs
2. Soil moisture levels are measured.
3. Utilization of irrigation technology (for example, soil moisture sensors and weather-based controls)
4. Water application techniques and consistency
5. Controlling waterlogging and salinity through drainage management

### 4. Rainwater harvesting

Rainwater harvesting in agriculture refers to the collection and storage of rainwater for subsequent use in irrigation and other agricultural activities. It involves capturing rainwater from rooftops, surface runoff, or other catchment areas and directing it into storage tanks, ponds, or underground reservoirs. This stored rainwater can then be utilized during dry periods to supplement or replace the need for freshwater sources.

**The benefits of rainwater harvesting in agriculture are numerous**

1. **Water availability:** In agriculture, rainwater harvesting refers to the gathering and storing of rainwater for later use in agriculture and other farming operations. Rainwater is collected from roofs, surface runoff, and other catchments and directed into storage containers, ponds, or subterranean reservoirs. During dry seasons, this saved rainwater can be used to complement or replace the requirement for freshwater supplies.
2. **Cost savings:** Rainwater collection assists farmers in overcoming water constraint by offering an extra supply of water. It lessens reliance on freshwater resources such as rivers, lakes, and underground aquifers, which might be restricted or declining. Rainwater may be used for irrigation, cattle watering, and other agricultural operations, providing water even during droughts.
3. **Sustainable water management:** Rainwater collection encourages environmentally sound water management practices. It contributes to water conservation by catching and using precipitation that would otherwise discharge or evaporate. It helps avoid excessive use of aquifers and ensures their long-term viability by minimising dependency on groundwater pumping.
4. **Improved crop health and productivity:** Rainwater is typically of good quality, devoid of salts, pollutants, and other toxins present in other sources of water. Irrigating with rainwater can improve soil health, minimize salinization, and increase crop output. Rainwater also restores soil moisture, ensuring that crops thrive in ideal circumstances.
5. **Environmental benefits:** Rainwater collection helps to protect the environment. It alleviates pressure on aquatic ecosystems by reducing the requirement for surface-level outflows or groundwater extraction. It also helps to prevent soil erosion and stormwater runoff by catching rainfall and enabling it to soak into the earth.

**Techniques for rainwater harvesting**

There are various rainwater gathering systems that may be used in agriculture. The volume and intricacy of these approaches varies, allowing farmers to select the best appropriate way depending on their unique demands and available resources. Here are some typical rainwater gathering techniques:

1. **Rooftop rainwater harvesting:** This method collects rainwater from the roofs of construction sites, sheds, or greenhouses. Rainwater is diverted into containers for storage such as drums or barrels via gutters and downspouts. The water collected can be utilised for small-scale watering or other agricultural purposes.
2. **Surface runoff harvesting:** Surface runoff can be collected and kept for agricultural use from fields, catchments, or footpaths. Small reservoirs, contour bunds, or agricultural ponds can be built to absorb and contain rainfall runoff. During dry spells, the saved water can be utilised for irrigation.
3. **Contour farming:** Contour farming is a method of planting crops following the line of contour of a slope. This reduces erosion and allows more water to soak into the soil by slowing the flow of precipitation. It encourages water saving while also increasing the moisture in the soil for crop development.
4. **Terracing:** Terracing is the process of building a sequence of levelled steps or platform on steep hillsides. This approach prevents soil erosion and improves precipitation infiltration into terraced beds, enhancing soil water storage and lowering runoff.
5. **Check dams:** Check dams are tiny structures constructed across slopes or drainage channels to limit water flow and capture debris. They aid in the retention of precipitation, letting it to soak into soil and replenish groundwater supplies.
6. **Percolation pits:** Shallow excavation or trenches filled with permeable soil such as gravel or sand are known as percolation pits. They are deliberately placed to absorb and store runoff rainfall. The stored water slowly seeps into the soil, recharging groundwater and promoting plant development.
7. **Bunding:** Bunding is the process of building tiny earthen banks or ridges across the contour of a field. This strategy aids in the retention of precipitation, the reduction of runoff, and the promotion of water penetration into the soil. Bunding is very beneficial in locations with undulating ground.

It should be noted that choosing and designing of rainwater harvesting systems should take into account local climate, soil characteristics, space available, and water requirements. Consultation with professionals or agricultural extension agencies can give information on the best practices to use in certain farming scenarios.

## 5. Water recycling and reuse

Water reuse and recycling in agriculture are critical to long-term water management and managing water shortage issues. Here are some of the most important reasons for water reuse and recycling in agriculture:

1. **Water conservation:** Agriculture is one of the most water-intensive sectors, and water scarcity is a significant concern in many regions. By recycling and reusing water, farmers can reduce their reliance on freshwater sources and conserve water resources. Instead of using freshwater for every irrigation cycle, treated wastewater or recycled water can be utilized, effectively extending the available water supply.
2. **Mitigating water stress:** Water recycling and reuse help alleviate water stress on existing water sources. By reusing water, the demand for freshwater extraction is reduced, relieving pressure on rivers, lakes, and groundwater reservoirs. This is particularly important in areas where water availability is limited or prone to drought conditions.
3. **Increased water availability:** Recycling and reusing water increase the overall availability of water for agricultural purposes. Wastewater, when properly treated and reclaimed, can be a valuable water resource. By treating and reusing wastewater, farmers have access to additional water supplies that can be used for irrigation, reducing the burden on freshwater sources.
4. **Nutrient management and soil fertility:** Treated wastewater or recycled water often contains nutrients that can benefit crop growth. When used for irrigation, these nutrients are applied directly to the soil, improving soil fertility and reducing the need for synthetic fertilizers. Proper water treatment processes can remove harmful contaminants and pathogens, ensuring that the reused water is safe for crop irrigation.
5. **Cost savings:** Recycling and reusing water can result in significant cost savings for farmers. Using treated wastewater or recycled water can eliminate or reduce the need for purchasing freshwater or pumping groundwater, which can be expensive. By utilizing alternative water sources, farmers can reduce their operational costs and achieve greater financial sustainability.

**Techniques for water recycling**

There are various water recycling techniques that may be used to cleanse and recycle water in agriculture. The intricacy and scope of these procedures varies, and their selection is influenced by the standard of the water supply, intended application, and available resources. Here are some typical water recycling techniques:

**1) Biological treatment**

Microorganisms are used in biological treatment techniques to break down organic materials and eliminate pollutants from water. Among the most common biological therapy approaches are:

- **Aerobic treatment:** To degrade organic substances, oxygen-consuming microorganisms are used. It effectively removes organic contaminants and improves water quality.
- **Anaerobic treatment:** Anaerobic digestion is the breakdown of organic materials by microorganisms in an oxygen-free environment. As a byproduct, it generates biogas, which may be used to generate electricity.

**2) Physical treatment**

Physical treatment techniques include the physical removal of solids and specific pollutants. Among the most prevalent physical therapy approaches are:

- **Screening:** To eliminate big particles such as bamboo sticks, leaves, and detritus, water is routed via screens or filters.
- **Sedimentation:** Solid particles sink at the bottom of sedimentation tanks or clarifiers, generating sludge, while cleared water is collected at the top.
- **Filtration:** To remove suspended matter and certain dissolved impurities, water is run through several types of filters such as screen filters or carbon filters.

**3) Chemical treatment**

Chemical treatment methods utilize chemicals to eliminate impurities and enhance water quality. The following are examples of common chemical treatment techniques:

- **Coagulation and flocculation:** Water is treated with chemical coagulants such as alum or ferric chloride to produce bigger particles that can be easily removed by sedimentation or filtration.

- **Disinfection:** Chemical disinfectants such as chloride or ozone are used to remove or inactivate viruses, bacteria, and other illnesses in water.

4) **Advanced treatment**

When higher purity criteria for water are necessary, sophisticated treatment processes are applied. Among these techniques are:

- **Reverse osmosis (RO):** RO is the process of moving water across a semipermeable membrane that eliminates salts that are dissolved, minerals, and other pollutants, resulting in high-quality treated water.

- **Ultraviolet (UV) disinfection:** UV radiation is used to destroy or kill microorganisms in water, adding an extra layer of sanitation.

5) **Combined treatment systems**

To obtain the required water quality, a mix of several treatment procedures is frequently utilised. A treatment system, for example, may involve physical procedures such as filtration and sedimentation, subsequently followed by biological treatments and disinfection.

It should be noted that the choice and layout of water recycling systems must depend on site-specific circumstances, water quality criteria, and the planned use of the recovered water. Furthermore, frequent inspections and servicing of the water recycle system are essential for ensuring the highest efficiency and security of the water recycled.

**Benefits and challenges of water reuse**

**Benefits of water reuse**

1. **Water conservation:** Water reuse contributes to the conservation of scarce freshwater resources by lowering the need for freshwater withdrawals. By reusing and recycling water, less water is needed from lakes, rivers, and aquifers beneath the ground, safeguarding these precious sources of water.
2. **Drought resilience:** Water reuse ensures a consistent supply of water even during times of drought or scarcity. Agricultural operations may continue without being overly reliant on fluctuating rainfall or restricted freshwater supplies by using reused water, which is often treated to meet certain quality requirements.
3. **Cost savings:** Reuse of water systems can result in cost reductions

for agricultural enterprises. Reusing and recycling water can lessen the need for expensive freshwater purchases or reliance on energy-intensive water extraction technologies such as pump groundwater. These cost savings can lead to better financial sustainability over time.

4. **Nutrient management:** Recycled water frequently includes elements that are useful to plant development, such as nitrogen as well as phosphorus. Farmers may minimize their dependency on synthetic fertilizers by utilizing reused water for irrigation, thereby saving money and reducing the flow of nutrients into water bodies.
5. **Environmental protection:** Water reuse is critical to minimizing the impact on aquatic ecosystems. Reduced freshwater withdrawals assist to sustain sustainable river flows and aquatic environments. Furthermore, appropriately cleaned recycled water may be utilized to restore wetlands and improve natural ecosystems.
6. **Climate change adaptation:** Droughts are predicted to become more common and severe as a result of climate change, making water reuse a vital method for adjusting to changing conditions. Agricultural businesses may increase resilience and minimize susceptibility by diversifying their water sources and using water reuse technologies.

**Challenges of water reuse**

1. **Water quality concerns:** The fundamental problem of water recycling is ensuring that recovered water fulfils the quality criteria necessary for the intended usage. To eliminate impurities and assure the safety of recycled water, proper treatment techniques and monitoring are required.
2. **Public acceptance:** Because of worries about the security and aesthetics of recovered water, water reuse initiatives may meet public opposition or scepticism. To address these issues and develop confidence among stakeholders, public outreach and educational programmes are essential.
3. **Infrastructure and implementation costs:** The initial expenses of establishing reused water facilities, such as treatment facilities and distribution lines, can be high. The cost of implementation varies based on the size of the undertaking, treatment needs, and the existing facilities.
4. **Regulatory frameworks:** Water recycling projects require

regulatory frameworks that are supportive of their development and execution. Setting norms, regulations, and permissions for water reuse may be difficult, as they may differ among regions or jurisdictions.

5. **Monitoring and maintenance:** Water reuse systems must be monitored and maintained on a regular basis to ensure their continuous effectiveness and the security of the reclaimed water. In order to carry out these efforts successfully, adequate assets and expertise are necessary.
6. **Water rights and legal considerations:** Water reuse initiatives may confront issues with rights to water and legal frameworks. Water allocation regulations and legal limits differ from area to region, and managing these complexity can be a substantial barrier to implementing water reuse efforts.

Despite these obstacles, the benefits associated with water recycling in agriculture provide a viable option for long-term water management, increasing agricultural production, and alleviating the effects of water shortage.

## 6. Crop selection and management

Crop choice and management are critical components of agricultural practices that entail selecting appropriate crops and applying effective tactics to maximize growth and output. Here are some recommendations regarding crop choice and management:

1. Crop selection for water-stressed environments
2. Crop diversity and rotation
3. Crop cultivars that need little water
4. Conservation tillage and mulching practices
5. Crop residue management to conserve soil moisture

## 7. Soil management

### Soil moisture conservation techniques

1. **Conservation tillage:** Conservation tillage is a farming practice that tries to reduce soil disturbance while preserving residue from crops on the soil surface. Conservation tillage, as opposed to typical tillage practices that require deep ploughing and soil inversion, helps prevent soil erosion, retain soil moisture, and increase organic matter content. Conservation tillage improves the soil's structure,

infiltration of water, and nutrient cycling by leaving agricultural waste on the soil surface. It also has the potential to save money on fuel and labour while improving soil health and the long-term viability of agricultural systems. Overall, conservation tillage is critical for encouraging conservation of soil, water efficiency, and environmentally friendly agriculture practices.

2. **Organic matter addition:** The addition of organic matter to soil offers various water-saving benefits. Organic material, such as crop waste, compost, or manure, improves the soil's structure and water-holding capacity when it is mixed into the soil. The incorporation of organic matter into soil is critical for water conservation. Organic matter functions as a sponge, increasing the soil's capacity to hold onto moisture and decreasing evaporation. It strengthens soil structure by forming an arrangement of pore spaces that enable water to permeate and be held in the soil. Furthermore, organic matter increases the water-holding capacity of the soil, allowing it to keep more water for plant usage during dry seasons. As a result, the organic matter addition reduces watering requirements while increasing plant tolerance to water stress. Organic matter addition leads to increased water usage and sustainable farming practices by boosting water retention and lowering water loss.

3. **Cover cropping:** Cover cropping is an agriculture practice in which certain plant species known as cover crops are grown during periods when the primary cash crop is not growing. These cover crop varieties are produced primarily for covering the soil surface while also providing a variety of advantages. Cover cropping has various advantages for agricultural systems. Cover crops, for starters, assist in preventing the soil against erosion by hiding it from water and wind. Cover crops' thick growth functions as an organic barrier, limiting soil erosion and maintaining important topsoil. Furthermore, cover crops improve soil overall wellness and fertility. Their vast root systems aid in soil structure improvement, aeration, and nutrient cycling. Cover crops absorb minerals from the soil as they grow, avoiding leaching and increasing soil fertility. Cover crops also help to decrease weeds by shading off weed development and minimizing competing with cash crops. In addition, cover crops can act as a source of nutrients by incorporating vital organic material into the soil. Overall, planting cover crops is a sustainable practice that promotes the health of the

soil, preserves soil moisture, controls weeds, and increases agricultural systems' overall resilience and production.

## 8. Integrated water resources management (IWRM)

IWRM is a comprehensive approach for water management that tries to balance the economic, social, and environmental elements of water use. It acknowledges that water sources are interdependent and finite, and that they must thus be managed in a planned and sustainable way. To ensure effective and fair water distribution, IWRM emphasizes the integration of diverse sectors, stakeholders, and organizations engaged in water management. The method takes into account the complete water cycle, from the source through use to treatment of waste water and reuse. It encourages participatory decision-making procedures that consider the needs and interests of many users while balancing conflicting water demands. Countries and regions may improve the availability of water, encourage water preservation and efficiency, preserve ecosystems, and minimize the effects of water-related disasters by implementing IWRM principles. Overall, IWRM provides an extensive framework for long-term, integrated water management that takes into consideration the complex relationships between water, civilization, and the environment.

## 9. Conclusion

Finally, good water management systems are critical for establishing effective irrigation in agricultural practices. Farmers may optimize water consumption and reduce waste by taking into account crop particular demands, the state of the soil, and water availability. Techniques like irrigation scheduling, water-saving irrigation technologies, and irrigation management practices all lead to increased water utilization and sustainable agriculture. Rainwater gathering, reuse of water, and the use of cultivation of cover crops also assist to preserve water resources and minimize dependency on freshwater supplies. Integrated methods, such as the application of Integrated Water Resources Management (IWRM) principles, improve the efficiency and sustainability of irrigation practices by taking into account the whole water cycle and balancing the requirements of diverse water users. We can assure the long-term accessibility of water sources for agriculture, increase food security, and limit the environmental implications of excessive water consumption by applying these water management measures. To achieve effective irrigation and water conservation in agriculture, farmers, governments, and stakeholders need to prioritize and make investments in these techniques.

## References


1. Agarwal, S. K., and R. De (1979). Effect of nitrogen rates, mulching and antitranspirants on water use and water use efficiency of barley (*Hordeum vulgare L.*) varieties grown under dryland conditions. Journal of *Agricultural Sciences.* 92: 263– 268.
2. Evans, G.R., Sadler, E.J. (2008). Methods and technologies to improve efficiency of water use.
3. *Water Resources Research.* 44(7): 1-15.

# Chapter - 12

# Vertical Farming: Revolutionizing Urban Agriculture


**Authors**

**Anuj Sohi**

PhD Scholar, Department of Vegetable Science, College of Horticulture, Dr. Yashwant Singh Parmar University of Horticulture and Forestry, Nauni, Solan, Himachal Pradesh, India

**Tarushi Attri**

PhD Scholar, Department of Plant Pathology, College of Horticulture, Dr. Yashwant Singh Parmar University of Horticulture and Forestry, Nauni, Solan, Himachal Pradesh, India

**Gurpreet Kaur**

MSc Scholar, Department of Vegetable Science, College of Horticulture, Dr. Yashwant Singh Parmar University of Horticulture and Forestry, Nauni, Solan, Himachal Pradesh, India

**Balbir S Dogra**

Professor, Department of Vegetable Science, College of Horticulture, Dr. Yashwant Singh Parmar University of Horticulture and Forestry, Nauni, Solan, Himachal Pradesh, India

# Chapter - 12

# Vertical Farming: Revolutionizing Urban Agriculture

**Anuj Sohi, Tarushi Attri, Gurpreet Kaur and Balbir S Dogra**

## Abstract

The vertical farming is a world-changing innovation which an upcoming technology in the future farming. Imagine a world where every town has their own local food source, grown in the safest way possible, where no drop of water or particle of light is wasted. Smart farming makes a tremendous contribution to food sustainability for 21$^{st}$ century. The cause is that plant development is directly impacted by environmental and water management. A contemporary method for feeding the growing global population by 2050 is vertical farming. It serves a farm close to the community by providing affordable, disease-free organic produce while preserving the finite natural resources.

**Keywords:** Vertical farming, sustainability, healthy food.

## Introduction

India has one of the world's strongest economies. The majority of Indians rely on farming as their main source of income. Due to urban culture and rising urban population, several countries today are experiencing issues with food production and scarcity. According to estimates, India has 1.37 billion people, or 18% of the world's population, but only 2.4% of the planet's land area and 4% of its fresh water, of which 80% is used for agricultural purposes (GFFA 2017). Growing crops in a precise indoor environment with controlled lighting, nutrients and temperatures is the goal of this method, known as vertical farming. In indoor farming methods, vertical farming technology is known as controlled-environment agriculture (CEA) technology. Vertical farming is a planting or growing plants vertically involves stacking them horizontally. Growing food and medicinal plants (herbs) in vertical layers and on vertically inclined surfaces is a common practice. Irrigation component, lighting, sensor, climate control, building material, glass greenhouse, plastic greenhouse and others are among the vertical farming components. The primary agricultural, residential, and small and medium company uses for vertical farming. Plants can be grown vertically at home using the hydroponics approach. With the growing industry of vertical farming, small and medium businesses will expand. However, smart use of vertical farming can make it possible to efficiently grow practically anything in a controlled environment. In light of ecological and societal sustainability, the current research aims to pinpoint justifications for India to take into account and embrace this cutting-edge farming technique.

**Need for vertical farming**

1. **To meet increasing food demand:** Because of this, there is a need to meet the growing population's food need, which can only be done by vertical farming. According to estimates, the world's population will reach 9.8 billion in 2050.
2. **To reduce pressure on arable land resource:** The world is experiencing a steady but persistent loss in the amount of arable land due to the rising urbanisation and industrialization of society. Numerous small and large concrete structures have been produced as a result of urbanisation, mostly to accommodate the constantly growing population at the expense of farmland. Therefore, soilless vertical farming is a potential means of producing food even in arid regions (Ali and Srivastava 2017).
3. **To produce clean, green and gourmet food:** The most alluring aspect of the vertical farming paradigm is without a doubt the potential for CGG food production. Fresh food can be produced year-round by crop production without seasonality in a climate-controlled setting (which includes both temperature and humidity). If hydroponic and aeroponic technologies are used, crop failures caused by drought or flooding would not occur (Benke and Tomkins 2017)
4. **Sustainable use of resources:** In addition to enabling efficient and ecological food production, vertical farming has the potential to improve the economy, reduce pollution, provide new job opportunities, rebuild ecosystems and increase access to wholesome foods (Rameshkumar *et al.* 2020)

**Brief history of vertical farming**

In 1915, Gilbert Ellis Bailey coined the term “Vertical farming” and wrote a book titled “Vertical Farming”. In the early 1930s, William Frederick Gerick pioneered hydroponics at the University of California at

Berkley. In the 1980s, Ake Olsson a Swedish ecological farmers, invented a spiral-shaped rail system for growing plants and suggested vertical farming as a means for producing vegetables in cities (Rameshkumar *et al.* 2020). The hanging garden of Babylon poses as the earliest method to grow plants vertically. In 1915, the American geologist Gilbert Ellis Bailey used the concept of the tall multi-story buildings for indoor cultivation Vertical farming as a concept was developed in the recent years 1999 through the advances in technology by Dickson Despommier, an Emeritus Professor of Microbiology at the Columbia University; He explained that hydroponic crops was grown on upper floor and the lower floors was suited for chickens and fish that eat plants' waste.

**Vertical farming can be done indoor or outdoor**

**How does indoor vertical farming works?**

1. **Physical layout:** Utilizing space to produce a high-quality crop is the fundamental goal of vertical farming. Growing on a horizontal plane with vertical stakes or growing the plant vertically and starching it upward is the most promising layout in vertical farming. With this design, the grower may make effective use of the vertically available space. In comparison to the same area on open land, it would enable the planter to obtain a higher yield in less time (Chaudhary *et al.* 2022).

2. **Lighting:** Light is the second-most crucial element. Regardless of where in an area vertical farming is practiced, either natural or artificial light is necessary to promote plant development. It must be kept in mind that in order to perform photosynthesis when artificial light is applied, it must emit electromagnetic spectrum. There are two choices: HPS (high-pressure sodium) or LED (light emitting diode). Because it releases little heat and has a wide spectrum, LED lighting is highly favoured because it promotes good plant germination and growth. Heat, which would impede plant growth in the summer, is the biggest problem with lighting. However, in order to keep the heat inside the nearby building or structure during the winter, a heater must be built (Chalabi 2015).

   i) **Natural lighting:** The structure is constructed to absorb as much light as possible. In areas with temperate weather, it is possible to build certain elements, such as the roof, to take full advantage of the sun's energy throughout the year.

   ii) **Solar cell:** Non-PAR waves (Photosynthetically Active

Radiation), which are not absorbed by chlorophyll, should be filtered in order to maximize the utilization of solar energy in vertical farms. Selected solar cells must filter these waves before using them to generate photovoltaic energy.

iii) **LED:** There is a need for artificial illumination in addition to natural lighting and one such source might be LEDs (Light Emitting Diodes). Long-lasting and less expensive LEDs make good alternatives. They have a long lifespan, are energy-efficient and can target certain light wavelengths to better control the photoperiod. The ability of LEDs to control the ratio of red and far-red colours to elicit the greatest responses from plants is another benefit. The stem elongation is also increased if this proportion is low. Red light is necessary for photosynthesis, while blue light helps to create the ideal conditions for plant morphogenesis. The ideal colours for LEDs are red and blue.

**Panasonic indoor VF farm, Singapore**

## 3. Growing medium

To grow crops without using chemicals and with a high-water usage efficiency, vertical farming can be done on soil-based medium or is frequently integrated with soil-less techniques like hydroponics, aeroponics, and aquaponics.

### 3.1 Hydroponics

This method of growing crops without soil is used. The crop plants' roots are immersed in a nutrition solution that comprises macronutrients and micronutrients in this system. To keep the quantity of nutrients in the flowing liquid solution constant, the nutrient solution is frequently fed and regenerated at predetermined times by an electric pumping system. It guarantees the produce is free of illnesses, pests and/or insects (Chaudhary *et al.* 2022).

**Vertical farming using hydroponics**

### 3.2 Aeroponics

When a nutrient-rich liquid is misted in an air chamber where plants' roots are hung, it is called liquid medium growing of plants. 90% less water is used in an aeroponics system than in other vertical approaches. It is an energy-saving technology since hydroponic systems requires a motor pump to often circulate the water throughout the entire system, whereas gravity drains away excess water. Consequently, hydroponic systems are not frequently used (Chaudhary *et al.* 2022).

**Vertical farming using aeroponics**

**Example: Aerofarms**

- It is the first sustainable indoor vertical farming agriculture company to be a Certified B Corporation based in **Newark, New Jersey**. It was founded in 2004 and since then it has been the commercial leader in indoor vertical farming.
- It is using patented aeroponics technology and using upto 95% less water and zero pesticides. The farms are optimized for year-round production and growing over 550 different varieties of plants including leafy greens, berries, tomatoes and more. (www.aerofarms.com)

### 3.3 Aquaponics

The word aquaponics or aquaculture, which refers to the farming of fish and hydroponics, which describes growing plants without using soil. In this approach, crop agriculture coexists with fish breeding in the same ecosystem. In this ecosystem, the fish excrete nutrient-rich wastes that are fed to the plant and then recycled back to where it all began after filtration. The balance of the ecosystem is maintained by repeating this cycle. Because of its intricacy, this system is rarely often utilised (Chaudhary *et al.* 2022)

**Example:** Aquaponics vertical farming system, Guru Angad Dev Veterinary and Animal Sciences University (GADVASU) - This system was established by the Centre for Development of Advanced Computing (C-DAC), Mohali, under the collaborative project funded by Ministry of Electronics and Information Technology (MEITY), GOI.

- Aquaponics involves synergistic combination of aquaculture and hydroponics to promote vertical farming with enhanced nutrient utilization efficiency through recycling of water between the two components to produce fish and vegetables together.
- It offers potential future solution to global warming, with only 10-12% water and land requirement compared to traditional fish and vegetable farming systems, fulfilling the concept of 'More crop per drop'. (https://www.gadvasu.in)

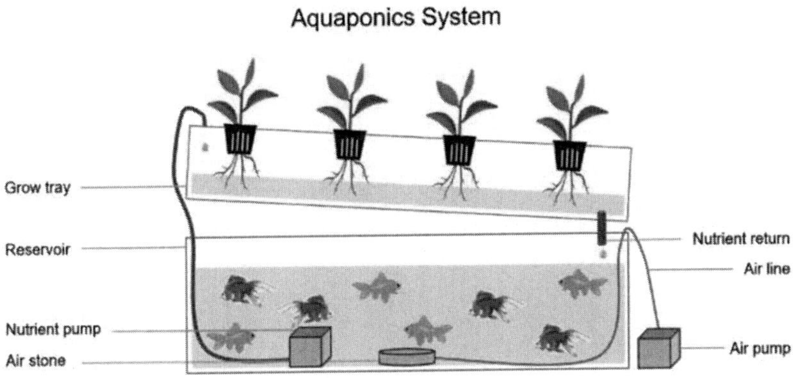


## 4. Controlled environment agriculture (CEA)

CEA systems are housed in buildings or enclosed structures like greenhouses where environmental elements like air, temperature, light, water, humidity carbon dioxide, and plant nutrition may be controlled. CEA is frequently used with soilless farming methods including hydroponics, aquaponics and aeroponics in vertical farming systems (Nassoori *et al.* 2021).

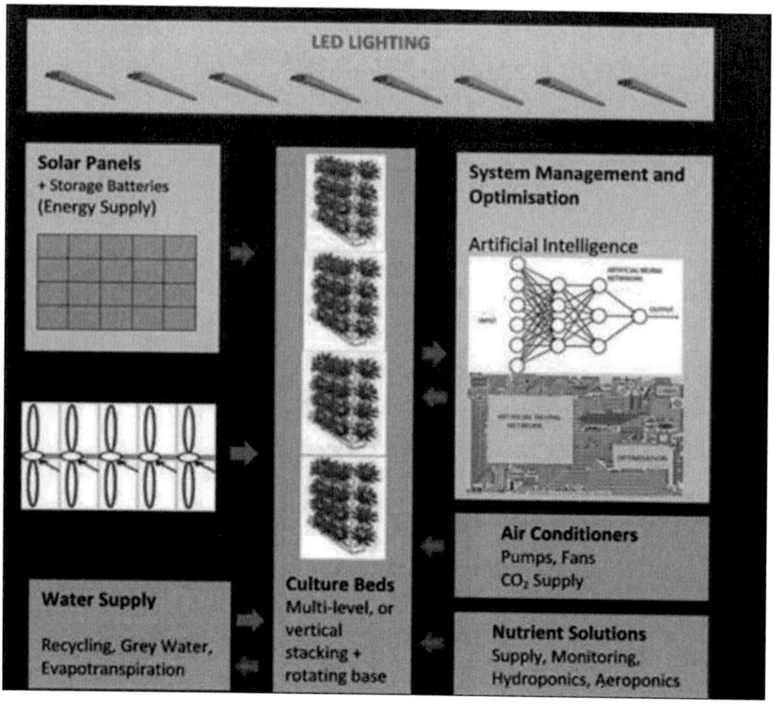


**Components of CE vertical farming**

**Estimated yield of a Vertical Farm compared to traditional agriculture**

| Crops | Yield in VF due to tech. (tons/ ha) | Field Yield (tons/ ha) | Factor increase due to Tech |
|---|---|---|---|
| Carrots | 58 | 30 | 1.9 |
| Radish | 23 | 15 | 1.5 |
| Potatoes | 150 | 28 | 5.4 |
| Tomatoes | 155 | 45 | 3.4 |
| Pepper | 133 | 30 | 4.4 |
| Strawberry | 69 | 30 | 2.3 |
| Peas | 9 | 6 | 1.5 |
| Cabbage | 67 | 50 | 1.3 |
| Lettuce | 37 | 25 | 1.5 |
| Spinach | 22 | 12 | 1.8 |
| Total (average) | 71 | 28 | 2.5 |

(Banerjee and Adenaeuer 2014)

**Vertical farming v/s traditional farming**

- Due to the closed environment and controlled lighting, the land productivity of Vertical Farming is twice as high as traditional agriculture.
- Taking into account that only 0.25 ha on which the farm is built are needed, the total yield increases 516-fold compared to traditional agriculture through stacking the production.
- In total this leads to an estimated production of 3,573 tons of edible fruits and vegetables.
- So, practicing vertical farming is the paramount solution to feed the ever-increasing urban population.

**Types of plants suitable for outdoor vertical farming (Fell 2011)**

1) **Climbers:** Vegetables with vine-like growth patterns are best suited for vertical planting. Most veggies require aid to climb, even with their natural predisposition to vine, so encourage their stems to grow upward using string or twist ties. e.g. sweet peas, cucumber, sweet potato, pole beans, squashes, melons, tomatoes etc.
2) **Foundation plants:** These are foundation plants low-growing, non-climbing plants that can be positioned at the base of climbers. Their foliage melds harmoniously with the stems and vines of these soaring neighbours and they coexist peacefully with the roots of climbing plants. e.g. beets, cabbage, carrot, eggplant, lettuce, peppers etc.
3) **Container and tower pot plants:** A central climbing plant, such an pole bean or cucumber, may usually be grown in the centre of most containers and cascade planters with bowl-shaped receptacles, with additional plants placed around the rim. Basil and miniature lettuce for example, are ideal for surrounding a central climber. For planting around the rim, also consider edibles that cascade such as strawberries, sweet potatoes and New Zealand spinach.
4) **Support plants:** Some of the climbing plants in a vertical garden might be supported by tall plants with sturdy stems. Sweet corn and okra plants, for instance, are so strong that they can hold up a variety of climbers, such as pole beans, cucumbers and single-serving-size melons like 'Tigger', which will wrap around the stem of the support plant and shoot upward.

**Planters and supporters for outdoor vertical farming (Fell 2011)**

1) **Bamboo supports:** Bamboo canes are frequently used to support the vertical growth of tomatoes, but they also form excellent towers and trellises for many other kinds of plants in a vertical garden. For instance, growing pole beans, tomatoes, and squash on a basic trellis made of bamboo canes is an excellent idea. A handcrafted bamboo trellis, perfect for 'Sugar Snap' peas, is made sturdy by the straightforward crisscross design of bamboo canes. For heavier vines like indeterminate tomatoes and pole lima beans, the diamond pattern trellis, which uses bamboo canes, provides extra sturdiness.
2) **Fan trellis:** Made of wood or more durable metal, plastic or vinyl, the prefabricated fan trellis is mostly used to grow plants against a wall. It is especially useful for vining.
3) **Garden netting:** There are different kinds of garden netting in different widths and lengths for rolling out and cutting to suit your needs. A larger mesh for supporting veggies with a 4- to 6-inch-wide grid that you can reach through to harvest is recommended. Depending on the plants, a certain length of garden netting is required. For instance, pole lima beans can reach a height of 15 feet in just one growing season.
4) **Builder's wire:** It is ideal for heavy vines like pole beans, vining squash, and to support individual tomato vines. Builder's wire, which is primarily used for reinforcing concrete floors, makes a strong, affordable trellis for mounting between two posts or attaching to a wall. It uses a heavy-gauge #10 mesh, creating a framework of 6 x 6-inch reach-through mesh that plants can climb or twine through.
5) **Skyscraper garden:** Vertical plant growth is made simple by the Skyscraper Garden, a commercial product. The majority of veggies with the potential to climb respond to it just as well. It includes plants that aren't often used for vertical gardening, like the gourmet French melon called "Charentais," the burpless (bitter-free) cucumber "Orient Express" and a climbing zucchini resembling a trombone ('Trombocino').
6) **Chain-link and other fence support:** It is an ideal support for climbing plants such as sweet peas etc.
7) **Upside-down planters:** These bottom-up planters are specially built to grow a column of vegetables (often tomatoes) upside-down.

They are normally composed of a flexible polypropylene fabric. Topsy Turvy is an interesting discussion point for guests in the garden and an excellent strategy to conserve space when cultivating a tomato vine. 'Sungold', 'Sweet 100' and other cherry tomatoes are quite productive; however, a medium-fruited variety like 'Early Cascade' yielded the highest results. Larger cultivars like "Beefsteak" did bear fruit, but there weren't many of them. Use indeterminate varieties (those with long vines) exclusively for best results; some indeterminates (such as 'Better Boy') can reach 15 feet. Avoid determinate tomatoes (those with short vines), like 'Patio'.

8) **Maypoles:** A maypole is simply a sturdy pole with strands of string that radiate down from the top, creating a tepee support for twining plants like pole beans and cucumber.

9) **Tower system:** Tower plantings create an exceedingly decorative accent that's more appealing than a collection of potted plants clustered at ground level, which allow to create a vertical display in a small space. Many of the stackable containers used for these tower systems are made of plastic, wire or Styrofoam, lightweight enough to work well in a tall tower grouping.

10) **Cascade planters:** The three-tier cascade planter is one type of tower system, consisting of three metal wire bowls or baskets held up by a central metal pillar. The size of each bowl planter increases from top to bottom, making it possible to produce a wide variety of flowers, herbs or vegetables (or a combination of all three) in a sunny spot on a deck or patio.

**List of vegetable cultivars suitable for vertical farming**

| S. No. | Vegetable | Cultivar |
|---|---|---|
| 1. | Pole beans | **Snap beans:** Blue Lake, Cherokee Trail of Tears, Lazy Wife, Purple Pod, Rattlesnake<br>**Lima beans:** King of the Garden, Christmas Lima<br>**Asparagus beans:** Chinese Red Noodle, Thai Black Seeded |
| 2. | Beets | Chioggia |
| 3. | Cabbage | Stonehead, Gonzales |
| 4. | Carrot | Royal Chantenay, Nantes |
| 5. | Swiss chard | Bright Lights |
| 6. | Chayote | Perlita |
| 7. | Cucumber | County Fair (The first seedless cucumber for outdoors), Poona Kheera, Miniature White (west Indian gherkin) |

| | | |
|---|---|---|
| 8. | Eggplant | Dusky Hybrid |
| 9. | Gourds | Snake gourd, 'Cuccuzi' |
| 10. | Lettuce | Buttercrunch, Sierra, Batavia |
| 11. | Melons | Ha'Ogen, Charentais, Minnesota Midget, Burpee's Ambrosia Hybrid, Burpee's Early Hybrid Crenshaw, Emerald Gem, Jenny Lind, Tigger |
| 12. | Okra | Annie Oakley, Clemson Spineless |
| 13. | Peas | Blue Podded, Golden Sweet<br>**Shell peas:** Green Arrow<br>**Snow peas:** Oregon Giant<br>**Snap peas:** Super Sugar Snap |
| 14. | Pepper | Gypsy Hybrid, Green Anaheim |
| 15. | Sweet potato | Beauregard |
| 16. | Pumpkin | Jack-be-little |
| 17. | Squash | **Winter squash:** Delicata, Buttercup, Table Ace, Waltham Butternut<br>**Summer squash:** Black Forest, Trombone, Vegetable Spaghetti |
| 18. | Tomato | Better Boy Hybrid, Big Rainbow, Black Krim, Brandywine, Early Cascade |
| 19. | Kale | Tuscan, Winter Boar, Dinosaur |

**Advantages of vertical farming (Naskoori *et al.* 2021)**

1. **Increased crop production:** Indoor farming can produce crops year-round. All-season farming multiplies the productivity of the farmed surface by a factor of 4 to 6 depending on the crop.
2. **Reduction in transportation cost:** Crops will be sold in the same infrastructures in which they are grown, they will not need to be transported between production and sale, resulting in less spoilage, infestation and energy required than conventional farming encounters.
3. **Protection from weather-related problems:** Vertical farming provides a controlled environment, the productivity would be mostly independent of weather-related problems such as undesirable temperatures or rainfall amounts, monsoons, hailstorms, tornadoes, flooding, wildfires and severe drought.
4. **Environment friendly:** reduction in transportation distance, thereby reducing cost, energy and carbon footprint.
5. **Higher quality produce:** Provides higher quality produce with

greater nutritional value and a longer shelf life as there is no need for the use of harmful herbicides or pesticides due to low incidence of disease and insect-pest.

6. **Conservation of resources:** Vertical farming reduce the need for new farm land due to overpopulation, thus saving many natural resources such as Vertical farming uses 70 to 95% less water compared to traditional farming.
7. **Vertical farming flexibility:** Able to grow over 80 varieties of leafy greens, micro greens and strawberries. It also works on non-arable lands and close to major markets or urban centers. Vertical farming creates sustainable environment for urban centres.

**Disadvantages of vertical farming**

- High initial investment cost
- High energy cost as growing plant is entirely with artificial lights
- excess nutrients used in vertical farming
- Lot of garbage, plant residues, etc.
- Skilled labor requirement
- High cost of LED lighting systems
- **Cost of production:** A skyscraper-sized vertical farm would cost hundreds of millions of dollars to build and equip for agricultural needs.
- **Cost of maintenance:** Urban property is much more costly than rural property and the cost to run thousands of LED lights, keeps the temperature perfectly ambient and supply water to plants will outweigh the cost to run a traditional farm unless they can find a way to make the energy renewable and self-sustaining.

**Conclusion**

Vertical farming is a best alternative for the city dwellers. It can deliver food in sustainable ways to improve global food security and solve the environment degradation problems. No harvest would fail by severe weather phenomenon. It has the benefit for easily minimize the cooling and heating water by indoor temperature. It helps to reduce poverty, increase food safety and well-being of human. Effectiveness of vertical gardening depends on the demand and supply of food, urban population and densities, technological development, water and energy supply and weather conditions. As the population is increasing at a high rate, this is a great challenge for agriculture to feed the large population when agriculture land is decreasing. So vertical farming is definitely a solution to critical problems in Indian farming like lack of supply or oversupply of farm produce, overuse of pesticides, overuse of fertilizers, deteriorating soils and even the un-employability.

**References**


1. Al-Chalabi M. 2015. Vertical farming: Skyscraper sustainability. Sustain. Cities Soc. 18: 74–77. doi: 10.1016/j.scs.2015.06.003

2. Ali F and Srivastava C. 2017. Futuristic Urbanism: An overview of vertical farming and urban agriculture for future cities in India. *Int. J. Adv. Res. Sci. Eng. Tech.* 4 (4): 3767-3775.

3. Balashova I, Sirota S and Pinchuk Y. 2019. Vertical vegetable growing: creating tomato varieties for multi-tiered hydroponic installations. *IOP Conf. Series: Earth Environ. Sci.* 395(1): 012079. doi:10.1088/1755-1315/395/1/012079.

4. Banerjee C and Adenaeuer L. 2014. Up, up and away! The economics of vertical farming. *J. Agric. Stud.* 2 (1): 40-60. doi:10.5296/jas.v2i1.4526

5. Benke K and Tomkins B. 2017. Future food-production systems: vertical farming and controlled-environment agriculture. Sustain.: Sci. Pract. Policy. 13(1): 13-26. DOI:10.1080/15487733.2017.1394054

6. Chaudhary PS, Kumawat SK and Baradwal H. 2022. Vertical farming in India: New era in Indian agriculture. The Agriculture Magazine. The Agriculture Publication, Jaipur. 1 (4): 61-64.

7. Despommier D. 2010. The Vertical Farm: Feeding the World in the 21$^{st}$ Century. Thomas Dunne Books, New York, USA.

8. Fell D. 2011. Vertical gardening: grow up, not out, for more vegetables and flowers in much less space. Rodale Books.

9. Naskoori K, Reddy KK, Reddy MV and Devi MR. 2021. To study the scope of vertical farming in India. *Pharma innov*. 10(12): 158-162.

10. Naskoori K, Reddy KK, Reddy MV and Devi MR. 2021. To study the scope of vertical farming in India. *Pharma innov*. 10(12): 158-162.

11. Nwosisi S, Nandwani D and Chowdhury S. 2016. Organic vertical gardening for urban communities. *VI International conference on landscape and urban horticulture. Acta Hortic.* 1189: 399-402.

12. Rameshkumar D, Jagathjothi N, Easwari S, Rajesh S, Muthuselvi S, Kumar PN, Krishnakumare P, Minithra R and Suresh R. 2020. Vertical Farming - Agriculture of the future. Indian Farmer. 7(11): 1013-1017

# Chapter - 13

## Automation and Robotics in Agriculture

**Authors**

**Reshma P**

Ph.D. Scholar, Department of Plantation, Spices, Medicinal and Aromatic Crops, Kerala Agricultural University, Trivandrum, Kerala, India

**Soumya Unnikrishnan**

Research Scholars, Department of Plantation, Spices, Medicinal and Aromatic Crops, Kerala Agricultural University, Trivandrum, Kerala, India

**Shanila**

Research Scholars, Department of Plantation, Spices, Medicinal and Aromatic Crops, Kerala Agricultural University, Trivandrum, Kerala, India

**Reddappa JB**

Research Scholars, Department of Plantation, Spices, Medicinal and Aromatic Crops, Kerala Agricultural University, Trivandrum, Kerala, India

# Chapter - 13

## Automation and Robotics in Agriculture

**Reshma P, Soumya Unnikrishnan, Shanila and Reddappa JB**

### Abstract

Agriculture serves as the foundation of society because it primarily produces food, fiber and other necessities for human survival. Precision agriculture aims to deliver low input, high efficiency, and sustainable agricultural production by applying sufficient treatments at the proper time and place. Automation and robots have emerged as one of key frameworks in precision agriculture, which focuses on reducing environmental effects while simultaneously maximizing agricultural output. The basic component of a robotic system includes sensors, end-effectors, control system, manipulators or arms and power supply. Utilizing contemporary technologies, automation and robots are primarily applied to exact farm management in precision agriculture. Research on the use of mobile robots for agricultural tasks like planting, inspecting, spraying, weeding, harvesting and post-harvest operations, has been heavily concentrated in recent decades. The development of an effective autonomous and robotic system for agricultural applications is fraught with difficulties. Because different operations face distinct obstacles, different solutions are required to handle each individual operational problem. As a result, to reduce system mistakes during implementation, the development process of an effective autonomous agricultural robotic system must take into account all potentialities and difficulties in various types of agricultural operations. For the farmers to invest their wealth as consumers, the development costs must also be completely taken into account. The autonomous agricultural robotic system will therefore have a very high chance of being extensively used in the future.

**Keywords:** Automated agriculture, precision farming, robotics, artificial intelligence, smart farming.

### 1. Introduction

The future of agricultural production systems intends to produce high quality and quantity agricultural products for a broad range of consumers.

The biggest obstacle facing agricultural experts worldwide is understanding sustainable agriculture through the careful use of ever-dwindling natural resources and environmental protection. One of the emerging agricultural fields, known as precision agriculture, aims to increase the efficiency with which agricultural inputs are used by optimizing site-specific input use based on plant, soil and environmental demands. In modern times, precision agriculture plays a vital role in ensuring future food security with less labour and energy while enhancing environmental management to ensure fruitful agricultural output. Precision agriculture focuses on controlling how seeds, fertilizers and agrochemicals are administered to the soil and how the harvesting process is carried out.

Precision agriculture has advanced in various areas, including technology, digitalization, societal impact, skills, the environment and production. The application of a wide range of technologies, such as geo-referencing, Global Navigation Satellite System (GNSS), data storage and analysis, advisory systems, and autonomous navigation, has been widely used in the development of precision farming (PF). Agriculture has been widely digitalized in tandem with the advancement of the Information and Communications Technology (ICT) revolution. Cloud computing and the Internet of Things (IoT) have enabled efficient and precise farming management in the context of massive data collection that can be readily shared and analysed. Precision agriculture is intended to cause cultural changes in rural communities, launch new social models and launch new businesses, in addition to making life easier by reducing workloads for the farmers, much like computers have transformed human existence (Mahmud *et al.,* 2020). Numerous researchers around the world have used precision agricultural techniques to raise crop output and quality while lowering input costs. Precision seeding and planting have improved crop productivity and accelerated crop maturation; timely fertigation for giving targeted water and nutrients has decreased agricultural input and accelerated crop maturation (Jyoti *et al.*, 2020). In addition to already-existing precision farming techniques, the deployment of agricultural robots and autonomous tractors has decreased farm operation drudgery and input cost of farming.

The term "automated agriculture" refers to any device or mechanism that is designed to eliminate manual intervention in agriculture. Automation in agriculture has recently become a trend in which agricultural tasks such as cultivation, inspection, spraying, trimming, and harvesting are performed autonomously to solve labour shortages in agriculture. Agricultural automation is primarily concerned with autonomous vehicle applications such as robots and tractors, which are used to reduce the harsh, deadly, risky and lengthy working conditions endured by farmers while also providing a precise and efficient operation and control system. Furthermore, the output quality and quantity must be maintained in order for the output to be of excellent quality and consumable for humans. As a result, the creation of an effective agriculture automation system has become a focus of contemporary agricultural research in order to assure the sustainability of food security in the near future (Mahmud *et al.,* 2020).

## 2. Agricultural robotics

Robotics is an intelligent machine developed via the technical marvels of computing, electronics and mechanical engineering that mirrors the human body's natural working patterns. Agricultural robots, according to Bechar and Vigneault (2017), are perceptive, programmable devices that carry out various agricultural operations like cultivating, transplanting, spraying and selective harvesting. The term "agrobot" is unquestionably a useful one for describing autonomous machines that may perform many repetitive agricultural chores on a farm, from harvesting to land preparation, without the direct involvement of humans. In agriculture, robotic technology is used for a variety of tasks including planting, spraying, weeding, harvesting and post-harvest operations (Xia *et al.*, 2015).

Agriculture deals with live produce (fruit, vegetables, grains, and flowers), which is extremely sensitive to environmental and physical factors (Eizicovits & Berman) such as temperature, humidity, gas, pressure, damage and acceleration. To retain acceptable quality over the distance and time separating its manufacturing site from customers, live produce requires delicate, accurate and frequently difficult handling processes. The majority of farming operations such as trellising, harvesting, sorting and packing chores associated with growing fruit, vegetables and flowers are still done manually. As a result, manual labour is a significant cost factor in field operations (Eben-Chaime *et al.*, 2011), accounting for up to 40% of the entire cost. Robotic technology not only eliminates tedium in farm operations but also shields farmers from working in hazardous environments. The use of robotics in agriculture is a difficult process that requires expert systems to deal with a variable and unstructured environment while performing the intended task. The development of robotic systems for various agricultural tasks, such as seeding, planting, weeding, spraying, harvesting and material handling by agricultural robots, has been attempted in recent decades by researchers from various universities. According to FAO (2019), over 90% of farmers globally operate on a small scale, and technology must become available to this huge population. In India, robots have been effectively used in the security, service and industrial sectors. Research organizations such as TNAU, IARI New Delhi and CIAE Bhopal have tried a few seeding and planting prototypes (Jyoti *et al.*, 2020).

## Components of robots

Robots usually have five parts: power supply, sensors, controller/ computer, arms/ manipulator and end effectors.

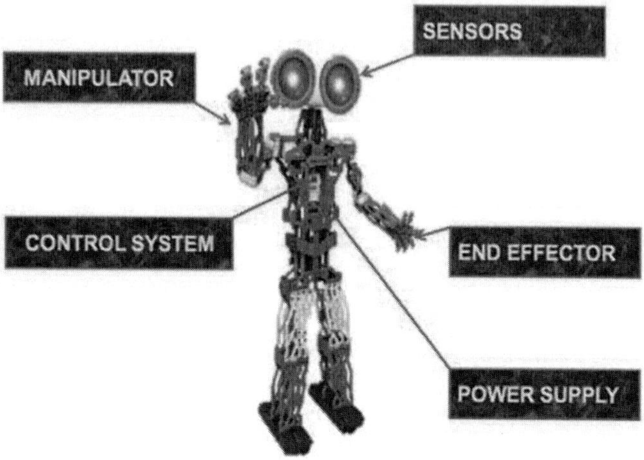


**Figure 1:** Components of a robot (Jyoti *et al.*, 2020)

End effectors are the robotic parts at the end of a robot, attached to its arm or its appendages. They are used to handle, grab or grasp objects to manipulate them. End effectors, in a broader sense, are the components of a robot that interact with the workplace. Robotic arms are also referred to as manipulators. Robotic arms have a shoulder, elbow, wrist and fingers, just like human arms. They are made of finite and non-rigid parts called links. The links are connected to each other by joints. Robotics frequently uses revolute, cylindrical, planar, spherical, spherical, screw and prismatic joints. The control system is used to regulate the behaviour of other subsystems; it requires information and knowledge about all the subsystems to be controlled, including their current and future stages. Controller acts as the brain of a robot. It is also called as computer. There are two types of control systems: open and closed loop. It is critical to have an effective control system to monitor and manage the robotic technology subsystems in order to complete a task with the planned purpose. Additionally, the controller enables the robot to be networked to other systems, enabling it to collaborate with other devices, processors or robots. The sensors are an essential part of assessing the environment and transforming the data into a form that can be read. The static and dynamic features of the sensors define them. In robotics, there are two different types of sensors: wheeled sensors and tactical sensors. Wheeled sensors are used to detect the motor position or speed, while tactical sensors are used to detect physical contact and proximity. Power sources are a crucial component when developing a robotic system, thus attention must be paid to consider how they will affect the system's mechanism, packaging, weight, and size. Batteries, solar cells, fuel cells, hybrids, generators and other power sources are frequently employed in robotics systems.

Autonomous robot systems (ARS) for agriculture are designed to carry out tasks, take actions and make judgments in the dynamic, unstructured environment of an agricultural production system in real-time. When building an ARS, there are several obstacles to overcome, such as the need for real-time reaction in the formulation of sensor controls and modelling the strategy to deal with difficulties (Ng & Trivedi, 1998). The ARS in agriculture is made up of numerous subsystems to carry out various tasks in the agricultural production system. Primary function of the system is to do routine agricultural tasks including seeding, planting, weeding, spraying and monitoring, while its secondary duties include mobility, steering, navigation, manipulator control and human-robot interaction. The primary is made to mimic the conventional agricultural unit operations with the usage of end effectors and manipulators. The mobility and steering subsystems engage with each other through soil contact in the secondary subsystem, which then steers the vehicle at the correct speed and in the desired orientation (Jyoti *et al.*, 2020).

## 3. Application of automation and robotics in agriculture

### 3.1 Planting

The act of planting involves putting seeds or young plants into the ground in preparation for their growing period. This method involves a higher degree of accuracy because different plants require particular spacing between one another in order to maximize output while also optimizing growth. According to the traditional planting procedure, each seed must be manually placed into the ground. This method takes a lot of time and effort because it requires great uniformity and accuracy over a large agricultural region. As a result, a planter machine the farmer controls the machine's motion while simultaneously planting the seed into the ground has been developed. Typically, the intended planter will be pulled behind a tractor and use repetitive motions to plant the seed. The consistency of the row will be impacted because the tractor and planter are operated manually by a human, which results in the row not being in a straight line and some locations where the planter is unable to or misses planting the seed. Therefore, an effective autonomous system is required, one that will guarantee the production of a straight plant row, and don't forget to plant any seeds.

The autonomous system has been introduced for several crops such as corn, wheat, sugarcane and vegetables. Several issues have become the key objectives in the design process to create an efficient autonomous system for the planting process. First, despite the uneven road in the farm field, the robot or vehicle must be able to proceed properly in a straight line. This is a critical requirement because the consistency of the straight seeding process is required to support the efficiency of the automation process in the subsequent planting phase, such as inspection or harvesting. The second necessity is to evaluate the degree of soil moisture, which may affect the soil digging process. Specified seed requires a specified digging depth throughout the soil digging process. As a result, the digging depth must be kept constant by adjusting the digging force for the cultivation process based on the soil moisture level and soil compaction. Finally, seeding detection is required, in which this system detects the presence of seed when the vehicle is ready to plant it. This technique is critical for ensuring that the planter plants the seed without missing any undesired planting areas.

The majority of planting process research is focused on the development of autonomous seeding systems with the main goal of ensuring that the seed is planted at a consistent distance and depth. The Agribot platform was used to construct an autonomous seeding robot. An Infrared (IR) sensor was utilized in this development. To ensure planting consistency, a seed metering unit control system has been devised. During the planting process, seed metering equipment is typically utilized to discharge metered seeds into the soil. It is typically used to separate the seed into a single or group of seeds before depositing it at the predetermined time with precise intervals. Different speeds and seed spacing are required to access the performance of the seed metering device, and planting quality is determined using plant spacing uniformity, variance among rows, fuel consumption, and negative slippage. As a result, it has been demonstrated that an efficient design of seed metering units can deliver higher planting quality and a 22% gain in fuel savings. Another method to ensure the consistency of the seed spacing in planting operations is by using a global positioning system. Using Global Navigation Satellite System (GNSS)/Inertial Measurement Unit (IMU) combined sensing technology, a seed and fertilizer control system was built. The vehicle will execute the operation in the designed system based on the target application rate of seed and fertilizer by comparing the target parameter to the time and space data collected from GNSS/IMU. At 5.8 km/h, this application has been shown to be efficient, with a maximum error of 4.91%. The GPS receiver is used to locate the real-time seeding position, travel speed, and course angle at a fixed frequency. Then, the data receives from GPS is further processed to obtain latitude, longitude, travel speed and course angle. This implementation is found to be a low-cost system for an efficient variable-seed control system with an average seeding accuracy of 97.64%. In addition to the global positioning system technique for an efficient seed planting operation, the use of a highspeed camera can be done to measure seed spacing consistency.

As planting is the first step in agricultural operations, the consistency of the automated system for the planting process is critical. If the operation is carried out inconsistently, it will have a detrimental impact on the subsequent activities, which are entirely dependent on the plant distribution generated during the planting procedure. The development of an efficient planter with seed detection will be able to ensure an optimized planting procedure with reduced operational costs while ensuring seed distribution consistency throughout the field. As a result, with superior planting quality, automation of the planting process would be much more efficient and convenient for farmers in the future.

**Figure 2:** Robot for planting

### 3.2 Inspection

In agriculture, inspection is the process of inspecting or observing plants for diseases or quality flaws. Plant diseases are the primary source of productivity losses in agriculture, resulting in economic losses. Because the agricultural environment is highly dynamic, a variety of unexpected and atypical stress events, such as changes in temperature, humidity, water levels, disease onset, and pests, have impacted plants and their products. If those abnormalities are not addressed promptly, significant and irreversible harm may result. Farmers have always used their human vision system to physically observe plant problems. As the quality of the human vision system degrades with age, the efficiency of inspection operations has decreased. Automation in agricultural inspection uses computer vision to replace the ability of human vision to carry out the inspection procedure. The majority of vision system deployments in agriculture are used for disease detection, with some being utilized for product quality monitoring. Artificial intelligence applications in the agricultural inspection are used to monitor plant diseases in real-time by capturing photos and gathering sensory data from the farm, such as moisture, temperature, humidity, and soil pH level, and displaying it on a website or mobile app. As a result, diseases like early blight, light blight, and powdery mildew can be spotted by the farmer before an outbreak breaks out. In addition to monitoring farm conditions, IoT technology allows farmers to control other agricultural systems such as irrigation and spraying, and it can also offer warning messages to farmers for any abnormalities in plant conditions using artificial intelligence. The inspection process aims to detect disease and maintain the quality of agricultural products. The automation of the inspection process will reduce the inefficiencies of manual inspection procedures. The action of autonomous inspection is typically carried out by positioning a camera in a fixed place, either on the mobile robot platform or on a drone.

**Figure 3:** Disease monitoring robot (Cubero *et al.,* 2020)

### 3.3 Spraying

In agriculture, spraying is a common way of administering pest-control chemicals, fertilizers, or growing media to plants as a fine mist for disease treatment and plant growth management. Pest control chemicals are typically administered consistently throughout the fields in most farming practices to control disease spread. The automated selective spraying system, which is typically carried out by highly automated equipment or mobile robots, allows for the selective administration of pesticides only where and when they are required. The primary goal of this selected operation is to limit pesticide usage while also preventing illness and its epidemic spread across the greenhouse. A variable rate spraying method is designed that allows farmers to automatically adjust the pesticide or herbicide volume rate to the target based on the canopy size and treatment requirement. The variable rate spraying method is proven to reduce pesticide use by up to 85% compared to the homogeneous spray or canopy method. As a result, farmers are strongly encouraged to apply and implement this strategy. With the huge automation and robotics technologies, this strategy will be able to cut pesticide use with great precision while exposing farmers to hazardous chemical impacts. To reduce operational costs, a multi objective algorithm called Non-dominated

Sorting Genetic Algorithm using Reference Point Based to optimize objectives such as travel time, distance and routing angle. UAV swarm is used for non-uniform crop spraying where a multi-agent area coverage method Heat Equation Driven Area Coverage (HEDAC) is used to reduce over-spraying. Although most autonomous spraying operations in agriculture use a land-based vehicle, some implementations propose the use of an Unmanned Aerial Vehicle (UAV) for this task. Spraying accuracy may be maintained with this solution by adjusting the behaviour of the UAV adaptively based on numerous aspects such as wind speed and direction and pesticide deposition. It is critical to substitute human labour in spraying operations in order to limit the risk of toxic chemical chemicals to farmers. With the rapid advancement of automation technology, the transition phase of human replacement in spraying operations will become considerably easier, with the primary goal of treating plant disease at the lowest possible operational cost. However, significant study is still being undertaken to assure that the application of automation in spraying operations would bring unlimited benefits to farmers in managing plant diseases and ensuring future food security. As a result, researchers can continue to improve current methods and strategies for producing an efficient spraying operation for agricultural applications.

**Figure 4:** Autonomous spraying

### 3.4 Weeding

Regular farm work involves challenging conditions like clearly defined field boundaries and loose, uneven, or sticky surfaces. As a result, new weeding instruments have been created, decreasing labour costs for naturally produced sugar cane and vegetable crops by up to 100% while reducing the usage of herbicides in high-value yields by 75% to 100%. They can be used with vegetation that is frequently removed. Robots will be used in place of additional people on third-class weeds on the farm during this process. The goal of the weed-eliminating gadget is to eradicate the weed. Weeds can be eliminated by employing a weed control tool between the rows of crops that are planted in rows. When weeds are perceived in a sophisticated manner that recognizes how harvests are organized and lead themselves specifically between them, the need for herbicides is significantly reduced.

**Figure 5:** Robot weeder

## 3.5 Harvesting

Harvesting is the process of gathering agricultural goods for processing or sale which is particularly labour and time intensive because it requires careful observation and repetitive procedures. As a result, throughout the past few decades, there has been active development of autonomous harvesting systems. Several implementations have been made in the past few years for different types of produce, including strawberries, apples, tomatoes, kiwis, capsicums, grapes, litchi, citrus, pumpkin, and heavyweight crop. The majority of implementations aim to increase harvesting accuracy.

There are various phases involved in autonomous harvesting. For the mobile robot to be able to recognize the object or area that needs to be harvested, it must first be able to locate the target site. The robotic arm will then carefully proceed in that direction, avoiding any obstructions on the way. The procedure will end with the cutting mechanism which generally begins with fruit grasping stem cutting, and harvesting the product, which will then be placed in a storage compartment built within the mobile robot structure. Therefore, each phase in the autonomous robotic harvesting process represents a unique set of challenges that agricultural experts must optimize and resolve in order to create an efficient harvesting robot. The majority of the works that have been implemented take advantage of visualization techniques to locate the fruits. The intended vision techniques were created to address two complex issues: the complicated and loosely organized workspace with a wide range of illumination circumstances and object occlusion levels, and the great variety of detected objects due to their natural characteristics. For target detection in agricultural harvesting, four separate vision techniques are used-monocular, binocular, spectral imaging, and laser.

**Figure 6:** Broccoli harvester and fruit picking robot

The success rate of harvesting with agricultural robots in dense crops depends on robust motion control and end-effector placement at the target fruit or vegetable. The primary obstacle to the replacement of manual labour by robotic systems is the ability to grasp the harvested product without affecting its quality. Therefore, a number of studies have been carried out utilizing various grasping mechanisms and designs to examine the factors that influenced product quality. Target positional and localization error, human factors (body height, shoulder tip, waist, and knee height), grasping velocity, input force, contact time, gripper stiffness, target position accuracy and grasping pressure, position and angle between plant parts, stalk orientation, harvesting posture, target orientation and compressive force are the factors taken into consideration when designing the end effector for an efficient grasping mechanism in maintaining the quality of the harvested products. Therefore, it is necessary to custom design the gripper designs and mechanisms for various applications in order to maintain product quality during the harvesting process. Even if the plants are given proper care during the growing season, there is no guarantee that the product will be of high quality because robotic and automated harvesting methods run the risk of causing product damage. Because of this, numerous ongoing studies are still being carried out to ensure that the efficiency of robotic and automation applications in the harvesting process will be comparable to or better than the way humans work to harvest agricultural product in a timely manner without affecting its quality.

## 4. Agricultural automation and robotics: challenges and next phase

It is anticipated that the global market for automation and robotics systems in agricultural applications would grow from USD 7.4 billion in 2020 to USD 20.6 billion by 2025. The growth of agricultural automation and robots has been made possible by factors including the reduction of labour, expanding population, and the need for high production. In order for the manufactured automation and robotics products to be dependable and strong enough to be utilised in diverse agricultural operations, the development opportunities demand a large technology maturity. As a result, a great deal of study is still being done to find ways to operate various agricultural activities in a variety of working circumstances. The majority of research activities are concentrating on seed uniformity and detection, which is one of the main objectives of the planting process. To keep the straight-line row, more attention should be paid to path correction. The research has become more challenging to track and maintain the robot structure in carrying out planting tasks because of issues like non uniform soil surface and soil stiffness for a variety of soil types like sand, loam, and clay with dry or muddy soil conditions.

The majority of currently active research projects focus on developing an effective spraying system with low operating costs. Volume rate, target location, airflow rate, weed species and size, sprayer travel speed and position (latitude, longitude), sprayer setting, nozzle types combination, airflow rates and the air direction, plant needs and requirements based on image features (entropy, energy, and local homogeneity), average leaf velocity, leaf RMS velocity, turbulence level, and spray cove are just a few of the parameters that have been identified to ensure the optimal execution of the spraying process. To ensure that a fully autonomous system can be realistically implemented in real spraying applications, the developed autonomous system must be able to identify an optimized route to execute the selective spraying operation by taking into account a number of spraying characteristics, such as spraying capacity and refill mechanism. An autonomous system is able to plan a specific strategy to harvest the most agricultural product in the shortest amount of time while also taking into account the limitations of storage capacity and distance to the depot. As a result, the harvesting procedure can be carried out effectively at a minimal operational expense.

Farmers are concerned about the expense of investing in agricultural robots and automation, despite numerous challenges in various agricultural operations. Some farmers are reluctant to spend money on technologies that may not be useful to them in the future. Therefore, agricultural experts need to develop ideas for creating a multipurpose agricultural robot that is also cost-effective. The modular robotic design with excellent resilience is one idea that could be applied to develop agricultural robots. It is advised that the robot's structure be readily detachable and can be changed by various structures based on the specific purpose and requirements of operation, as different agricultural operations have distinct operating mechanisms and methods. Therefore, to perform all agricultural tasks from planting to harvesting, farmers only need to invest in one primary robotic system with several types of detachable structure. As a result, the farmers are not required to buy various types of robots for various agricultural operations, which will result in a significantly lower investment cost.

## 5. Indian scenario

The development of several types of agricultural robots has also been attempted by Indian experts.

- Tamil Nadu agricultural university (TNAU) - developed a pneumatic actuated gripper for grasping and releasing the plug type seedling (transplanting rate 20 – 25 seedlings/minute); robotic rice pellet seeder, which can be remotely controlled from the telemetry controller. The basic features of the controller are the two Joy sticks for right and left moment 4b.
- Indian Agricultural and Research Institute (IARI), New Delhi has developed a robotic precision planter with wireless control through a microprocessor using a Wi-Fi module.
- A Bangalore-based robot company has created the GRobomac commercial robot to locate and harvest cotton balls.
- IIT Kharagpur has developed a robotic arm for cotton picking based on 3D machine vision techniques.

## 6. Conclusion

Precision agricultural technologies have resulted in a paradigm shift in the agricultural production system over the last few decades. Future food security could be greatly improved by the use of robotics and automation in agriculture. Farmers may now execute agricultural activities rapidly with the help of robot equipment thanks to the excellent technology offered by the established system. Planting, inspecting, spraying, weeding, harvesting and post-harvest operations will be completed efficiently with the least amount of operational costs and human labour because the development of robotic systems in agriculture is generally focused on simulating the behaviour of human labour in the completion of agricultural operations. A systematic autonomous agricultural robotic system may be created in the future by combining all the technologies created for each operation, resulting in a reliable and effective agricultural robotic system that will be widely adopted by farmers around the world with the primary goal of producing a high level of agricultural output in order to maintain food security in the future.

## References


1. Bechar, A. & Vigneault, C. (2017). Agricultural robots for field operations. Part 2: Operations and systems. Biosystems Engineering, 153: 110–128.

2. Cubero, S., Marco-Noales, E., Aleixos, N., Barbé, S., & Blasco, J. (2020). Robhortic: A field robot to detect pests and diseases in horticultural crops by proximal sensing. *Agriculture*, *10*(7), 276.

3. Eben-Chaime, M., Bechar, A., & Baron, A. (2011). Economical evaluation of greenhouse layout design. International Journal of Production Economics, 134(1): 246-254.

4. Eizicovits, D., & Berman, S. (2014). Efficient sensory-grounded grasp pose quality mapping for gripper design and online grasp planning. Robotics and Autonomous Systems, 62(8): 1208-1219.

5. FAO. (2019). Smallholders and family farming. In: Family Farming Knowledge Platform [online]. Rome. [Cited 23 May 2023]. http://www.fao.org/family-farming/themes/small-family-farmers/en/

6. Jyoti, B., Chandel, N. S., & Agarwal, K. N. (2020). Application of robotics in agriculture: an Indian perspective. In Proceedings of the 8th Asian-Australasian Conference on Precision Agriculture.

7. Mahmud, M. S. A., Abidin, M. S. Z., Emmanuel, A. A., & Hasan, H. S. (2020). Robotics and automation in agriculture: present and future applications. Applications of Modelling and Simulation, 4, 130-140.

8. Ng, K. C., & Trivedi, M. M. (1998). A neuro-fuzzy controller for mobile robot navigation and multirobot convoying. IEEE Transactions on Systems, Man, and Cybernetics, Part B: Cybernetics, 28(6), 829-840.

9. Xia, C., Wang, L., Chung, B.-K., & Lee, J.-M. (2015). In situ 3D segmentation of individual plant leaves using a RGB-D camera for agricultural automation. Sensors, 15(8), 20463- 20479.

# Chapter - 14

## Crop Protection Strategies: Integrated Pest Management


**Authors**

**Diksha Abrol**

Assistant Professor, Baddi University of Emerging Science and Technology, Baddi, Himachal Pradesh, India

**Vijaya**

Assistant Professor, Baddi University of Emerging Science and Technology, Baddi, Himachal Pradesh, India

**Isha Sharma**

Research Associate, Dr. Yashwant Singh Parmar University of Horticulture and Forestry, Nauni, Solan, Himachal Pradesh, India

# Chapter - 14

## Crop Protection Strategies: Integrated Pest Management

**Diksha Abrol, Vijaya and Isha Sharma**

### Abstract

By using pesticides, cropping systems may be made simpler, yields could be raised, and more intricate crop protection techniques could be avoided. However, an excessive reliance on chemical management is linked to ecosystem contaminating and unfavourable health effects. The development of insect resistance and the dwindling supply of active ingredients both pose threats to crop output in the future. Therefore, agricultural methods need to be created that are less reliant on synthetic pesticides. Here, we provide farmers, advisors, and researchers a dynamic and adaptable strategy that takes into consideration the variety of agricultural circumstances and the complexity of agroecosystems and can increase cropping systems' resilience and our ability to customize crop protection to local conditions. We propose that the key to prevention is the construction of fundamentally robust cropping systems utilizing a mix of agronomic levers. It is necessary to deal with the local availability of monitoring, warning, and forecasting systems. Cropping system elements can be incorporated into the decision-making process to create longer-term strategies. Synergies can be produced by combining non-chemical techniques that may be individually less effective than pesticides. There are choices for choosing goods that have minimal negative effects on health, the environment, and biological pest control thanks to the development of novel biological agents and products as well as the utilization of existing databases. Reduced pesticide use can be paired with other strategies to good effect. The greatest method to develop sustainable crop systems is to address the underlying reasons of pesticide resistance.

**Keywords:** Integrated pest management, resistance, pesticides, sustainable agriculture.

### Introduction

IPM, or integrated pest management, is a sustainable method of crop protection that chooses pest control measures and incorporates them into a management strategy after conducting a cost-benefit analysis that takes into account the impacts on the economy, society, and environment (Kogan, 1998). IPM is defined as "all available plant protection methods and subsequent integration of appropriate measures that discourage the development of populations of harmful organisms, keep the use of plant protection products and other forms of intervention to levels that are economically and ecologically justified, and reduce or minimize risks to human health" by the Food and Agriculture Organisation of the United Nations (2005) and the European Union (EU Framework Directive 2009/128/EC 2009b). In this context, the term "pest" refers to "any species, strain or biotype of plant, animal or pathogenic agent injurious to plants or plant products," in accordance with the International Plant Protection Convention, (2010).

Entomologists at the University of California are credited with creating integrated pest management (IPM) (Stern *et al.*, 1959). They did this in order to address the serious issues (such as secondary pest outbreaks brought on by the eradication of natural enemies and pesticide resistance) resulting from the widespread and occasionally indiscriminate use of broad-spectrum insecticides that were noticed after the discovery and success of these products. Chemical control should only be used as a supplementary measure to biological control, according to Stern *et al.* (1959) innovative approach to pest management that "combines and integrates biological and chemical control". The essential components of integrated control were comprised by the concepts of "economic injury level" and "economic threshold" (Stern *et al.*, 1959), which were regarded as the axioms of decision-making in pest management. All available control measures, including cultural and mechanical ones, were included in the notion of integrated control (Franz, 1961; van den Bosch and Stern, 1962; Smith and Huffaker, 1973). The following ten years saw a major increase in the importance of biological control in this new method of pest management (Perkins, 1985).

Australian entomologists introduced the idea of "protective population management" a few years after Stern *et al.* (1959) defined integrated control (Geier and Clark, 1960). Later, "pest management" (Geier, 1966) was used to describe this and was described as "intelligent manipulation of nature for humans' lasting benefit, as in wildlife management." A. J. Nicholson's theories on insect population dynamics and the ecological aspects of pest control, as well as the work of other ecologists like C. S. Elton, A. G. Tansley, M. E. Solomon, H. G. Andrewartha, L. C. Birch, A. Milne, C. B. Huffaker, and D. Chitty, served as the foundation for this new concept of crop protection (Clark *et al.*, 1967).

IPM is an amalgamation of the principles of integrated control and pest management (Smith *et al.*, 1976; Poston *et al.*, 1983; Pedigo *et al.*, 1986; Van Emden and Peakall, 1996). IPM as a word was originally used by Apple and Smith, 1976. However, Kogan (1998) makes reference to US President Nixon, who in his message to the US Congress in February 1972, included a paragraph on IPM in a programme for environmental preservation. The predominant paradigm for a sustainable approach to crop protection is now IPM.

**Conceptual aspects**

Three questions are frequently asked while making decisions in integrated pest management:

1) IF (Is it required to deploy curative control measures?),
2) While (while should action be taken?),
3) HOW (What kind of control measures should be chosen?), which can be addressed using the ideas of economic harm level, economic threshold, and IPM tactics and strategies, respectively.

**EIL concept**

The term "economic injury level" (EIL) (Fig. 1) refers to the lowest pest population density at which economic harm will result (i.e., the degree of harm needed to make artificial control techniques economically viable) by Stern *et al.*, 1959. Later definitions of economic damage included the amount of harm for which the expense of preventing pest damage matched the economic value of the potential agricultural losses brought on by the pest population (Pedigo, 1996).

EIL was defined as a function of pest density because it is typically challenging to obtain direct measures of injury in field conditions (Pedigo *et al.*, 1986; Funderburk *et al.*, 1993; Pedigo, 1996). Although EIL is related to injury level, i.e., the level of injury to justify the application of control measures, it was defined as a function of pest density. The following is a representation of the general model of economic injury level (Mumford and Norton, 1984; Pedigo *et al.*, 1986; Funderburk *et al.*, 1993; Pedigo, 1996):

EIL=C/VID

Where

C = cost of management per hectare,

V = market value per production unit (Rs/kg)

I = injury units per insect per production unit (stated as a percentage, for example, percent defoliation/insect/ha)

D = damage for every injured unit (for example, kg lost/ha/percent defoliation)

In case, some loss of the insect is avoided then

EIL=C/VIDK

Where K= porportinate reduction in damage

EIL=Gain threshold/ damage per insect in kg × Amount of loss avoided =Insects/ha

Secondary variables like host damage/injury and injury/insect-density interactions, which are controlled by tertiary variables like weather, soil, biotic, and social aspects, have an impact on the components of the models C, V, I, and D.

The application of EIL has a number of restrictions. For instance, it cannot be used when (1) specific pest types are present, such as insect vectors of plant pathogens, (2) sampling techniques are impractical, (3) only preventive control methods are available, and/or (4) there are several pests present (Poston *et al.*, 1983; Mumford and Norton, 1984; Pedigo *et al.*, 1986; Ruesink and Onstad, 1994; Pedigo, 1996).However, despite criticism, the concept has persisted to this day for a number of reasons, one of which is possibly its simplicity.

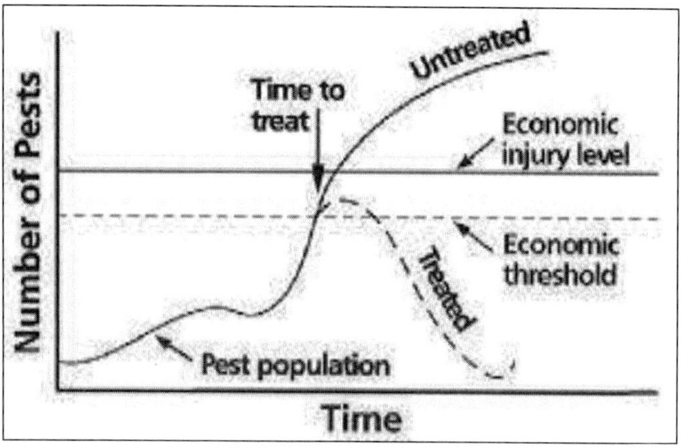


**Fig 1:** ETL and EIL

**Economic threshold (ET)** is a threshold action point that designates the point at which a management strategy should be implemented. According to

Stern *et al.* (1959), the ET is the "density at which control measures should be determined to prevent an increasing pest population from reaching the EIL." If pest numbers reach the ET, it is deemed likely that the economic harm level will be exceeded (Pedigo *et al.*, 1986; Funderburk *et al.*, 1993; Pedigo, 1996).

ETL= EIL - Daily reproduction rate of insect

Table 1 lists the four stages of evolution for the ET implementation: no, nominal, simple, and complete thresholds (Poston *et al.*, 1983; Pedigo *et al.*, 1986; Metcalf and Luckmann, 1994; Pedigo, 1996).

**Table 1:** Implementation of the economic threshold

| Stage | Description |
|---|---|
| **No threshold** | Instances in which there are no clearly defined ETs or in which they do not apply, such as when only preventative measures are at hand. It can involve making decisions in accordance with suggestions made by local plant protection services. |
| **Nominal thresholds** | Technical advisors with area experience typically serve as the foundation for empirical ETs. These arbitrary ETs, which are still prevalent, have helped to reduce the use of pesticides. |
| **Simple thresholds** | Scientific ETs based on the EIL model |
| **Comprehensive thresholds** | These ETs, which are still in development, will be based on how a pest community affects the overall production system. Systems for delivering computer-based information will assist with implementation. |

An integrated pest management strategy is a "overall plan to eliminate or alleviate a real or perceived pest problem". This approach, or the pest management methods used to implement the strategy, often combines one or, preferred, several tactics, such as chemical and biological control (Pedigo 1996). The type of IPM method depends on the four types of pest status: noneconomic, sporadic, perennial, and severe (Pedigo 1996). Perennial and severe pests, also known as key-pests, are often the source of the most serious and difficult pest problems, which are frequently associated with high-value crops and/or high pest population densities. Even the biggest variations in the size of the noneconomic pest population do not surpass the EIL, preventing them from producing economic harm. The population of noneconomic pests has a general equilibrium position (GEP) that is much below the level of economic injury. The occasional pests are the ones that are most common. The EIL is occasionally approached by population fluctuations even while their GEP is likewise below that level. Perennial pests frequently result in financial loss, even if they have a GEP below the EIL.

Based on the economic factors and pest characteristics, various IPM techniques may be categorized and put into practice using various methods (Table. 2). The prophylactic (or preventative) measures seek to avert prospective pest issues. If the ET has been reached, the therapeutic, curative, or responsive techniques should only be used after the occurrence of pest injury in order to effectively address a pest problem. While curative control methods, including the use of pesticides, should only be employed if there are no other options, integrated pest management plans should be environmentally responsible and primarily based on preventive control techniques. The biological control of crop pests by their natural enemies is one such control strategy that should receive special consideration (Tscharntke *et al.*, 2007; Bommarco *et al.*, 2013; Holland *et al.*, 2016; Hatt *et al.*, 2018). In structurally complex settings, natural adversaries are typically more numerous and diversified. These landscapes exhibit higher functional biodiversity and may better offer the hosts, prey, or overwintering refuges that crop pest predators and parasitoids require to complete their life cycles (Rusch *et al.*, 2010; Wäkers *et al.*, 2013; Gardarin *et al.*, 2018; Pollier *et al.*, 2019).

**Table 2:** Integrated pest management prevention and therapeutics (Pedigo, 1996)

| Tactic | Target | Objective | Examples |
|---|---|---|---|
| **Prevention** | Pest | Reducing the GEP | Some biological control tactics<br>Crop<br>Rotation<br>Sanitation<br>Tillage<br>Trap cropping<br>Mating disruption<br>Sterile insect technique |
| | Host plant | Reducing Plant susceptibility | Rational irrigation and fertilization<br>Disrupting crop and pest synchrony (e.g., modifying planting dates)<br>Selecting tolerant/resistant plant varieties |
| **Therapeutic** | Pest | To stop pest population growth and the corresponding crop injury | Selective pesticides<br>Inundative biological control<br>Modifying harvest dates<br>Pruning of infested branches in fruit trees |

## Elements of IPM

IPM aims to increase socioeconomic and ecological benefits while minimising pesticide-related drawbacks. In order to get the desired result from IPM, one needs keep the following in mind:

### a) Knowledge of the agricultural ecosystem

A natural ecosystem like a forest contains a greater diversity of animal and plant species than an agroecosystem. Only one to four major crop species and six to ten major pest species are present in a typical agroecosystem. An agroecosystem is heavily managed by man and is subjected to abrupt changes, such as tillage, intercropping, and pesticide application. Since insect populations are increasing, these procedures are essential for pest management. Due to the lack of diversity in plant and insect species and abrupt changes brought on by the weather and humans, agroecosystems can be more vulnerable to pest damage and catastrophic outbreaks. An agroecosystem, on the other hand, is a complex of food chains and food webs that interact to form a stable unit.

### b) Agricultural ecosystem planning

By integrating crop protection with the crop production system, an IPM programme allows the agricultural system to be planned in terms of anticipating pest issues as well as measures to decrease them. It is advisable to avoid growing sensitive types and allied crops. The prevalence of the spotted borer is increased when okra is followed by cotton. The frequency of the leaf miner increases when soybean is followed by groundnut.

### c) Cost-benefit ratio

The cost-benefit ratio should be prioritized based on the likelihood of pest harm by anticipating the pest issue and determining the economic threshold level. To give reliable information analysis of pest damage and the cost-benefit ratio in pest management, the crop life table must be established. When a chemical pesticide is used in an agroecosystem, a benefit-risk analysis is conducted to evaluate how it will affect society and the environment in relation to the benefits it will provide.

### d) Pest damage tolerance

Most of the time, neither high yields nor insect pest control call for the pest-free crop. Any bug can be tolerated to some extent. Castor plants are able to withstand up to 25% of defoliation with defoliators. In the case of plant disease transmission by vectors, there are exceptions. Through threshold values like economic threshold and economic harm levels, the relationship between pest population density and the profitability of control techniques is expressed.

**e) Leaving pest remains behind**

In addition to losing their individual insect hosts, natural enemies are gradually wiped out due to indiscriminate use of broad-spectrum pesticides, which in turn wipes out other natural enemies. Leaving a persistent pest residue below the economic threshold level in order to allow for the survival of natural enemies is crucial to pest management.

**f) Treatment timing**

Pesticide application should be based on need, with a minimum number of applications, prompt scheduling, and enhanced methods of pest monitoring, such as the use of pheromone traps for pest population monitoring.

**Tools for IPM**

**Monitoring:** The cornerstone of IPM is crop monitoring, which keeps tabs on pests and any possible damage they may cause. This information about the state of the crops and pests now is useful for choosing the most effective pest management strategies. When compared to other monitoring devices like light and sticky traps, pheromone traps offer an advantage. They have demonstrated their value in extensive IPM validations in cotton, basmati rice, chickpea, and pigeonpea due to their selectiveness to certain pests Birthal and Sharma, 2004.

**Resistant pests varieties:** The process of breeding for pest resistance is ongoing. Additionally, pests co-evolve with their hosts, particularly plant diseases. Gene transfer technique is therefore helpful in creating cultivars that are resistant to pesticides, plant diseases, and insects. An illustration of this is the insertion of genetic material from the naturally occurring bacterium *Bacillus thuringiensis* (Bt) in cotton, maize and potatoes, which renders the plant tissues poisonous to the insect pests. The scientific community is awed by its enormous potential for controlling pests, but it is equally worried about the prospect of increasing selection pressure for resistance against it and its consequences on non-target natural species. This possible technology has, however, generated debate due to ethical, scientific, and social issues.

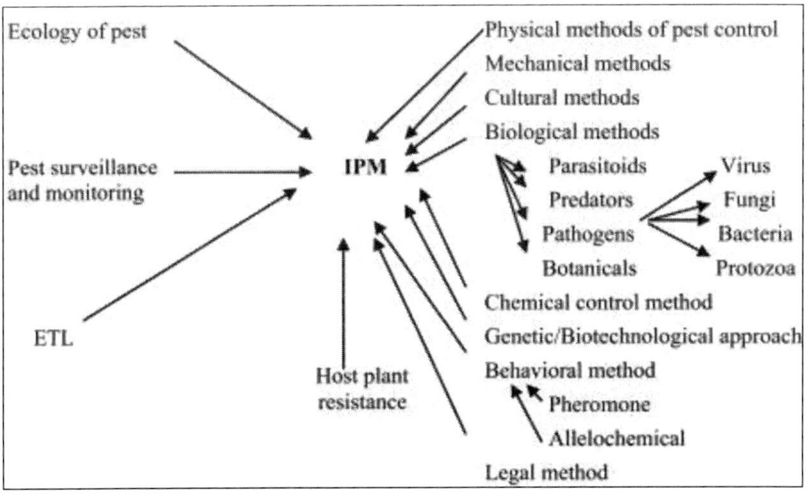


**Fig 2:** Tools of integrated pest management

**Cultural pest management:** It consists of agricultural production techniques that reduce the susceptibility of the crop environment to pests. A few cultural approaches for managing pests include crop rotation, fallowing, altering planting and harvesting dates, altering plant and row spacing, and destroying old crop detritus. Important management approaches include the interplanting of various crops, the planting of cover crops, and the planting of plants that produce nectar. Cover crops, which are frequently legume or grass species, stop soil erosion and control weed growth. A cover crop can also be utilised as green manure, which is added to the soil to provide the succeeding crop with nitrogen and organic matter. Some cover crops from the Brassicaceae family can control nematode pests and wilt diseases when they are introduced into the soil. Rye and wheat, when left in the field as residues, effectively reduce weeds by more than 90%. Based on an understanding of the biology and development of the pest, cultural controls are chosen.

**Mechanical or physical control:** These are based on an understanding of how pests behave. One instance of physical control is digging holes coated with plastic in potato fields to trap Colorado potato bugs that are migrating. In locations where pigeonpeas are grown, shaking the plant to get rid of *Helicoverpa* larvae is a frequent practice. The simplest form of pest control may be hand plucking insects. The bollworm infestation in cotton and chickpea crops has been successfully controlled by installing both live and dead bird perches. Other examples include installing row coverings to shield plants from insects and using mulch to smother weeds.

**Biological controls:** These involve preserving and enhancing the pest natural predatory insects, parasitoids, parasitic nematodes, fungus, and bacteria. Native natural enemy populations are preserved in IPM programmes, and non-native agents may be discharged with extreme caution. The most widely used parasitoids on a variety of host crops are *Trichogramma* spp. Numerous microorganisms that target and suppress plant infections, including *Trichoderma* spp., *Verticillium* spp., *Aspergillus* spp., *Bacillus* spp., and *Pseudomonas* spp., have been used as biological control agents.

**Controls with the chemical:** When pest populations cannot be controlled by other measures, pesticides are employed to keep the populations below levels that are economically destructive. Both plant-derived and manmade insecticides are classified as pesticides. Numerous synthetic compounds that are produced by humans are used as insecticides. These are quick-acting, simple to use, and reasonably priced. IPM programmes should ideally only use pesticides as a last resort because of the possible harm they may cause to the environment. The best pesticides are those with minimal side effects on the environment and non-target creatures. Fortunately, new generation pesticides are being created and approved for usage. These chemicals have novel mechanisms of action and little environmental consequences. This category includes pesticides with a short half-life or those that only affect one or a few particular organisms.

Economic threshold analysis is predicated on the idea that the majority of plants can withstand at least some insect damage. Despite extensive research, there is still no consensus over the damage thresholds for a range of crops and pest circumstances. Chemical controls are only used in an IPM programme when the economic threshold is known when the pest's ability to cause damage is getting close to it, even while other alternative management practices are being used (Fig. 2.).

Different methods can be used to create botanical insecticides. They can be as straightforward as unprocessed, crushed plant leaves, plant extracts, and compounds made with only the best plant materials. Botanicals include, among others, pyrethrum, neem, tobbaco, garlic, and pongamia compositions. Some plants have broad-spectrum insecticide properties. Because of their rapid degradation, botanicals tend to be less damaging to the environment. They are easier to move safely. The main benefit is that farmers themselves may create these on the farm.

**Legal methods:** Plant quarantine is the legal use of measures intended to stop the spread of pests or their continued multiplication in the event that they have already achieved admission and have established in new restricted regions. When the grapevine phylloxera was introduced into France from America around 1860 and the San Jose scale spread into the US in the latter part of the eighteenth century and caused significant damage, it became clear how important it was to place restrictions on the movement of pest-infested plants or plant materials between nations. While India passed the "Destructive Insect and Pests Act of 1914" to forbid the importation of any insect, fungus, or other pest, the United States passed its first quarantine law in 1905. Later, a more comprehensive statute from 1917 augmented this.

**Techniques for implementing IPM**

The IPM packages outperform farmers practices, according to tests conducted at several research facilities. IPM techniques made it possible to use less chemical sprays. According to Dhaliwal and Arora (1996), the IPM system also caused a three-fold rise in natural enemies and a decrease in insecticide use and environmental degradation. By

i) Developing effective pest control techniques through pest surveys and monitoring,

ii) Preserving and enhancing natural enemies like parasites, predators, and insect pathogens,

iii) Biological control of pests, it is possible to develop an integrated strategy for the management of major pests and diseases.

Major pests in rice, cotton, legumes, sugarcane, and other crops have been controlled using economically sound integrated pest management systems. The discharge of biocontrol agents has been successful in controlling the *Pyrilla* and top borer of sugarcane, mealy bug of coffee, lepidopterous pests affecting cotton, tobacco, coconut, sugarcane, etc. The invention of mass rearing techniques for biocontrol agents including *Trichogramma* spp., *Chrysoperla* spp., and nuclear polyhedrosis viruses (NPV) of *Heliothis* and *Spodoptera* has been a significant accomplishment.

The concept of economic thresholds and the negative externalities of pesticides are both understood by Indian scientists and extension personnel. The State Agricultural Universities and other research organisations receive financial support from the Department of Biotechnology, Government of India, for the development and production of biopesticides and biocontrol agents. In recent years, a number of biopesticide production facilities and clinical plant protection centres have been built and strengthened. As a result, although it has not yet achieved the required level, India is using more biopesticides and biocontrol chemicals. Less expensive than chemical insecticides are biopesticides. They are not only eco-friendly but also do not run the risk of developing resistance.

**Major challenges**

Although IPM has been recognised as the most appealing approach for protecting crops from pests, farmer adoption has been rather low. The biggest threat to IPM is the continued dominance of pesticides and their careless use. The following are some of the barriers to its dissemination that must be identified for an effective implementation strategy:

- Low awareness and innovativeness of extension personnel and target groups;
- Inadequate interaction between research and extension agencies;
- Problem of timely and adequate supply of quality inputs, including biocontrol agents and biopesticides;
- Complexity of IPM vs. Simplicity of Chemical Pesticides;
- Dominant Influence of Pesticide

**Requirements for implementation**

- Accessibility to site-specific IPM modules that are ethically sound, financially viable, and socially acceptable;
- High target group participation;
- Area-wide dissemination strategy;
- Removal of barriers to IPM dissemination;
- Measurement, evaluation, and publicization of IPM impacts.

The preservation and enhancement of pests' natural adversaries are crucial. Additionally, botanicals and biopesticides are the most reliable instruments for sustainable IPM due to their inherent renewability, reversibility, and resilience. Because of this, the usage of bio-agents and biopesticides/botanicals must be given top importance in order to maintain ecological balance and manage pests.

**IPM misunderstood**

- **IPM is not brand-new:** Since the beginning of agriculture, it has existed in some form or another. However, many programmes are only a few decades old and are founded on scientific research.
- **IPM implementation takes time:** Years of research may be

required to build an IPM programme, with participants including university researchers, extension agents, pest control specialists, business scientists, and most importantly farmers.

- **IPM is not organic agriculture:** Organic farming is a way of thinking about agricultural production that doesn't use any synthetic fertilisers or pesticides. Due to the fact that they are synthetic, some low-risk procedures and technologies that are available to growers utilizing IPM cannot be used by organic farmers.
- **IPM is not a method to get rid of or cut back on pesticide use:** IPM programmes that are well-developed and scientifically based consistently lower the need for pesticides because they make use of a larger variety of pest control methods. By design, IPM programmes lead to a safer and more responsible use of pesticides. IPM is not a strict set of management practices. IPM, which combines all appropriate strategies, gives growers alternatives for managing pests within a specific crop production system.
- **Not all organizations use IPM programmes:** Programmes for the same crop in different geographies may alter significantly depending on the pest complex and the area.

**Conclusions**

There is a growing consensus that industrial farming based on petrochemicals is unsustainable and that ecological approaches to food production need to be developed and promoted. The potential for doing this in biotechnology is enormous. The most straightforward and ostensibly eco-friendly substitute for pesticides is to use naturally existing biological methods. Numerous plant species have been shown to have pesticidal and growth-inhibiting capabilities, but the industry has yet to fully utilise these abilities.

Farmers are given the management tools they need via holistic planning to efficiently manage biologically complex farming systems. IPM programmes need time, money, patience, short- and long-term planning, adaptability, and commitment to be successful. The research managers must invest time in their own education and networking with extension and research staff to talk about farming practices, which differ greatly. This would make it easier to create integrated plans. Government policies could be developed to support IPM promotion. The federal and state governments must take the lead in transforming the pest control landscape by implementing legislative, regulatory, and fiscal measures that would reduce the appeal of chemical pest control.

The development and promotion of IPM throughout the nation is a priority for the Indian Council of Agricultural Research (ICAR) and the Department of Agricultural Research and Education of the Ministry of Agriculture, Government of India. The ICAR and the Government of India place a high premium on offering safe and efficient solutions to guard against unacceptably high losses caused by weeds, illnesses, and insect pests.

**References**


1. Apple, J. L., & Smith, R. F. (1976). Progress, problems, and prospects for integrated pest management. *Integrated Pest Management*, 179-196.
2. Birthal, P. S., & Sharma, O. P. (2004). Integrated pest management in Indian agriculture.
3. Bommarco, R., Kleijn, D., & Potts, S. G. (2013). Ecological intensification: harnessing ecosystem services for food security. *Trends in ecology & evolution*, *28*(4), 230-238.
4. Clark, L. R., Geier, P. W., Hughes, R. D., & Morris, R. F. (1967). The ecology of insect populations in theory and practice.
5. Dhaliwal, G. S., & Ramesh, A. (1996). *Principles of insect pest management*. National Agricultural Technology Information Centre.
6. Franz, J. M. (1961). Biological control of pest insects in Europe. *Annual Review of Entomology*, *6*(1), 183-200.
7. Funderburk, J., Higley, L., & Buntin, G. D. (1993). Concepts and directions in arthropod pest management. *Advances in agronomy (USA)*.
8. Gardarin, A., Plantegenest, M., Bischoff, A., & Valantin-Morison, M. (2018). Understanding plant–arthropod interactions in multitrophic communities to improve conservation biological control: useful traits and metrics. *Journal of pest science*, *91*, 943-955.
9. Geier, P. W. (1966) Management of insect pests. *Annual Review of Entomology* 11:471–490
10. Geier, P. W., & Clark, L. R. (1960). An ecological approach to pest control. In *Proceedings of the eighth technical meeting. International Union of Conservation of Nature and Natural Resources, Warsaw* (pp. 10-18).
11. Hatt, S., Boeraeve, F., Artru, S., Dufrêne, M., & Francis, F. (2018). Spatial diversification of agroecosystems to enhance biological control and other regulating services: An agroecological perspective. *Science of the Total Environment*, *621*, 600-611.

12. Holland, J. M., Bianchi, F. J., Entling, M. H., Moonen, A. C., Smith, B. M., & Jeanneret, P. (2016). Structure, function and management of semi-natural habitats for conservation biological control: a review of European studies. *Pest management science*, *72*(9), 1638-1651.

13. International Plant Protection Convention (2010) International standards for phytosanitary measures No. 5.Food and Agriculture Organization of the United Nations, Rome

14. Kogan, M. (1998). Integrated pest management: historical perspectives and contemporary developments. *Annual review of entomology*, *43*(1), 243-270.

15. Metcalf, R.L., & Luckmann, W. H. (Eds.). (1994). *Introduction to insect pest management* (Vol. 101). John Wiley & Sons.

16. Mumford, J. D., & Norton, G. A. (1984). Economics of decision making in pest management. *Annual review of entomology*, *29*(1), 157-174.

17. Pedigo, L. P. (1996). *Entomology and pest management* (No. Ed. 2). Prentice-Hall Inc..

18. Pedigo, L. P., Hutchins, S. H., & Higley, L. G. (1986). Economic injury levels in theory and practice. *Annual review of entomology*, *31*(1), 341-368.

19. Perkins, J. H. (1985). Naturally occurring pesticides and the pesticide crisis, 1945 to 1980. In *CRC Handbook of Natural Pesticides: Methods* (pp. 297-325). CRC Press.

20. Pollier A., Tricault, Y., Plantegenest, M., & Bischoff, A. (2019). Sowing of margin strips rich in floral resources improves herbivore control in adjacent crop fields. *Agricultural and Forest Entomology*, *21*(1), 119-129.

21. Poston, F. L., Pedigo, L. P., & Welch, S. M. (1983). Economic injury levels: reality and practicality. *Bulletin of the ESA*, *29*(1), 49-53.

22. Ruesink W.G., & Onstad, D. W. (1994). Systems analysis and modeling in pest management. *Introduction to insect pest management. Wiley, New York*, 393.

23. Rusch A., Valantin-Morison, M., Sarthou, J. P., & Roger-Estrade, J. (2010). Biological control of insect pests in agroecosystems: effects of crop management, farming systems, and seminatural habitats at the landscape scale: a review. *Advances in agronomy*, *109*, 219-259.

24. Smith, R. F., & Huffaker, C. B. (1973). Integrated Control Strategy in the United States and its Practical Implementation1. *EPPO Bulletin*, *3*(3), 31-46.

25. Smith, R. F., Apple, J. L., & Bottrell, D. G. (1976). The origins of integrated pest management concepts for agricultural crops. *Integrated pest management*, 1-16.

26. Stern, V.M., Smith, R.F., van den Bosch. R., Hagen, K.S. (1959) The integrated control concept. *Hilgardia* 29:81–101

27. Tscharntke T., Bommarco R., Clough, Y., Crist, T. O., Kleijn, D., Rand, T. A.,... & Vidal, S. (2008). Reprint of "Conservation biological control and enemy diversity on a landscape scale"[Biol. Control 43 (2007) 294–309]. *Biological control*, *45*(2), 238-253.

28. Van den Bosch, R., & Stern, V. M. (1962). The integration of chemical and biological control of arthropod pests. *Annual review of entomology*, *7*(1), 367-386.

29. Van Emden, H. F., & Peakall, D. B. (1996). *Beyond silent spring: integrated pest management and chemical safety*. London: Chapman & Hall.

30. Wäkers F.L., van Rijn, P. C. J., & Bruin, J. (Eds.). (2013). *Plant-provided food for carnivorous insects: a protective mutualism and its applications*. Cambridge University Press.

# Chapter - 15

## Role of Agricultural Extension in Knowledge Transfer

**Authors**

**Mupparapu Deepika**

Ph.D. Research Scholar, Dept. of Agricultural Extension Education, S. V. Agricultural College, Tirupati, ANGRAU, Andhra Pradesh, India

# Chapter - 15

## Role of Agricultural Extension in Knowledge Transfer

**Mupparapu Deepika**


### Abstract

Agricultural extension plays a crucial role in the transfer of knowledge and information between researchers, experts and farmers. It serves as a vital link in the agricultural knowledge system, bridging the gap between scientific research and practical on-the-ground application. Agricultural extension agents facilitate the dissemination of research findings, new technologies, best practices, and relevant information to farmers and other stakeholders. They act as intermediaries, translating complex scientific knowledge into accessible and understandable language for farmers. Extension services promote the adoption of new agricultural technologies and practices by providing farmers with information about their benefits, demonstrating their effectiveness, and guiding them through the process of implementation. This helps farmers improve their productivity, efficiency, and sustainability.


**Keywords:** Agricultural extension, knowledge transfer, farmers, technology promotion.

### Introduction

In the dynamic and ever-evolving field of agriculture, agricultural extension plays a key role in knowledge transfer and is crucial for the sustainable development of farming practices. Agricultural extension acts as a bridge between researchers, farmers, and rural communities, facilitating the exchange of information, skills, and technologies. Agricultural extension is essential for enhancing agricultural productivity, strengthening food security, enhancing rural livelihoods, and promoting agriculture as a crucial factor of economic growth. Besides, it provides a critical support service for rural producers to face the new challenges confronting agriculture (*IFPRI*).

Extension is basically education, although it takes place outside formal education systems, and as such, aims at bringing positive changes in the behaviour among the targeted. The extension function is crucial for the welfare of the farming community, just like many other key roles in daily life, like health and education, regardless of who performs it in a successful manner. The stakeholders in the extension function, includes government extension departments, private extension service companies, NGOs, private extension advisors, farmer's associations, research institutes, universities, and others as seen in the recent years. Extension organisation refers to how the department or agency in charge of the function organises itself to carry out the assignment (FAO, 2005).

**Agricultural extension: A knowledge system (FAO)**

Agricultural extension operates within an extensive knowledge system which includes research and agricultural education. FAO and the World Bank refers this larger system as Agricultural Knowledge and Information Systems for Rural Development (AKIS/RD). The OECD countries refer Agricultural extension as the Agricultural Knowledge System (AKS) whereas, others identify the three pillars of this system as being research, extension and agricultural higher education – as "the agricultural knowledge triangle" and suggest that as the three pillars include complementary investments, they should be planned and executed as a system instead of as separate entities.

Agricultural extension receives pertinent information from the agricultural education system and shares back field observations to this system. Extension is also professionally linked to the agricultural vocational and higher education systems as these systems also produce the extension agents. The relationship between agricultural extension and agricultural research is much closer, as the knowledge, the agricultural extension transfers is usually generated from agricultural research through adaptive and applied agricultural research development.

Agricultural extension is the process of carrying the scientific agriculture technology to the farmers in order to enable them to utilize the knowledge for a better economy (Altalb, *et al.,* 2015)

**Agricultural extension - An essential pillar in Agricultural research and development (FAO, 2005)**

Before being examined by farmers for potential adoption, the innovations must first pass via appropriate research institutions and the farmer's fields for critical scrutiny. Then, they must be demonstrated in simple language to the farmers regarding positive aspects of the technology in comparison to traditional method in a persuading manner. Followed by which, the prerequisites for trying the new technology, like cost and risk factors must be explained. Several adult education methods and participatory decision-making approaches must be used in order to promote discussion on the information presented regarding the new technology for assessment. Assuming that some progressive farmers might be willing to try the recently developed technology, arrangements must be made for other farmers so that they can benefit from this limited "sample adoption" opportunity. The regular monitoring of the trial of the technology, is mandatory including discussion with the progressive farmers, for any trouble-shooting and eventually to measure the performance of the new technology under *in situ* conditions. The results along with possible reasons must be carried back to those particular researchers who recommended that specific technology. If the advantages outweigh the drawbacks, there is a larger possibility that other farmers will adopt the technology; otherwise, it will be deemed inappropriate and will be moved down the list of extension recommendations.

Agricultural research institutes in some of the developing countries initiated small extension units as part of their organizational structure in order to publicize their technologies, mostly through publications. While the publications were a source of gratification for the researchers, such publicity hardly went beyond the institutes. Such units were eventually disbanded. Several research projects were also carried out, that involved research trials in farmer's fields. This kind of approach gave more favourable outcome only in those places where both farmers and extension workers were actively involved in the entire stages. However, the practice cannot be replaced for a viable extension service.

**Extension in developing countries (FAO, 2005)**

Agricultural extension was formally introduced in the 1950s by visiting American professors, technical experts in agricultural and rural sciences, national government officials, and university faculty who had studied or been trained in the United States of America, despite the fact that the function of extension had existed in developing countries for centuries. They were motivated by the exceptionally structured agricultural research, extension, and education efforts of the American Cooperative Extension Service and the associated Land Grant College system. The agricultural extension organizations that are meant to deliver the improved technologies from agricultural research institutes were established in developing countries. The delivery methods included oral method to farmers (individual/group method) at their farms or at homes augmented through demonstrations of recommended technologies in farmer's field. This was further augmented by communication channels like printed material, radio, video and television. In most of the countries, the extension agents were entrusted with disbursement of agricultural inputs, particularly insecticides, herbicides, chemical fertilizers and pesticides.

**Alternative extension approaches (FAO)**

Most of the approaches, Axinn referred, were supported by FAO. The main approaches along with respective success criteria are briefed below.

1) **The general agricultural extension approach:** The rate of implementation of the recommendations and the rise in national production serve as the primary indicators of success.
2) **The commodity specialized approach:** The overall crop production is typically used as a measure of success.
3) **The training and visit approach:** Increases in the output of the specific crops covered by the programme are used to measure success.
4) **The agricultural extension participatory approach:** The number of farmers actively engaging and gaining from the programme, and the longevity of local extension organisations, are used as indicators of success.
5) **The project approach:** Success is determined by short-run change.
6) **The farming systems development approach:** The extent to which farming people embrace and adopt the technologies and continue using them over time is the yardstick for success.
7) **The cost sharing approach:** The ability and willingness of farmers to contribute financially, either on their own or through their local governments, is a key metric for success.
8) **The educational institution approach:** The level of involvement and attendance by farmers in the school's agricultural extension programmes serves as a gauge of success.

Agricultural extension involves different approaches and various methodologies. It is managed and delivered through a variety of institutional arrangements. Therefore, it is reasonable to claim that no single approach is appropriate for extension development in all situations, just as there is no one method that works best for development.

**Global developments demanding reforms in extension (FAO, 2005)**

Globalization, liberalization, privatization, decentralization and devolution, pluralism, client participation in decision-making, emphasis on integrated, multi-disciplinary, holistic and sustainable development along with natural and man-made disasters, rural poverty, food insecurity, HIV/AIDS epidemic, COVID pandemic necessitated the new learning approaches for both subsistence and commercial farmers in developing countries. These criteria pose challenges to decades-old mandates and operations within conventional extension systems, particularly when seen in the present era of information technology revolution.

**Guidelines for modernization of national extension systems (FAO, 2005)**

- Determine whether to improve or revamp the current extension organisation in view of farmer demands.
- Decentralize extension only after capacity-building of the staff and training of relevant elected officials.
- Broaden the technical mandate of extension to focus at broader development of human resources.
- Design national policy on extension which guarantees financial and political commitment.
- Accord the extension profession the standing it deserves on par with other agricultural disciplines.
- Align agricultural extension pre-service education with the modernization of the national extension system.
- Encourage the involvement of institutions from the public, corporate, and civil society to further pluralism.
- Privatize extension services partially or completely if it is feasible - socially and economically.
- Develop information and communication technology (ICT) tools to aid the work of extension workers.
- Develop location-specific, gender-sensitive, participatory, and least expensive extension materials and methodologies instead of applying those that are promoted as universally suitable.
- Orient extension staff towards issues related to food security that could eventually enhance rural livelihoods.
- Encourage the extension services which involves empowering farmers by organizing them into groups in order to form a strong lobby for farmers and for extension.
- Encourage bottom-up *i.e.,* grassroots extension programme planning by the farmers in order to make extension not only demand-driven, but also supply-driven.

- If the extension function is likely to be performed with relatively smaller number of extension personnel, appropriate strategies must be followed in order to get maximum output.
- Ensure effective operational linkages among research and extension and other relevant institutions.

**Extension division in India - An overview**

Extension Division is accountable for conducting Agricultural Extension activities through various schemes and programmes. It aids farmers in learning about recent advances in agricultural knowledge and scientific research. It assists and motivates state governments in establishing, sustaining, and running various agricultural extension programmes. Extension Division implements Sub Mission on Agricultural Extension (SMAE), a Submission of Krishonnati Yojana. The Extension Division also oversees the well-known ATMA Scheme, which aids the extension activities of the State Governments.

The Division also offers information Support to farmers with regard to agricultural activities by coordinating with All India Radio (AIR) and Prasar Bharati for smooth running of Krishi Vani and DD-Kisan channel programmes respectively. Kisan Call Centres (KCCs) were setup at twenty one locations all over the nation covering about all states and UTs, to where the farmers can freely call on toll free number (1800-180-1551) in order to get information related to agriculture. Additionally, the Extension Division publishes four magazines, two in Hindi and two in English, that provide information on multiple aspects of Extension endeavours and recent advancements.

The Division is headed by Joint Secretary who is assisted by a Director and two Additional Commissioners. The Division subordinate office is situated in Pusa, New Delhi. Directorate of Extension offers technical support to the Extension Division in conducting various schemes / programmes and other activities.

National Institute of Agricultural Extension Management (MANAGE), an autonomous institute is also associated with establishing connections between eminent State, Regional, National, and international entities involved in Agricultural Extension Management. Besides it also undertakes research and trainings related to agricultural extension.

Extension Division has established four Extension Education Institutes (EEIs), located in Nilokheri (Haryana), Anand (Gujarat), Hyderabad

(Telangana), and Jorhat (Assam) which serve to the training requirements of field extension functionaries under agriculture and allied departments of States/UTs in their regions. The entire team of Extension Division is committed towards effective execution of extension activities through which agricultural productivity and farm income in the country can be improved (DA&FW,2022)

**Agricultural extension: Challenges (Extension digest, MANAGE)**

Axinn highlighted five major challenges to agricultural extension in the twenty-first century which are as follows:

**1. Control and accountability**

Farmers are unlikely to have control over a government extension system or an NGO if there is minimal accountability to them.

**2. Sustainability**

Ecological sustainability is a problem for the future survival of humanity. It is important to think about how sustainability may affect agricultural extension from a technical and economic perspective as well as from a social, political, cultural, and other perspective.

**3. Role of women in farming and extension**

Gender issue impacts all aspects of extension activity. There are implications for women-farmer led extension for the timing, location and language of extension activities.

**4. Participation**

Strategies for achieving maximum participation through farmer-led approach. Greed and corruption if left ignored, can even defeat the best extension strategies.

**Programmes and Schemes towards revamping of extension services in India (*agricoop.nic.in*)**

The Sub Mission on Agricultural Extension (SMAE) was recently placed under the umbrella of Krishonnati Yojana. It focuses on creating awareness and enhanced utilization of appropriate technologies in agriculture and allied sectors. It has four components, specifically:

i) Support to State Extension Programmes for Extension Reforms

ii) Mass Media Support to Agricultural Extension

iii) Establishment of Agri-Clinic and Agri-Business Centres by Agriculture Graduates (AC&ABC), and

iv) Extension Support to Central Institutions.

These components of Sub Mission on Agricultural Extension operate as separate schemes of Extension Division. Some details about these are discussed below.

**Support to state extension programmes for extension reforms**

This scheme aids the state governments in the Extension Activities. By using new institutional arrangements for technology dissemination in the form of an Agricultural Technology Management Agency (ATMA) at the district level to operationalize the extension reforms, it intends to make extension systems farmer driven. Under this Scheme Grant-in-Aid is released to the State Governments in compliance with the scheme guidelines (in the ratio of 60:40 (centre: state) to General States, 90:10 for North-Eastern & 3 Himalayan States & 100% for UTs) in order to support State Government's to revamp the extension system and to make available the recent agricultural technologies and modern agricultural practices in agriculture and allied sector. The extension activities include farmers training, farmer's exposure visit, demonstration, mobilization of farmers groups, kisan mela, organizing farm schools etc.

**Mass media support to agricultural extension**

The scheme intends to make it possible to modernize the extension services offered in the nation by transferring technology and knowledge to farmers through electronic and print media. Five tier publicity and information support are provided to the farmers and extension functionaries through the following:

- First tier information support through All India Radio, Doordarshan, and private TV channels Programmes. The purpose is to make the farmers aware of new technologies and recent advancements related to agriculture and allied sectors. Success-stories and new farm-practices are promoted through advertisements, audio-video spots etc.
- Second tier information support through print media *viz.*, newspaper ads, leaflets and pamphlets, magazines etc.
- Third tier information support through supporting/ organizing kisan melas, exhibitions, fairs, at regional / national level.

- Fourth tier information support through SMS / inward voice calls in Kisan Call Centres (KCCs) and internet and
- Fifth tier information support through coordinating and facilitating delivery of information at the lowest level *i.e.,* blocks level through Common Service Centres (CSCs).

An emphasis on using social media platforms to connect with farmers and provide them with information is also being made with vigour in order to leverage the popularity and reach of social media for the advantage of agriculture extension.

**Establishment of agri-clinics and agri-business centres by the agriculture graduates (AC&ABC)**

This scheme intends to utilise the expertise available among the pool of Agriculture Graduates. They can establish Agri-Clinic or Agri-Business Centre as a profession on their own and can offer extension services to farmers. These centres would provide free/paid services to the farmers and can contribute to increase the agriculture production and farmer's income. This is beneficial as it opens up gainful self-employment opportunities to unemployed agricultural graduates besides complementing the efforts of public extension by providing extension and advisory services to the farmers at grass root level. Under this AC&ABC, 45 days specific training is provided by the selected institutes across the country at free-of-cost to AgriGraduates, who are interested in setting up of such centre. Subsidized loans and credit-facilities are also provided to the Agri-Clinic entrepreneurs by the commercial banks.

**Extension support to central institutions**

This scheme aims towards capacity building of extension functionaries while simultaneously working towards skill development of rural youth, farmers and farm-women. The prominent institutions involved in this initiative are – Directorate of Extension (a subordinate office under Extension Division), MANAGE, Hyderabad, four Extension Education Institutes (EEIs) at the regional level and the State Agricultural Management & Extension Training Institutes (SAMETIs) at the State level.

The main programmes under this scheme are as below:

a) **Programmes for capacity building**

   i) **Extension Education Institutes (EEIs):** Ministry of Agriculture & Farmers Welfare (MOA&FW) has established four Extension

Education Institutes at Nilokheri (Haryana), Anand (Gujarat), Hyderabad (Telangana), and Jorhat (Assam). These EEIs meet the training requirements of middle-level field extension functionaries working under agriculture and allied departments of states/UTs of respective regions.

The activities of EEIs include organization of on-campus/off-campus trainings, conferences, workshops, etc. in the areas of communication technology, training management, extension methodology, Agriculture Knowledge Information System (AKIS) and Information Technology.

ii) **Model Training Courses (MTCs):** Directorate of Extension (DOE), DAC&FW sponsors the organization of Model Training Courses (MTCs) of 8 days duration through reputed national/ central institutes, ICAR institutes and State Agricultural Universities (SAUs) in specialized areas for capacity building of middle level extension functionaries working under agriculture & allied departments of states/ UTs. These training courses focus on subjects including Agriculture, Horticulture, Animal Husbandry, and Fisheries Extension.

**Kisan Call Center (KCC):** The project's goal is to respond to farmer's phone calls in their own dialect. These call centres are currently operating in 21 different locations throughout the nation, serving all of the States and UTs. For Kisan Call Centre, a standard eleven-digit toll-free number with nationwide coverage, 1800-180-1551, has been assigned. All landlines and mobile devices connected to telecom networks, including those operated by commercial service providers, can dial this number. Answers to the farmer's questions can be obtained on all days of the week from 6:00 am to 10:00 pm. Kisan Call Center agents are known as Farm Tele Advisors (FTAs), who are graduates or above in agriculture or allied areas and possess excellent communication skills in their respective local languages. In call conferencing mode, questions that FTAs are unable to answer are routed to higher level specialists. These experts are Subject Matter Specialists (SMSs) of state agricultural departments, ICAR & State Agricultural Universities (SAUs). The restructured KCC has a number of unique features, including 100% call recording, call barging, voice mail service, customized IVRs, call conferencing with experts, playing state-specific advisories during call wait time, SMS to caller farmers summarizing responses from FTAs, and registration of farmers for receiving SMS from experts on the subject area provided by them for receiving regular updates on selected crops.

b) **Programmes for skill development**

i) **Skill Training of Rural Youth (STRY):** This aims at training rural youths, rural artisans including farm women across the country. The training under this component focuses on specific vocational areas in agriculture & allied sectors. Both public and private/non-Governmental institutions including vocational training organisations, youth organisations like Nehru Yuva Kendras are actively involved in implementation of this programme.

ii) **Diploma in Agricultural Extension Services for Input Dealers (DAESI):** Agri input dealers in the nation, in addition to providing inputs and finance, are a major source of farm information to the farming community. Most of the dealers do not have formal education in agriculture. In order to build their technical competency and to facilitate them in serving the farmers better and to act as Para–Extension professionals, a self-financed one year "Diploma in Agricultural Extension Services for Input Dealers (DAESI) Program" was launched during the year 2003 with a course fee of Rs.20000/- to the input dealers. National Institute of Agricultural Extension Management (MANAGE) is the nodal agency to implement this program.

## Conclusion

Agricultural extension embraces a participatory approach by involving farmers in the knowledge transfer process. Extension agents engage with farmers, understand their needs, and encourage their active participation in decision-making. This participatory approach helps to tailor information and technologies to the local context, ensuring their relevance and effectiveness. Extension services collect feedback from farmers, aggregating their experiences, observations and challenges. This information is valuable for researchers, policymakers and other stakeholders in refining agricultural technologies and policies. It helps to identify research gaps, improve existing practices, and align agricultural development strategies with the needs of farmers and local communities.

Overall, agricultural extension serves as a vital conduit for the transfer of knowledge, fostering innovation, empowering farmers, and promoting sustainable agricultural practices. By strengthening the link between research and practice, extension services contribute to the development and transformation of agriculture, ultimately improving livelihood and food security.

## References

1. Altalb, A. A. T., Filipek, T., & Skowron, P. (2015). The role of Agricultural extension in the transfer and adoption of agricultural technologies. *Asian Journal of Agriculture and Food Sciences*, 3(5), 500-507.

2. Chikaire, J. U., Emerhirhi, E., Anyoha, N. P., & Onoh, P. A. (2018). Perceived competencies of Agricultural Extension and Advisory services providers in building rural farmer capability in Imo state Nigeria. *International Journal of Research in Agriculture and Forestry*, 5(6), 25 -32.

3. https://www.fao.org/

4. https://www.ifpri.org/topic/agricultural-extension

5. https://www.fao.org/publications/card/en/c/f2c836cf-a527-5915-898e-fc0001c2cc6a/

6. https://www.manage.gov.in/publications/edigest/edigest-1999.asp

7. https://pib.gov.in/

8. https://agricoop.nic.in/en/Extenson#:~:text=Through%20its%20various%20programmes%2C%20schemes,and%20operating%20professional%20Extension%20Services.

9. https://agricoop.nic.in/sites/default/files/Programme-Schemes_new.pdf

10. https://krishivistar.gov.in/Home.aspx

11. http://www.agriclinics.net/

# Chapter - 16

# Value-Added Agri-Processing for Horticultural Crops

**Authors**

**Jyoti Sengar**
Ph.D. Scholar, Department of Horticulture (Fruit Science), Jawaharlal Nehru Krishi Vishwa Vidyalaya, Jabalpur, Madhya Pradesh, India

**Anjali Mishra**
Ph.D. Scholar, Department of Horticulture (Vegetable Science), Jawaharlal Nehru Krishi Vishwa Vidyalaya, Jabalpur, Madhya Pradesh, India

**Pragya Uikey**
Ph.D. Scholar, Department of Vegetable Science, Odisha University of Agriculture and Technology, Bhubaneswar, Odisha, India

# Chapter - 16

## Value-Added Agri-Processing for Horticultural Crops

**Jyoti Sengar, Anjali Mishra and Pragya Uikey**


### Abstract

Value added describes the extra value generated with a product to increase its marketing shelf life. The procedure used on diverse farm products to gain the advantages of supplying a processed version of the raw material is known as value-adding in agricultural products. Additionally, it is a method of raising a commodity's economic value and consumer attractiveness. Value-added goods have the ability to open up prospects for local economic development. The exponential rise in global population and the preference for using eco-friendly and practical agricultural byproducts provide a stable foundation for the advancement of innovation.


**Keywords:** Value addition, Horticultural crops, fruits & vegetables, flowers, spices.

### Introduction

A horticulture revolution is currently taking place in the Indian agricultural sector. With only 13.1% of the total gross cultivated land, horticulture contributes 30.4% of the agricultural GDP. The nation's overall horticulture production has outpaced its total food grain production over the years, setting new records for production each year. According to the Second Advance forecasts, the production of horticulture crops is expected to reach a record 329.86 million tonnes (MT) in 2020–21, a 2.9% increase over the final forecasts for 2019–20 (320.47 MT). With a huge increase in tomato, onion, and potato (TOP) output over the years, India is now the second-largest producer of all three veggies worldwide (Tiwari 2021). India has emerged as world leader in the production of a variety of fruits like mango, banana, guava, papaya, sapota, pomegranate, Lime & aonla and is the second largest producer of fruits and vegetables. Besides, India has maintained its dominance in the production of spices, coconut and cashewnut. Among the new crops, kiwi, gherkins, kinnow, date palm and oil palm have been successfully introduced for commercial cultivation in the country (Horticulture 2022).

Despite having an abundance of excellent horticulture products, India lags behind in the processing and value-adding of vegetables and fruits. India makes up just 2% of the global horticultural trade. The Indian government has championed the idea of agri-export zones and mega food parks to support the food processing industry in India as well as its various subsectors, such as the processing of fruits and vegetables (Sharangi 2015). India, a major producer of fruits and vegetables, has a huge potential to take advantage of this agricultural wealth by giving its produce more value and raising the value of the food. In this aspect, adding value to horticulture products provides several benefits for all parties involved. Value addition increases the financial return for the food and beverage business while also extending the product's shelf life and lowering food waste.

## Value addition

The "Process of changing or transforming a product from its original state to a more valuable state" is known as value addition. Better nutrition for mothers and children, higher producer revenue, access to new markets, and innovative procedures for packing and storage that increase food safety and waste reduction are all advantages of value-added products.

In order for farmers to be profitable, for farmers to be empowered, for farmers to provide safe, high-quality, and branded food to consumers, for farmers to decrease post-harvest losses, for farmers to decrease their reliance on imports while increasing their reliance on exports, and for farmers to support the growth of ancillary industries, value addition in agriculture is necessary (Tripathi, *et al.* 2022).

In horticulture, value addition refers to the process of realising a high price for the same volume of a primary product through methods like processing, packaging, improving the quality, or other similar techniques. Making strawberry and pomegranate jam is an example of value addition

## Importance of fruits and vegetables (Kulshrestha 2018)

Root and tuber crops, fruits, and vegetables are some of the richest sources of calories, natural vitamins, and minerals needed for a healthy lifestyle. Amaranth, spinach, fenugreek leaves, chenopodium album (bathua), mint, and other green leafy vegetables, as well as roots and tubers like carrots, are all excellent sources of beta carotene, the most crucial precursor to vitamin A in human nutrition. An essential antioxidant component of beta carotene inhibits the activity of oxygen and free radicals, protecting against cancer. Normal growth and vision, reproduction, epithelial cell maintenance, and immunological function all depend on vitamin A, whose inadequate consumption lowers blood levels and serum levels, indicating vitamin A deficiency. In order to improve the nutrition of rural and urban poor, it is necessary to increase the production and processing of fruits and vegetables. It has been noticed that the amount of fruits and vegetables now available only provides about half of the requirements for various vitamins and minerals. Therefore, it becomes vital to expand vegetable processing by creating such methods that would not only be practical but also be sufficient to generate goods of economic quality. As a result, veggies that are out of season are available all year long. In India, less than 2% of the total veggies produced are processed, compared to 70% in Brazil and 83 percent in Malaysia. The dehydration method is the most popular way to preserve fruits and vegetables. For small-scale operations, the vegetables can be dried using the hot air-drying method, as well as the traditional tray or vacuum drying methods, and at the household level, the sun drying method. Vegetables that have been dehydrated can be consumed in a variety of ways without losing any of their nutritional value or flavour.

Simple technologies can be used to make vegetable powders like carrot, tomato, and fenugreek leaf powder, which can then be added to traditional food preparations to add value and achieve both food and nutrition security. Due to inadequate post-harvest management, there have been significant losses in farm output. According to numerous studies, food commodities suffer annual post-production losses of between Rs. 75,000 and Rs. 100,000 crore.

On adequate transmission and implementation of agro processing technology, the extent of losses is also predicted to be reduced to less than 50% of the current level.

Need for value addition in Horticulture (Post harvest, processing and value addition n.d.)

1. To increase farmers' profitability.
2. To provide chances for productive work to farmers and other underprivileged groups in society, particularly women, and to revitalize rural areas.
3. To give consumers with higher-quality, safer, and branded foods.
4. To draw attention to both main and secondary processing
5. To minimize losses following harvest.
6. Import reduction and satisfaction of export requirements.
7. A strategy for boosting foreign exchange.

8. Promote expansion of ancillary industries.
9. Lower the financial risk involved in marketing.
10. Expand market prospects for smaller farms and businesses;
11. Diversify the rural communities' economic foundation.
12. Overall, strengthen the financial security of farmers.

Importance of value addition in horticulture (Post harvest, processing and value addition n.d.)

1. India is the second-largest producer of fruits and vegetables, but only 10% of that produce is processed, compared to other developed and developing countries where 40–80% of produce is added value.
2. Horticulture deals with a large group of crops that have great medicinal, nutritional, and health-promoting values.
3. Horticultural crops offer a variety of components, such as pigment, amino acids, oleoresins, antioxidants, flavours, and scent, that can be profitably and successfully used for value addition.
4. Horticultural product experiences post-harvest losses of 5 to 30 percent, or more than 8000 crore rupees annually. The losses can be monitored if we add value to our produce.
5. Horticultural crops are right material for value addition because they are more profitable, has high degree of process ability and richness in health promoting compounds and higher potential for export.

**Value addition in horticultural crops (Kulshrestha 2018)**

India is becoming an emerging major producer of fruits and vegetables in the globe thanks to the combined efforts of R&D organisations, farmers, governmental agencies, and the trade.

About 45 million tonnes of fruits and 80 million tonnes of vegetables were produced in the nation in 2000–2001. It ranked first in the production of fruits and was second only to China in the production of veggies. However, the expansion of the post-harvest industry has lagged behind production. Even in 2000–2001, the number of fruit and vegetable production units in the nation had increased from around 1,000 in 1950–1951. Although the installed capacity of the processing sector increased gradually from 0.27 Mt in 1980 to around 3 Mt in 2000–2001, less than 1% of the total produce was processed.

Better knowledge of the ripening phase of fruits, the ideal time for harvesting, pre-cooling newly picked product, cold storage of the raw fruits and vegetables, sorting, cleaning, waxing, and fruit packing technologies are just a few examples of significant technological advancements. Numerous technologies have been created at the following institutions: CFTRI, DFRL, IIHR, Bangalore; IARI, New Delhi; GBPUA&T, Pantnagar; IIVR, Varanasi; and HPKV, Palampur. The technique for controlled fruit ripening has been the focus of the most significant research to date. A commercial success is the production of juices and value-added goods like jams, jellies, pickles, canned goods, etc. The industry employs indigenous technology in the extraction of juices, concentration of juices, canning, and production of a number of goods, including jams, jellies, canned fruits, dried vegetables, and other things. Technology is still being imported in order to set up large-scale export-oriented facilities for the production of goods like banana paste, concentrations of different fruit juices, and raw fruit and vegetable sorting, cleaning, washing, waxing, and packaging.

**Commodity-wise value addition**

**Fruits and vegetables (Post harvest, processing and value addition n.d.)**

About 77.7 million tonnes of fruits and 159.5 million tonnes of vegetables were produced in the country in 2012–2013, accounting for 25% of global production (NHB 2013). However, the development in the post-harvest industry has not kept up with demand. In contrast to nations like the USA (65%), Malaysia (83%), the Philippines (78%), France and Brazil (70% each), etc., only around 2.2% of the total fruits and vegetables produced are processed. Fruit and vegetable processing in India is largely decentralized, with cottage/household and small-scale businesses accounting for 33%, organised 25%, and unorganized 42% of total production. These businesses have annual production capabilities of up to 250 tonnes. There are around 5,000 organised units in India, and thousands of people are employed in the unorganized sector of the country's fruit and vegetable processing industry.

**Value added products of different fruits**

Mango: Mango has mostly been used as a raw material for tasty pulp and as a value-added product in beverages. Additionally, there are mango juices, jams, pickles, chutney, vines, and more.

In order to add value to mangoes, they must be processed properly after harvest, together with all of their waste, and exported for high returns. Children adore mango ice cream and mango toffees. Vitamins A and C are both abundant in mangos. This value-adding process can be carried out at any point between harvest and finished product. Green mangoes are used to make the popular summertime beverage mango pana. Sugar, salt, citric acid, and spices are all mixed together in a thick mixture. Mango leather, which is comprised of mango pulp combined with deep sugar, is well recognised for being extremely dry (Jayamali 2021).

Banana: Bananas are used in numerous goods with value added. Some of them are banana chips, banana juice, banana wine, and banana jam. Making flour is likewise seen as a banana manufacturing byproduct. For stomach discomfort and pain, banana flour is helpful. Banana jam is highly well-liked everywhere. Other processed goods include banana sauce, banana powder, and banana drinks.

Pineapple: Pine apples are often eaten fresh or tinned. In today's world, pineapple is consumed in a variety of drinks, including banana wine, cocktail-style cocktails, dried chips, dehydrated sugar, and canned slices of the fruit.

Canned pineapple, dried pineapple, nectar, juice, sauce, jam, vinegar, and toffees are some examples of pineapple value-added goods.

Guava: This is primarily eaten as fresh fruit. Moreover, several different varieties of guava are used in value-added goods. Guava juice, jelly, concentrate, cheese, toffee, and other value-added products are a few examples.

Other value-added goods include guava powder, ice cream toppings, and guava wine. Guava has a terrific flavour and strong nutritional value in addition to being a significant source of vitamin C.

Avocado: The fame of avocado oil is global. It is employed in cooking and beauty care. Ripe avocado flesh is used to make oil. Briquettes can be used in place of charcoal. They are produced from a variety of compressed waste materials. Avocado peels are really beneficial. Avocado seeds are yet another waste item that can be useful. They are capable of being powdered. The cosmetic industry employs this powder. Avocado paste can be used to create juices and smoothies.

Durian: Despite having a wide range of value-added items. Some of them include durian cake, frozen durian, chips, paste, and powder. Moreover, durian powder can be added as an essence to ice cream, coconut milk, and other foods.

Strawberry: Fruits can be consumed raw, tinned, or processed into milkshakes, jams, juices, and cake decorations. This is employed in the cosmetics sector as a flavour agent.

Given that they include keratin, they are also utilised in face masks and creams. Jams can be made using the fruits. Strawberry milkshakes and smoothies are two typical examples of value added products of strawberry.

Oranges: Oranges are one of many citrus fruits with added value that have numerous advantages. An important role is played by cold press orange oil. By pressing the orange peel, this is removed. Orange peel that has been dried is yet another byproduct of juice extraction. The peel is dried and used in sauces, baked goods, and cereals.

A by-product of the juice extraction process is frozen orange pulp. These can be made into goods used in animal feed.

Cherry: Cherries can be enhanced in a variety of ways. Fresh, frozen, canned, wine, and juices are all options. Over 75% of food is consumed uncooked and 25% are consumed as products with value added. Cakes are decorated with cherries.

Value added products of passion fruit: Some value-added products of passion fruit include juice, chocolate, tea, and sweets such ice cream cordial, jam, and cake.

Grapes: Grapes are used to make raisins, wine, juices, vinegar jelly, jam, butter, marmalade, and dried goods, which are considered value-added products. These added-value products help manage diabetes, treat anaemia, support bone and dental health, and maintain skin health.

Value added products of wood apple: Jujubes, candy, juice, leather, sauce, and other products with additional value are examples. Even raw versions of these are edible.

Papaya: Papaya pulp is used to create blended drinks and to make nectar. Pulp is preserved, frozen, or concentrated for later use. Papaya immature fruit dried latex is used to extract papain. Additionally, papain is utilised in the manufacturing of chewing gum, dietary supplements, meat tenderizers, cosmetics, and pharmaceutical products.

Some value-added goods include marmalade, sweets, pickles, chocolate, canned papaya, and freeze-dried papaya.

Additionally, cosmetics are created. Papaya removes damaged and dead skin cells from the body.

**Value added products of different vegetables (A. and T. 2020)**

**Onion:** Onions that have been dehydrated are thought of as prospective trade goods. India is one of the world's two biggest producers of dried onions. As one of the key elements in the preparation of numerous food products is onion. Low-acid foods like onions must be pressure canned in order to destroy any potentially hazardous microorganisms. So, canned and dehydrated onion are 2 major value added products of onion.

**Carrot:** Carrot is eaten all across the world in a variety of ways, including fresh, shredded, sliced, sticks, and processed goods. Some value added products are as follows:

**Canned Carrot:** Small carrots are typically used in canning procedures. There are several ways to can food, including whole, in halves, quarters, or diced form. Better quality canned goods are produced when the product is blanched at 71°C for 6 to 8 minutes.

**Dehydrated Carrot:** According to a report, blanched pre-treated dried carrot has a higher concentration of beta-carotene and a lower ascorbic acid content than unblanched dried carrot (S and K. 2001).

**Carrot pickle:** Pickles are said to make a tasty appetizer and enhance the meal's flavour. The commercial manufacturing of pickles uses potassium metabisulfite or sodium chloride (NaCl) brine (Sultana, Iqbal and Islam 2014).

**Carrot juice:** In addition to being used in yoghurt fermentation, carrot juice is frequently combined with other juices including orange juice and pineapple juice in non-alcoholic beverages. Carotene and vitamin A are abundant in carrot juice.

**Carrot Candy:** It is a sweet food item made from carrots that has been submerged in sugar syrup, dried of extra syrup, and dried to a stable form.

**Carrot jam:** It is made by boiling the pulp with a specific amount of sugar and pectin until it takes on the consistency of jelly.

**Cabbage:** Although cabbage is typically eaten raw, it can also be prepared into processed foods like fermented cabbage, commonly known as sauerkraut. It is uncooked, small-cut cabbage that has been fermented with lactic acid bacteria (Holzapfel, Schillinger and Buckenhuskes 2003). It extends cabbage's shelf life and gives it a sour flavour as a result of the lactic acid produced when bacteria digest the carbohydrates in the cabbage leaves.

**Cucumber:** Although cucumbers are frequently eaten raw and uncooked, they are also well-known when fermented into pickles. The gherkin, also known as pickled cucumber, is used to make pickles out of cucumbers. It has a significant export potential. Pickling is the process of fermenting cucumber with the use of salt, vinegar, spices, and other ingredients. To make pickles, the immature fruits are employed (Purseglove 1969).

**Ash gourd:** It is a highly significant vegetable that has a very long shelf life and is used to make products with added value like Petha Badi Sweet, also known as Pethamethai in India. It is a well-known confection worldwide. The most popular sun-dried vegetable is pethabadi. This badi can either be eaten straight away or added to various dishes. India's Petha is a very well-liked confection. Its structure is soft, chewy, and candy-like.

**Bottle gourd:** The fames product produce from it is known as Tutti frutti. Tutti frutti is a confection made from bottle gourds that can be made in a variety of ways, including slow syruping, single operation (boiling in sugar syrup), and slow syruping with 1% $CaCl_2$ (Babar 1996).

**Pumpkin:** Fresh pumpkin vegetables are consumed, as well as pumpkin puree, jam, and pickles, which are given below:

**Fruit bar:** When pumpkin pulp or puree is dehydrated, it takes on the form of a bar or a piece of leather, which is an intermediate food product (FSSAI 2011).

**Pumpkin Jam:** Fruit and sugar are the main sources of James' production. Jam is sweetened by the addition of sugar, which is made from sugar cane or sugar beets.

**Pumpkin Pickle:** Pickling is a crucial step in food preservation for long-term storage of pumpkin. In addition to being incredibly nutritious, pumpkin pickles have a unique flavour from other pickles available on the market.

**Potato:** Potato chips are a potato product that is used commercially as a popular meal. Value added products of potato are as follows:

**Potato Chips:** The Kufri Chipsona-1 and Kufri Chipsona-2 potato varieties are the most popular for making chips. This cultivar is perfect for making potato chips. Thin, fried, and baked potato chips are common ready-to-eat snacks used in both residential settings and as food in restaurants.

**Canned Potatoes:** Potatoes that are young and tiny are preferred for canning purposes. For larger tubers, they first chop into little pieces before canning.

**Tomato:** Potato products which are made through value addition are as follows:

**Tomato powder:** Country tomato (Namdhaari) is the variety that is utilised to make the powder. Powdered dried tomatoes are used.

**Tomato juice:** Tomatoes that are fully ripe and have developed colours are used to make tomato juice. They gave them a thorough wash before trimming, steaming, and cutting them into small pieces with knives.

**Tomato puree:** To make tomato puree, the juice must be vacuum-concentrated to a total solid content of between 9 and 12%. The product is then placed in bottles, which are then placed in boiling water for 30 minutes before being allowed to cool (Reimers. 2011).

**Tomato Ketchup:** Tomato juice is utilised in the making of ketchup. Salt, sugar, spices, and other ingredients in the juice concentrate. Cloves, cardamom, pepper, cinnamon, and other spices are used, and they are knotted loosely in a muslin cloth before being poured in boiling liquid in a steam jacketed kettle.

**Green leafy vegetables:** GLVs are a crucial part of the world's food supply. They give the body the right amount of vitamins and minerals. Vegetable leaves must first be washed twice or more to remove any remaining dirt or contaminants before being dried and ground into a powder. When they are crisp and their moisture level drops to 6-7%, they are adequately sun dried for 6-7 days. Leaves were then broken up in a mortar and put through a sieve (Joshi 2015).

**Value addition in spices**

India is a large supplier of spices and spice-related goods. Spices are high value export-oriented crops. India exported 1,77,625 tonnes of spices for Rs. 2,71,147 in 2012–13, a volume that increased by 13% and decreased by 9%. The nation offers oleoresins and spice oils to the world market for spices. India produced roughly 6 million tonnes of spices and condiments in the years 2012–2013 (Mani and Arpan 2019). Some value added products of spices are as follows:

**a) Ready-made spice powder and paste**

There is a high demand for airtight packaging for powdered spices. The amount of time spent cooking has decreased due to growing urbanisation and an increase in the number of working women. As a result, homemakers have begun to ask for ready-made spice powders such as turmeric, black pepper, fennel, cumin, and chilli powder. Also popular are packets of ready-made onion, garlic, and ginger paste.

**b) Spices extractives**

i) Essential oil: Major flavouring components of spices are essential oils, which are 75–100 times more concentrated than the fresh spice.

Coriander: Major component of volatile oil is linalool (67.7%) followed by alphapinene (10.5%), amma-terpinene (9.0%), geranyl acetate (4.0%), camphor (3.0%) and geraniol (1.9%). Minor components include beta-pinene, camphene, myrcene, limonene, p-cymene, borneoletc. Indian coriander oil differs from European oil in Possessing a lower linalool contents and comparatively higher linalyl acetate contents.

Cumin: Cuminaldehyde, along with three other aldehydes, makes up the majority of the compound in cumin. B-pinene, g-terpinene, and p-cymene are the most abundant monoterpene hydrocarbons (totaling around 50–55%), followed by myrcene, b-phellandrene, and limonene. Sesquiterpene hydrocarbons are present in smaller levels.

Fennel: With a yield of 1.5–6.5%, fennel seed oil, which is typically sold as fennel oil, is primarily produced by steam distilling whole or broken fruit. More recently, supercritical carbon dioxide extraction has also been used.

Fenugreek: Dihydroactinidiolide, 2,3-dihydrobenzofuran, and 1-heanol are the primary components, making about 7-9% of the mixture, while 20 other components make up less than 3% and the rest less than 1%. The distinctive fenugreek fragrance is said to be mostly caused by the furanone derivative sotolon.

ii) Oleoresin: Oleoresin is a term used to describe the entire flavour and non-volatile resinous component found in spices. The 'real essence' of the natural spice is created by appropriately blending the scent and flavour components.

iii) Derivatives of essential oil and oleoresins: They include seasonings in dry carriers like dextrose, salt, or rusk powder, as well as plate-encapsulated forms of spice extractives. They are simple to use and add the flavour of freshly ground, high-quality spices to dishes.

**4. Dried whole spice**

The final rates for dry ginger and kalaunji per quintal were Rs 15,200–18,200 and Rs 9,700–10,200, respectively.

**Black pepper:** Black pepper is used to create a wide range of goods. This also applies to products made with green pepper, black pepper, white pepper, and products made with pepper. Canned green pepper, green pepper in brine, bulk-packaged green pepper in brine, cured green pepper, frozen green pepper, freeze dried green pepper, dehydrated green pepper, green pepper pickle, mixed green pepper pickle, green pepper sauce, and green pepper-flavored goods are the main green pepper-based products.

**Cardamom:** The four main cardamom products are cardamom volatile oil, cardamom oleoresin, decorticated seeds and seed powder, and bleached cardamom. Additionally, the CFTRI in Mysore has created the following goods: Cardamom extract, tea, coffee, and soft drink mix in capsule form.

Ginger: The valuc-added products from ginger include ginger powder, ginger oil, ginger oleoresins, encapsulated ginger, ginger preserves, and salted ginger.

Turmeric: Turmeric powder, turmeric oil, turmeric oleoresin, and curcuminoids are the main value-added products.

Chilli: Oleoresin is the main value added product that can be obtained from chilli. Other products from chillies are dehydrated chilli, canned chilli, brined/pickled chilli and fermented chilli.

Nutmegandmace: Nutmeg oil and mace oil, Nutmeg oleoresin, Nutmeg butter are the main value added products. By utilizing nutmeg pericarp (rind), many value-added products have been developed *viz.*, Nutmeg (rind) pickle, Nutmeg (rind) preserve from slices, and Nutmeg (rind) preserve from shreds, Nutmeg (rind) candy, Nutmeg (rind) sweet chutney and Nutmeg (rind) powder.

Cinnamon: In addition to cinnamon bark, the tree also yields a number of additional goods, such as bark oil, leaf oil, bark oleoresin, etc.

Seed spices: Seed spices can be used to create value-added goods including ground spices, spice extracts, curry powder, consumer-packaged spices, and organic spices.

Curry Powders / blends and mixes: Curry powder is an indigenous flavouring created from a variety of spices (typical ingredients include coriander, cumin, fennel, and fenugreek). Curry powder is in high demand all around the world. India now dominates the curry powder export industry. Depending on the intended purpose, curry powder may contain as few as five or as many as twenty different spices.

**Value addition in flowers (L.C. De 2020)**

Two different flower arrangements are included in the value-added category of flowers: cut flower arrangements such bouquets, baskets, bunches, buttonaires, and corsages, and loose flower goods including garland, floral strings, pomanders, wreaths, and floral jewellery.

Cut Flower arrangements: Roses, carnations, chrysanthemums, lilies, and other high-value flowers are appropriate for creating various floral arrangements. The floral arrangements can be made in English or Japanese styles, such as erect, slanting, regular, irregular, curved shape, Hogarth S form, etc.

Flower Bouquets: A flower bouquet is a grouping of flowers that have been artistically arranged. Flower arrangements can be carried or used as wall art in homes and public spaces. Numerous common forms and patterns, like as nosegay, crescent, and cascade bouquets, can be used to categorize handheld flowers.

Rangoli: Rangoli is typically used to create colourful patterns at entrances, however considering how eco-friendly they are, loose flowers can also be utilised to create rangoli. Different coloured flowers are utilised in place of artificial colouring; turf grass clippings or foliage clippings are used to provide green hue.

Garlands: Garlands are prepared by using only one type of flower or a combination of different flowers. Generally, sweet scented flowers such as jasmine, tuberose, gardenia etc. are preferred for making a garland (L. De 2011)

Floral bangles: Female dancers use floral bracelets made of non-fragrant flowers like Tabernaemontana, marigold, and Thevetia peruviana or fragrant flowers like jasmine or tuberose.

Floral Crowns: Floral crowns consisting of fragrant tuberose flowers are typically worn by a monarch or a princess in a play or ceremony like "Annaprasna."

Floral earrings and Bajubandh: In dancing ceremonies, women wear floral earrings and bajubandh to honour epic heroes and heroines like Rama and Sita.

For this, flowers like tuberose, kadamba, jasmine, etc. are utilised.

Veni: In South India, long plaits of hair are frequently decorated with special flower arrangements during rituals like marriage or Bharat Natyam dance performances.

Wreath: It is a ring-shaped band made of flowers or leaves that is typically worn as a crown or a badge of honour or laid on graves as a tribute.

Buttonhole/ boutonniere: In general, both men and women wear this. It is a tiny floral arrangement made with a single, small flower, such a rose or an orchid, and filler, like a thuja leaf.

Corsage: A flower or small floral arrangement that someone wears as a personal decoration. Typically, it is larger or more ornate than a boutonniere.

**Dry flower value added products**

Dry flower value added products includes dried flower arrangements, products of press dried flowers and pot pourries etc.

Dry flower arrangement: Just like fresh cut flowers, dried flowers are presented in dry vases. Once they have been secured to decorative bands, they can be placed in bouquets or wall displays. Different drying techniques can be used on flowers including *Helichrysum*, *Delphinium*, *Helipterum*, *Amaranthus*, *Nigella*, *Carathmus*, *Gypsophilla*, and *Rosa* to create dried flower goods and flower arrangements.

Floral Jewelry: Orchid flowers are widely used in Thailand, Singapore and Malaysia to electroplate with gold and platinum to make exquisite jewelry. India has enormous potential with orchid species and hybrids like *Dendrobium*, *Phalaenopsis*, *Cattleya*, *Oncidium*, *Arnada*, *Mokara*, *Vanda*, *Pholidota* etc. to diversify into such avenues to exploit the vast genetic resources and expert craftsmanship available in the country.

Floral gifts: The development of the internet and the availability of inexpensive air travel have created new opportunities for placing and delivering flower arrangements as unique gifts across the nation.

Pot pourri: It is a blend of dried, aromatic plant material that is used to give homes a soft, natural scent. It is typically put in a gorgeous wooden bowl or knotted in little sheer fabric bags.

Candle making: The candles can be enhanced with dry flowers as well. It is possible to do this by simply adding dried flowers to the exterior of simple candles or by arranging crushed dried flowers on wax paper, pouring a little molten wax over the flowers, and then rolling the candle in the flowers.

Press dried flower products: It consists of things like cards, table tops, table mats, paper weights, wall hangings, and bookmarks. The flowers and foliage are herbarium-dried before being artistically pasted with fevicol or adhesive.

Petal embedded handmade paper: Real flower petals are inserted in the paper throughout the creation process, and all wood-containing elements were used in its creation. This gives the paper its upscale appearance. Peonies are woven into exquisitely textured handmade sheets, adding elegance and charm.

**Processed flower products**

Essential oils, absolutes, concrete, petal jam (rose, rhododendron), jelly, ready-to-drink wine, floral tea, rose hip juice, poultry feed, insect repellent, floral dyes, handmade paper with embedded petals, and cosmetics like calendula cream, rose water, and rose cream are examples of processed goods.

**Conclusion**

Nearly 90% of the nation's overall horticultural production is made up of fruits and vegetables. India is currently the world's second-largest producer of fruits and vegetables and a pioneer in a number of horticulture crops. However, in India today, a significant portion of the primary processing of fruits and vegetables is part of the agriculture value chain; there is relatively little value addition. Value addition in horticultural crops has a huge potential to raise the income of the farmer and economy of the country as well.

**References**


1. A., Nagil, and Chamroy T. "Processing and Value addition of Vegetable Crops-A Review." *International Journal of Current Microbiology and Applied Sciences* 9, no. 11 (2020).
2. Babar, V.D. "Preparation of Tutti-Fruity from Bottle Gourd." *M.Sc. Thesis, Mahatma Phule Agricultural University, Rahuri, India.*, 1996.
3. De, L. C. "Value Addition in Flowers." *VIGYAN VARTA* 48, no. 51 (2020): 10.
4. De, L.C. "Value Addition in Flowers and Orchids." (New India Publishing Agency, Pitampura, New Delhi-110088) 2011: 294.
5. FSSAI. *Food safety standard (Food products standards and food additives).* Ministry of Health and Family Welfare, 2011.
6. Holzapfel, W., U. Schillinger, and H. Buckenhuskes. "Sauerkraut." In *Handbook of Fermented Functional Foods*, by E. R. Farnworth, 395-410. Boca Raton: CRC., 2003.
7. *Horticulture.* November 18, 2022. http://agricoop.gov.in/en/Hirticulture#gsc.tab=0 (accessed may 29, 2023).

8. Jayamali, Thisara, S. "value added products of fruits." *International journal of processing and post harvest technology*, 2021.

9. Joshi, Pallavi and Mathur, Beena. "Development of value added products from the leaf powders of dehydrated less utilized green leafy vegetables." *Nutrition & Food Science* 45 (2015): 302-309.

10. Kulshrestha, K. "Horticultural crops value addition for nutritional security." *International Journal of Research-Granthaalayah* 6, no. 10 (2018): 110-120.

11. Mani, Arghya., and Kabiraj. Arpan. "Export potential of spices and its value added products." *Agriculture and Food: e-Newsletter* 1, no. 8 (2019): 339-347.

12. *Post harvest, processing and value addition.* n.d. https://dirhorti.assam.gov.in/portlets/post-harvest-processing-value-addition#:~:text=Horticultural%20crops%20provide%20varied%20type, antioxidants%2C%20flavors%2C%20aroma%20etc. (accessed may 28, 2023).

13. Purseglove, J. W. In *Tropiocal Crops Dicotyledons-1.*, 109-110. Longmans Green and Co.Ltd, 1969.

14. Reimers., Freeman. B. B and K. "Tomato consumption and health: emerging benefits." *American Journal of Lifestyle Medicine* 5, no. 2 (2011): 182–191.

15. S, Negi P, and Roy S K. "The effect of blanching on quality attributes of dehydrated carrots during long term storage." *European J Food Res Technol.* 212 (2001): 445-448.

16. Sharangi, A. B., & Datta, S. *Value addition of horticultural crops: recent trends and future directions.* new delhi: New Delhi: Springer India., 2015.

17. Sultana, S., A. Iqbal, and M. N. Islam. "Preservation of carrot, green chilli and brinjal by fermentation and pickling." *Int Food Res J.* 21, no. 6 (2014): 2405-2412.

18. Tiwari, A., Afroz, S. B., & Kumar, V. "Market vulnerabilities and potential of horticulture crops in India: With special reference to top crops." *Indian Journal of Agricultural Marketing* 35, no. 3 (2021): 1-20.

19. Tripathi V, Giftson S.P. Jeshwin, M. Pooja, S. Pranao, and GS.. Goutham. "Value addition in horticultural crops." *The Pharma Innovation Journal* 11, no. 5 (2022): 1058-1067.

# Chapter - 17

# Innovative Breeding Techniques to Enhance Quality Traits in Solanaceous Crops

**Authors**

**Pallavi Wani**

Ph.D. Scholar, Department of Horticulture, University of Agricultural Sciences, GKVK, Bangalore, Karnataka, India

**Dr. Srinivasappa, K.N.**

Professor, Department of Horticulture, COA, University of Agricultural Sciences, GKVK, Bangalore, Karnataka, India

# Chapter - 17

## Innovative Breeding Techniques to Enhance Quality Traits in Solanaceous Crops

Pallavi Wani and Dr. Srinivasappa, K.N.

### Abstract

World Population during 2020 is 7.8 billion, is growing steadily, and it is predicted to reach 9.1 billion by the year 2050. (United nation population division). Per capita calorie consumption of Fruit and Vegetable is expected to more than triple by the year 2050 as compared to 2010. To feed growing population FAO projects the need to increase agricultural output by at least 60 percent in the next decade. Over the coming decades a changing climate, growing World Population, rising food prices, and environmental stresses will have significant yet impact on food security. India is ranked 94$^{th}$ position out of 104 countries. And it is also leading producer of many vegetables, but it has estimated around 15 percent population is under nourished and lack of inadequate food intake both in quality and quantity. So thus improvement of nutritive value of crop is the major gole of breeding programme. By use of conventional breeding method to get the product will take more number of years and labours, so if we start by modern breeding technologies like, Marker assisted selection, Molecular Breeding, Antisense RNA and CRIPSR/cas-9 *etc*., along with conventional method of breeding we can save the lot of time to develop the product.

**Keywords:** Quality traits, Modern breeding techniques, solanaceous crops.

### Introduction

Vegetable crops are rich in dietary micronutrients, antioxidants and functional phytochemicals. Among these, solanaceous vegetables are key sources of health protective nutritional constituents such as minerals, vitamins and antioxidants in human diet. Some of the elements are of public health concern, such as β-carotene, lycopene, anthocyanin, calcium and folic acid, which attract breeders' interest to improve their content in vegetable crops so that their dietary intake could be increased. These solanaceous family includes Tomato, Brinjal, Chilli and Potato. The attempts through conventional breeding methods have been slow and inadequate to fulfill the above-mentioned traits. Hence newly developments in molecular and biochemical diagnostic tools and techniques have been used for improving the other quality traits (Kumari *et al.*, 2017).

**Quality-** It refers to "Degree of excellence" for a specific use. Quality refers to many aspects like colour, size, nutrient content, shelf life and suitability for processing.

**Classification of quality traits**

"A trait defines aspects of quality of produce". Each crop has specific and different set of quality traits. These differ from species to species depending upon plant part used as economic product. There are mainly 5 categories-

1. Morphological traits - Related to product appearance which includes fruit size, shape, color and thickness. These traits are usually plays in the main role in determining consumer acceptance of the produce.
2. Organoleptic traits - Concerned with palatability of the produce, examples are taste, aroma, flavor *etc.*, these traits are easily detected. These traits are play very important role in influencing consumer preference.
3. Nutritional traits (Neutraceuticals) - These are determine value of the product such as Beta carotene, ascorbic acid, TSS, Protein, vitamins and minerals *etc.*,
4. Undesirable traits - Includes Solanine content in brinjal and browning in potato
5. Other traits - These includes all other quality traits that are not included in the previous categories such as keeping quality, cooking quality *etc.*,

**Advantages of quality breeding**

- It enables grower to get the good product
- Helps in consumer acceptability
- Makes the product more suitable for processing
- Provides Food and health security

**Table 1:** Gene action of important quality traits

| Crop | Character | Gene action |
|---|---|---|
| Tomato | Fruit Weight | Non-additive |
| | TSS, Vit C, Carotenoids, Lycopene | Non-additive |
| | Pericarp thickness | Additive |
| | Locule number | Additive |
| Brinjal | Fruit Weight | Additive, also Non-additive |
| | Fruit size index | Additive |
| | Fruit length | Additive |
| | Fruit width | Additive |
| Chilli | Fruit Weight | Additive, also Non-additive |
| | Pungency | Additive genes |

Additive gene action refers to average effect of alleles. Non additive gene action includes both additive and dominant epistasis. Gene action of some quality traits are listed in Table 1.

## Quality traits of solanaceous vegetables

### 1. Tomato

Tomato is the second highest produced and consumed vegetable in the world today.

Tomato fruits are an important source of nourishment for the whole world's population. And it is botanically called *Solanum lycopersicum* L. having its chromosome number is 24. Breeding objectives for quality traits are-

- ✓ For fresh market- quality purpose - Breeding Objectives includes size, shape, color, Appearance and smoothness
- ✓ If we want to breed for processing quality traits some of important breeding objectives such as high Tss (5.5 Brix), Low pH (< 4.5) and low acidity.
- ✓ And also to develop variety rich in lycopene, vitamin C and Tomatine content.

### 2. Brinjal

There is wide range of diversity in India based on fruit size, shape, color as well as preference for consumption. Brinjal botanically called *Solanum melongena* L. It is an important source of plant-derived nutrients like vitamins, minerals available throughout the year and popular among the poor.

Breeding objectives for quality traits are - Low glyco-alkaloid content. And to develop variety rich in vitamin A and anthocyanin content.

### 3. Chilli

Chilli is the most important spice in Indian dishes. It is botanically called *Capsicum annum* L. chromosome number is 24.

The breeding objectives like, we need to breed variety rich in high sugar to acid ratio, High vitamin C in bell pepper, high capsaicin, capsanthin and oleoresin content in hot pepper fruits.

### 4. Potato

Potato is considered as the staple food crop, it has good source of CHO, vitamin C, amino acids and minerals. It is botanically called *Solanum tuberosum* L. having its chromosome number is 48.

The breeding objectives for quality traits like include - To breed variety which in Low glycoalkoloid content that is solanine, and if we want to breed for processing purposes for example, French fries and chips the most important breeding objectives includes tuber shape should be oval or round, dry matter content (21%), starch (21%) and reducing sugars *etc.,*

## Advanced breeding methods

Advance quality breeding needs a combination of genetic source, modern genomic tools and biochemical diagnostic techniques. (Singh *et al.*, 2021)

The Recent breeding methods includes,

- ➢ Mutation breeding
- ➢ Polyploidy breeding
- ➢ Molecular breeding
- ➢ SNP's discovery
- ➢ Transgenic approach
- ➢ RNA interference
- ➢ Targeted genome editing

### a) Mutation breeding

- ➢ Mutation is sudden heritable change in DNA sequence which leads to change in characteristics of an organism. The entire operation of mutation induction and isolation of mutants for crop improvement is called mutation breeding. The term mutation was introduced by Hugo Devries in 1900.

In vegetable there are several quality mutant genes are reported for developing neutral rich hybrids. For example, in Tomato two mutant gene *hp-1* and *hp-2* have been reported for carotenoid biosynthesis and used for developing carotene rich tomato hybrid.

The physical or chemical agents which enhance the mutation frequency are commonly called as mutagens. Physical mutagens include ionizing radiations like α, β, γ, X-rays and non-ionizing radiations like UV and infrared rays. The chemical mutagens consist of ethyl methyl sulphonate, dimethyl sulphonate, bromouracil, *etc*., Singh *et al.* (2021).

**Chlorogenic acid**

- ✓ Chlorogenic acid is the one of the most important phenolic compound found in brinjal. And it ranges from 50- 70 percent.
- ✓ It has Health promoting activities like Anti-diabetic, Cardio protective, Anti-carcinogenic, Anti-oxidant, Anti-inflammatory and analgesic, Neuroprotective, Anti-obesity properties Chada *et al.,* 1998.

Swathy *et al.,* 2021 studied the response of Helium-Neon (He-Ne) laser irradiation on chlorogenic acid content of *Solanum melongena* L. var. Mattu Gulla, seeds were irradiated with different doses (20, 25 and 30 $J/cm^2$) of He-Ne laser showed significant enhancement on the chlorogenic acid. Seeds irradiated with dosage of 20 $J/cm^2$ resulted with maximum chlorogenic acid content as compared to other treatments.

**b) Molecular breeding for quality traits**

- ➢ Molecular breeding is refers to use of DNA markers that are tightly linked to phenotypic traits to assist in selection scheme for particular breeding objective.
- ➢ A marker is usually a DNA sequence located close to the gene of interest that can be used to identify individuals or species.
- ➢ Molecular markers are widely used in breeding programme it includes randomly amplified polymorphic DNA (RAPD), simple sequence repeat (SSR), amplified fragment length polymorphism (AFLP), sequence characterized amplified region (SCAR), cleaved amplified polymorphic sequence (CAPS), single nucleotide

polymorphism (SNP) and insertion-deletion (InDel) markers. Among these markers SNP are widely used due to their abundance, stability and cost effectiveness.

- A SNP is defined as a single base change in a DNA sequence at a given locus that occurs in a significant proportion (>1%) of a large population. Single nucleotide polymorphisms (SNPs) are single nucleotide base variations, caused by transitions (C/T or G/A) or transversions (C/G, C/A, or T/A, T/G).

**Features of DNA/Molecular Markers**

- Marker must be polymorphic in nature.
- It should be codominant in nature (so that it can differentiate between heterozygote's allele from homozygous allele.
- It should be even distribution across the whole genome Easy, fast and cheap to detect and
- It should be reproducible

Manoharan *et al.* (2017) studied on developing a new tomato fruit color that is enriched with β-carotene and chlorophyll content. In Figure 1 we can clearly see that level of *CYC-B* gene expression was significantly higher at the turning stage in both orange and orange brown phenotypes as compared to other phenotypes.

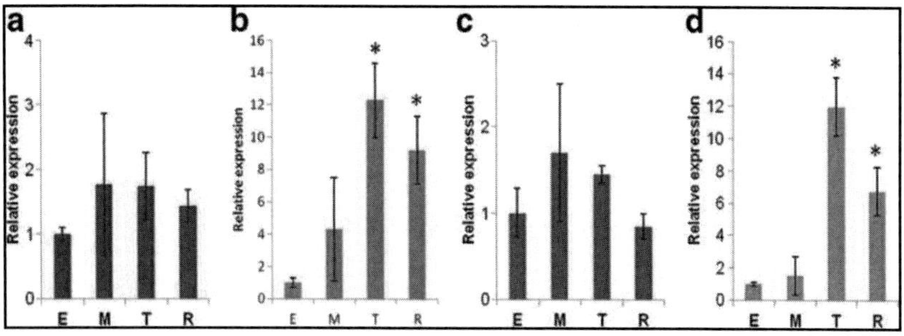


**Fig 1:** Expression of the *CYC-B* gene during fruit development in $F_2$ generation plants

The expression of *CYC-B* mRNA coincided with the accumulation of β-carotene and loss of *SGR* protein function leads to inhibit chlorophyll degradation. They revealed that highest amount of β-carotene and chlorophyll content was found in orange brown and brown phenotypes respectively.

**Table 2:** Molecular breeding for quality traits

| Reference | Marker | Trait | Crop |
|---|---|---|---|
| Lixia *et al.* 2011 | SCAR | high-lycopene | Tomato |
| Tanaka *et al.*, 2016 | CAPS | High Pungency in *C. Chinense* | Chilli |
| Fischer *et al.*, 2013 | SSR | Cold induced sweetness | Potato |

### c) Marker assisted selection

MAS refers to the indirect selection for a desired plant phenotype based on the banding pattern of linked molecular markers (Singh *et al.*, 2021).

**Steps in marker assisted selection**

1. **Selection of parents:** The parents selected for the mapping population will differ for one or more traits of interest. It means parents should be distantly related so that it provides adequate polymorphism.
2. **Development of mapping populations:** The mapping population should be larger in size. Several different populations may be utilized for mapping such as $F_2$ populations, Backcross (BC) populations, Recombinant inbred (RI) lines, Doubled haploid (DH) populations.
3. **Identification of polymorphism**
4. **The construction of Linkage Map with Polymorphic Markers:** Linkage map will be constructed by using the computer software's.
5. **Identification of molecular markers linked to the Trait of interest:** Based on the Linkage maps constructed, the tightly linked markers to the gene of interest are selected.
6. **Marker validation:** Markers should also be validated by testing for the presence of the marker on a range of cultivars and other important genotype.

### d) Marker-assisted backcrossing (MABC)

Marker assisted selection refers to indirect selection for a phenotype based on banding pattern of linked molecular markers. Molecular breeding is effective in introgression of desirable genes from wild into cultivated genotypes (Singh *et al.*, 2021). There are mainly two stages in marker assisted selection, those are

**Foreground selection:** Is used to identify our target locus with help of linked marker

**Background selection:** Is used to test recovery of recurrent parent

**Advantages of MABC over conventional method of breeding-are**

1) **Effective selection of target loci:** In Conventional breeding selection is based on phenotypic, that's why there was no effective selection in case of backcross breeding here with help of marker we can select our target loci means gene of interest which gene we want to transfer from donor to recurrent parent.
2) **Minimize linkage drag:** Actually, back crossing we are using two parents (one is recurrent parent and another one is donor parent) donor parent is any wild variety have poor economic importance so that's why along with gene of interest some undesirable genes are transferred these is called linkage drag, so with help of MABC we can minimize linkage drag.
3) **Accelerated recovery of recurrent parent:** In Conventional breeding method we need more number of back cross population to get recovery of recurrent parent but in marker assisted selection we using marker than we need only two back cross population for complete recovery of genome up to 99.99 percent.

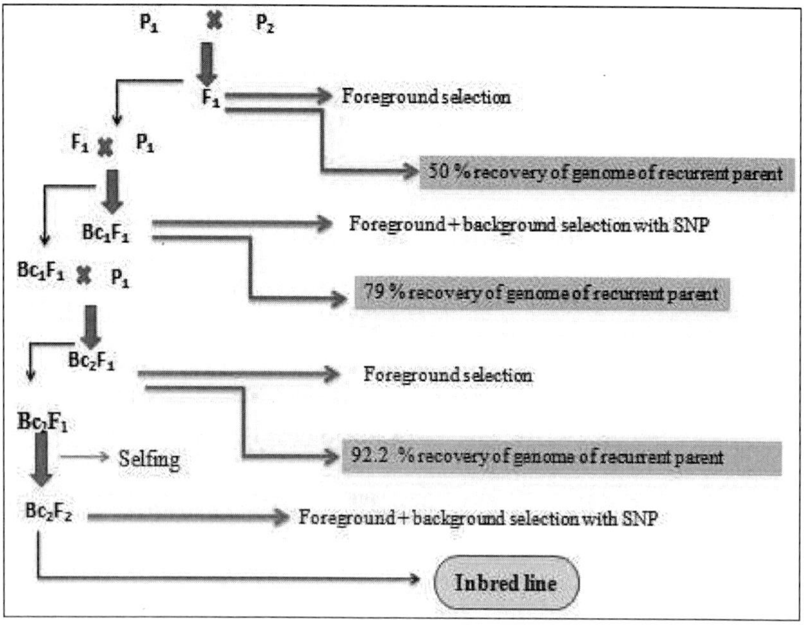


**Figure 2:** Schematic representation of marker-assisted back cross breeding in tomato by transferring *alcobaca* gene

**Population development between recurrent and donor parent:** Here recurrent parent will be used as female and donor parent as male. Each recurrent parents are crossed with donor parent which is carrying gene of interest that is *alc* to generate $F_1$ seeds. (Fig.2). Next as we can see $F_1$ plants will be subjected to true hybridity gene test by using foreground selection. By this will get to know about $F_1$ plants which are in heterozygous nature. After the conformation of target gene present in $F_1$ plants they will use these $F_1$ for backcross with RP to generate $BC_1F_1$ seeds. In each generation they will go for foreground and background selection. In $BC_1F_1$ generation they got 79 percent of recovery of RP. The selected $BC_1F_1$ plants again backcross with RP to generate $BC_2F_1$ seeds. Here they got 92.2 percent recovery of genome of RP. And after these $BC_2F_1$ plants carrying gene of interest so they will go for continue selfing for two generation to produce $BC_2F_2$ and $BC_2F_3$ population. The final $BC_2F_3$ family called inbred line or near isogenic lines these will be evaluated for agronomic performance (Maxwell *et al.*, 2022).

Lee *et al.,* (2021) studied on Breeding *Capsicum chinense* Lines with High Levels of Capsaicinoids and Capsinoids in the Fruit. The spicy taste of the pepper is mainly due to the compound called capsaicin, this capsaicin and its anologs are called as capsaicinoids. There is another compound called capsinoid which are smiliar to capsacin but mild pungent which are present mostly in sweet peppers.

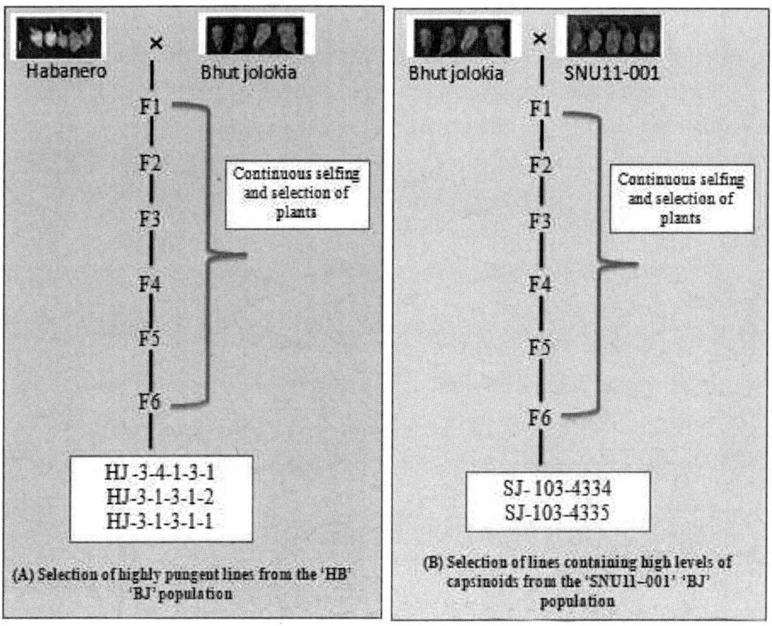


(A) Selection of highly pungent lines from the 'HB' 'BJ' population

(B) Selection of lines containing high levels of capsinoids from the 'SNU11-001' 'BJ' population

**Figure 3:** Diagram of the novel capsicum variety breeding program

In this study they used three parental lines of Capsicum *spp.* that is "SNU11–001, Habanero and Bhut Jolokia" these parents were used for HJ and SJ population development. The $F_2$ population are taken for QTL mapping through genotyping by sequencing, for QTL they used QTL cartographer programme. Then the quantification of capsicinoids and capinoids on the selected line were done by HPLC method (Fig. 3)

In Figure A showes 'HJ' population it was developed by crossing 'Habanero' and 'Bhut jolokia'. In this HB parent which has normal pAMT enzyme this popultion is majorly used to increase capascionoids content. In fig B the SJ population was obtained by crossing 'Bhut jolokia' with 'SNU11001' line, in this SNU11-001 parent has mutant pMAT enzyme. This population is majorly used to increase the capsinoid contents. Both the populations are advanced up to $F_6$ generation and from $F_6$ three superior lines were selected from HJ population and two superior lines were selected from SJ population which having high capasicinoid and capsinoid content respectively.

### e) Genetic engineering

Paul berg is considered as "Father of genetic engineering". Genetic engineering is the manipulation of plant genome through recombinant DNA technology to alter plant characteristics. It can also be defined as the controlled introduction of specific DNA segments into recipient plant genome. Gene transfer can be done by various methods like *Agrobacterium* mediated gene transfer, particle bombardment, glass bead method, ultrasonication, electroporation and shock wave method.

Ex: Golden rice, Flavr savr tomato *etc.*

**Steps involved in genetic engineering are**

1. Extraction of plasmid from bacterial cell which is followed by isolation of our gene of interest
2. After that cutting DNA at specific location with help of certain enzymes called restriction enzymes or molecular scissors
3. After the second step that is cutting DNA segment followed by the ligases process it is nothing but joining of DNA fragments of desired sequences along with plasmid molecule to form a specific type of hybrid DNA is called recombinant DNA technology
4. In next step the rDNA inserted into the host cell to form a transformed bacterial cell these transformed bacterial cell allowed

them to grow inside a media

5. Last step is selection and screening of transformed cells with desired character.

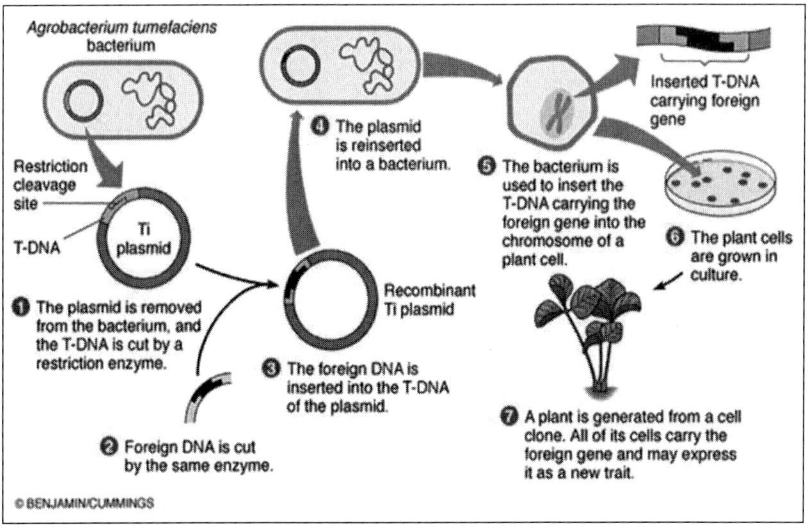


**Figure 4:** Steps followed in genetic engineering

Nagar and Mekawi, 2018 studied on Genetic engineering to improve β-carotene content in pepper. Pepper fruits an often be a major dietary source of cartenoids especially beta-carotene there is a considerable variation in Beta-carotene content in pepper fruits. With main objective is to introduce the *Arabidopsis lycopene beta-cyclase (β-Lcy)* gene *via Agrobacterium* mediated vector transmission method. They used two different pepper genotype Balady, and Topepo rosso, were analyzed with regard to their efficiency for regeneration and genetic transformation *in vitro*. Cotyledons were used *in vitro* from young seedlings as target tissue for regeneration and transformation of pepper. Over expression of *β-Lcy* gene *via Agrobacterium*-mediated transformation was used. The presence of the transgene was assayed on leaf DNA *via* PCR and biochemical analysis using HPLC. Results showed that the transgenic pepper plants showed a significant increase in β-carotene content reaching 7 to 10 folds than the control.

## f) RNA interference technology (RNAI)

**RNA interference technology (RNAi):** sequence specific gene silencing phenomenon caused by the presence of double stranded RNA. These RNAi allows down regulation in gene in a more precise manner without effecting expression of other genes. It was first observed by Richard in 1990.

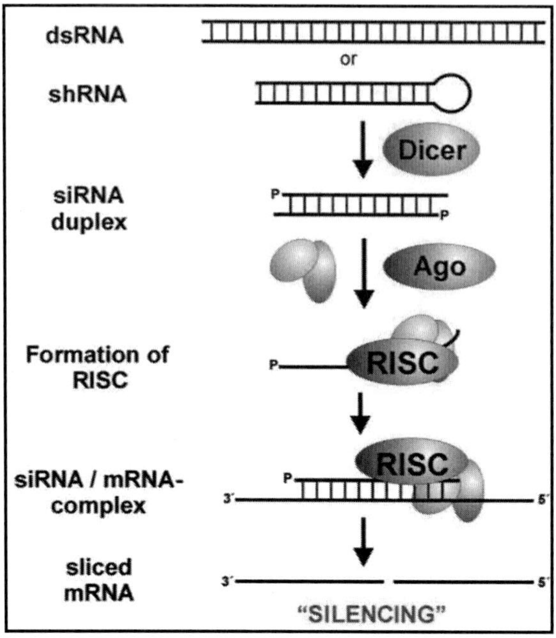


**Figure 5:** Mechanism of the RNAi pathway

Due to action of dicer enzyme on dsRNA is chopped and converted into fragments called small interfering RNAs (siRNAs). With the help of argonout enzyme siRNA forms the RISC (RNA induced silencing complex). The complex of SiRNAs and RISC binds to the target mRNA. Later this will inhibit the mRNA synthesis and silence the gene (Fig.5).

**Table 3:** Status of RNAi for quality improvement in vegetable crops

| Crop | Targeted genes silencing | Targeted traits |
|---|---|---|
| Tomato | 1-aminocyclopropane-1carboxylate (ACC) oxidase | Suppressed Ethylene production and Improve shelf life |
| | α-Man and βHex 9 (Ethylene ripening inducer) | Enhanced shelf life |
| | *MiR156* | Enhanced shelf life |
| | Colorless never ripe *(CNR) gene* | Enhanced shelf life |
| Potato | β-carotene hydroxylase gene | Enhanced β-carotene content |

## g) CRISPR CAS9- A genome editor

It has been successfully applied to various plants, including Rice,

Wheat, Maize and Tomato (clustered regularly interspaced short palindromic repeats) is types of DNA sequences found in the genomes of prokaryotic organisms such as bacteria. It is nothing but acquired immune system present in prokaryotes, which is identified in early 1987 in *E. coli*. Later, scientists identified and modified as a powerful gene editing tool and got a broad application in genome research field due to its ease of use and cost-effectiveness. Emmanuelle Charpentier and Jennifer Doudna got Nobel Prize for genome editing discovery.

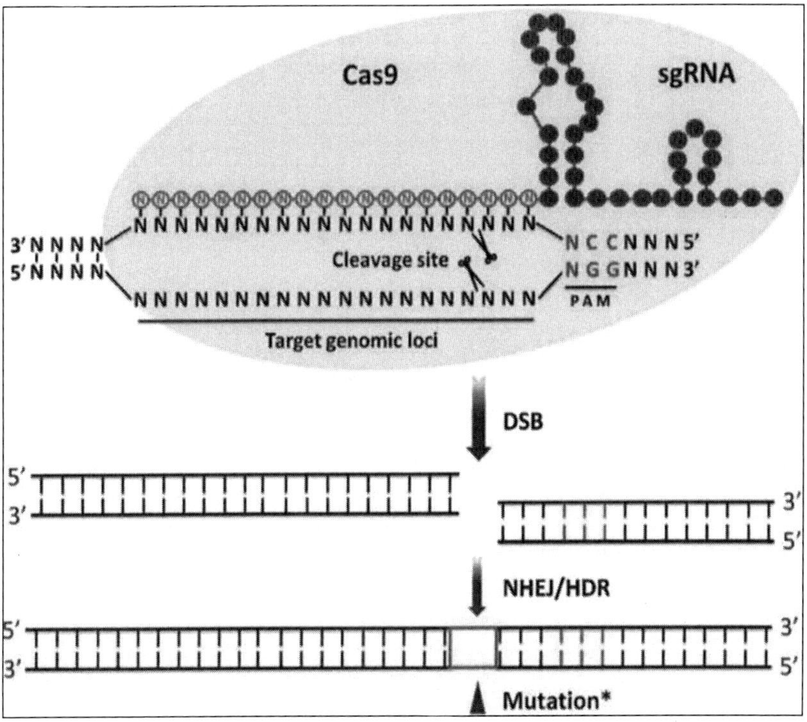


**Figure 6:** Schematic view of gene silencing by RNAi technology in plants

The synthetic guide RNA (sgRNA) contains a region (usually 20 bp in length) complementary to the target site on the genomic loci and stem loops that mediate the binding of the Cas9 protein. The protospacer adjacent motif (PAM, NGG) required for cleavage. Induced DSBs of the target DNA are repaired either by NHEJ (non-homologous end joining) or HDR (homology-directed repair) resulting in gene mutations that include nucleotide insertion, deletion, or substitution around the cleavage site.

Andersson *et al.* (2020) studied the application of the CRISPR Cas-9 system to induce mutations in the *StPPO2* gene in the tetraploid potato cultivar Desiree. They revealed that relative enzymatic browning was significantly lower in all the edited lines in comparison to the control line. The mutant edited lines showed reduction of up to 69 and 73 percent in tuber PPO activity and enzymatic browning respectively as compared to that of control.

**Achievements**

**Table 4:** Wild sources for quality traits

| CROP | Wild species | Trait |
|---|---|---|
|  | *Solanum pimpinellifolium, Solanum peruvianum* | Ascorbic Acid |
| | *Solanum chmielewskii* | Total soluble solids |
| | *Solanum cheesmaniae* | β-Carotene |
|  | *Solanum torvum*<br>*Solanum khasianum*<br>*Solanum aviculare* | Phenols |
|  | *S. vernei* | Starch |
| | *S. phureja spp. phureja* | Carotene |
|  | *Capsicum frutescens* | Capsaicin |

(Kumar *et al.,* 2020)

**Transgenic approach for bio fortification in vegetables crops**

Biofortification refers to process of improving nutritional quality of food crops, this can be achieved through agronomic practices, conventional breeding or biotechnological approaches like genetic engineering and genome editing by Kumar *et al.*, 2020

Some of examples of transgenic plants are in tomato Flavr savr, was th first commercially grown genetically engineered food for human consumation. Followed by Introgression of *AmA1* gene from *Amaranthus hypochondriacus* tend to increase 60 percent protein content in potato. And the scientists identified two genes (*SlFdAT1* and *Solyc12*) from snapdragon that induce the production of anthocyanins concentration in purple tomato Singh *et al.* (2021).

**Organizations involved in biofortification**

- ➢ Consortium of international Agricultural Research (CGIAR) International Food Policy Research Institute (IFPRI)

➢ International Crops Research Institute for semi-Arid Tropics (ICRISAT)

➢ International Potato Centre (CIP)

**Special initiatives in India**

✓ Balwadi nutrition programme

✓ Pradhan Mantri Matru Vandana Yojana

✓ Integrated Child Development Services (ICDS)

✓ Mid-day meal

**Anoop pur: India's first biofortified village**

➢ Recently, in Uttar Pradesh Chandra Shekhar Azad University of Agriculture and Technology, Kanpur has developed a completely biofortified village named Anoop pur in district Kanpur.

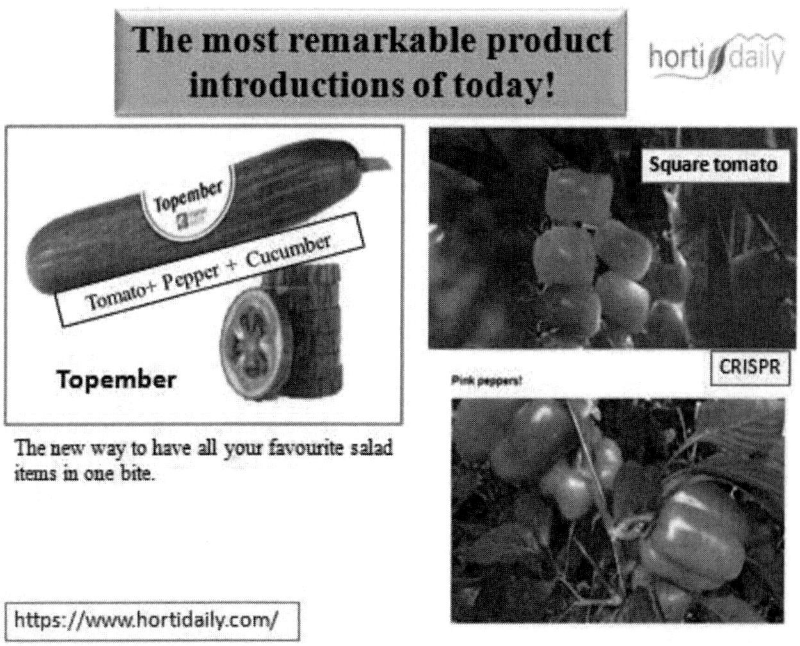


**Research regulation**

- The departments of biotechnology and the ICAR have reconginsed the potential use of gene-edting technology. The DBT's national Agri food biotechnology institute Punjab was one among the first in India to use CRIPSR/cas 9 to carry out a change in the *phytoene desaturase* (fruit ripening) gene of banana cv Rasthalli.

- ICAR institute are involved in the application of CRIPSR technology for enhancing stress tolerance and nutritional quality in a number of crops.
- Scientist He jianukui created shock to the world in the year 2018 with the stunning claim that he had altered the genetic makeup of IVF embryos and implanted them into a women's uterus, leading to the birth of twin girls name called Lulu and Nana.
- '*Topember vegetable is the newly created with* the sweetness of tomatoes, the crunchiness of peppers and the tastiness of cucumbers. And also, the square tomato is not only special in terms of shape, but it will also improve your quality of life. Because it contains more than 50 percent lycopene, it will protect you from many diseases. Ready to release for commercial purpose.

**Conclusion**

Nutritional qualities and associated health benefits of vegetable crops is becoming important criteria for their increase in consumer preference. Molecular tools and technologies showed effectiveness to speed up the classical breeding methods for quality trait. Utilization of advanced breeding methods such as molecular markers, Marker assisted selection, genetic engineering and CRISPR-cas-9 along with conventional breeding method have great promise in reducing time and cost involved in breeding for improvement of quality traits in solanceous vegetables.

**References**


1. Anderson, M., Gonzalez, M. N., Massa, G. A., Turesson, H., Olsson, N., Falt, A. S. & Feingold, S. E., (2020). Reduced enzymatic browning in potato tubers by specific editing of a polyphenol oxidase gene *via* ribonucleoprotein complexes delivery of the CRISPR Cas-9 system. *Frontiers in Plant Science, 10*(1): 1-12.
2. Behera, T. K. & Singh, S., (2019). Advances in vegetable breeding for nutraceuticals and quality traits. *Indian Journal of Gene*tics, *79*(1): 216-226.
3. Chadha, M. L., Hedge, R. K. & Bajaj, K. L., (1998), Heterosis and combining ability studies of pigmentation in brinjal (*Solanum melongeona* L.). *Vegetable Science, 15*(4): 64-71.
4. Kalloo, G., (1986), Tomato (*Lycopersicon esculentum* Miller). *Allied Publishers Private Ltd.* New Delhi-110002.

5. Kumar, A., Kardile, H. B., Raigond, P., Changan, S. S., Thakur, N., LaL, M. K. & Singh, B., (2020), Biofortification of vegetables. *Advances in agri-food biotechnology,* 105-129.

6. Kumari, S., Anokhe, A. & Kumar, R. (2017), Nutritional quality and health benefits of solanaceous vegetables. *Indian Journal of Genetics and plant breeding, 29*(1): 200-218.

7. Lee, D. G., Jang, S., Park, M., Lim, J. H., Jung, J. W. & Kang, B. C., (2021), Breeding *Capsicum chinense* lines with high levels of capsaicinoids and capsinoids in the fruit. *Agri*culture, *11*(9): 1-12.

8. Manoharan, R. K., Jung, H. J., Hwang, I., Jeong, N., Kho, K. H., Chung, M. Y. & Nou, I. S., (2017), Molecular breeding of a novel orange-brown tomato fruit with enhanced beta-carotene and chlorophyll accumulation. *Hereditas*, *154*(1): 1-8.

9. Maxwell, D. P., Kwabena Osei, M., Danquah, A., Danquah, E. Y., Blay, E. & Massoudi, M., (2022), Marker Assisted Backcrossing of *alcobaca* gene into two elite tomato breeding lines. *Frontiers in Horticulture, 6*(1): 1-10.

10. Nagar, M. M. & Mekawi, E. M., (2018), Genetic engineering to improve β-carotene content in pepper. *Plant Biotechnology, 56*(1): 119-126.

11. Singh, S., Yadav, S. & Singh, A. (2021), Recent approaches for breeding vegetable crops for quality, rich in nutrients and nutraceuticals. *Journal of Pharmacognosy & Photochemistry, 10*(1): 2061-2071.

12. Swathy, S. P., Kiran, K. R., Rao, M. S., Mahato, K. K. & Radhakrishna, M., (2021), Responses of He-Ne laser irradiation on chlorogenic acid content of brinjal (*Solanum melongena* L.) var. Mattu Gulla. *Journal of Photochemistry & Photobiology*, *164*(57): 182-190.

# Chapter - 18

## Food Security and Nutrition: Challenges and Solutions

**Authors**

**Pundlik Waghmare**
Assistant Professor (Agronomy), Vasantrao Naik Marathwada Krishi Vidyapeeth, Parbhani, Maharashtra, India

**Mamta Patange**
Assistant Professor (Agronomy), Vasantrao Naik Marathwada Krishi Vidyapeeth, Parbhani, Maharashtra, India

**Sharad Gosavi**
Assistant Professor (Agronomy), Vasantrao Naik Marathwada Krishi Vidyapeeth, Parbhani, Maharashtra, India

# Chapter - 18

## Food Security and Nutrition: Challenges and Solutions

**Pundlik Waghmare, Mamta Patange and Sharad Gosavi**

*Food security "exists when all people, at all times, have physical, social and economic access to sufficient safe and nutritious food that meets their dietary needs and food preferences for an active and healthy life." – FAO, 1996.*

### Abstract

The principal concept of food security and nutrition conveys the admittance to healthy food and optimal nutrition for all. Household food security is the application of this concept to the family level with individuals within households as the focus of concern. Food insecurity refers to a lack of access to enough good, healthy and culturally appropriate food. India has experienced notable economic growth in recent years and remains one of the fastest growing economies in the world. However, poverty and food insecurity in India are still areas of concern in spite of many strides. Food is considered as a basic amenity essential for the nourishment, development and growth of an individual. India has ranked 101 among the 116 countries on the Global Hunger Index, 2021. According to the Food and Agriculture Organisation, the Food Price Index has increased by 30% in the year 2021-22. Although the Government of India has been actively addressing food security at households for a long time through the Public Distribution System and the National Food Security Act (NFSA) 2013, there are still concerns related to Food Security in India amidst increasing population, climate change and global supply chain disruption (Russia-Ukraine War) that need to be addressed.

**Keywords:** Food, nutrition, poverty, hunger, economic growth, food security.

### Introduction

In the past half-century, the world has become increasingly aware of the challenges and threats to food security. This heightened awareness has been encouraged by a range of well exposed humanitarian disasters and food price crises on the one hand and the expanding growth of the world's population and the changes in its dietary patterns on the other. Development organizations are increasingly concerned about food security and have focused their efforts on helping those who are not able to feed themselves adequately and satisfactorily. Over the course of decades, these bodies have received the support of governments, private foundations and United Nations organizations. Many of their efforts have focused on provisioning food in situations of crisis or emergency and increasingly on providing cash or food for development. Securing an adequate supply of food, however, is by no means the same thing as securing adequate nutrition.

The distinction between food and nutrition security is critically important because it affects what can be expected from the large and in some cases growing, investments in improving agriculture productivity and encouraging global food security worldwide.

**Food security**

Food security has been defined by the Food and Agriculture Organization of the United Nations (FAO) as existing *"when all people, at all times, [*have] physical, social and economic access to sufficient, safe and nutritious food to meet their dietary needs and food preferences for an active and healthy life" (FAO 1996, par. 1). Importantly, this definition stipulates that food should be available in sufficient quantity as well as in sufficient quality, should be culturally acceptable, and should be available at all times throughout the year.

The concept of Food Security is multifaceted. Food is as essential for living as air is for breathing. But food security means something more than getting two square meals. It has following dimensions:

- **Availability:** It means food production within the country, food imports and the stock stored in government granaries.
- **Accessibility:** It means food is within reach of every person without any discrimination.
- **Affordability:** It implies that having enough money to buy sufficient, safe and nutritious food to meet one's dietary needs.

Thus, Food security is ensured in a country only when sufficient food is available for everyone, if everyone has the means to purchase food of acceptable quality and if there are no barriers to access.

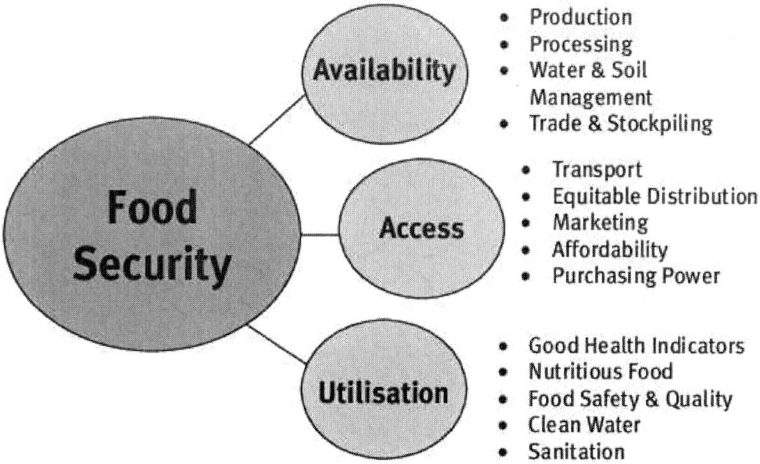


**Food security scenario in India**

**Background on food security in India**

Food Security means that every member of society must always have access to enough food to meet their dietary needs and lead healthy, active lifestyles.

- The ultimate goal of Food Security in India should be ensuring that everyone always has physical and financial access to the food they need.
- The three pillars of food security access are: Money (having enough money or other resources to buy food), consumption (having enough of the correct kinds of food on hand), and availability (having enough of the right kinds of food on hand).
- However, the challenges of food security create a hurdle in the implementation of these pillars.
- Also, there have been certain instances in the past in India when the issue of food security has been of prime concern.
- For example, during the Indo-China War of 1962 or the India-Pakistan War of 1971.
- Also, during the occurrence of the National Emergency in 1977, the issue of food security gained importance.
- Thus, many policies have been formulated and implemented thereafter to ensure Food Security is maintained in India.

**Characteristics of being food secured**

Concept of food security also includes the presence of following characteristics:

- **Enough food:** If there is adequate food as per one's need and requirements.
- **Good food:** It should be with desirable taste, texture and other attributes.
- **Healthy food:** It should be nutritious and sustaining with proper vitamins, minerals, fibre, energy, and all the other things that our bodies need.
- **Culturally appropriate food:** Having access to culturally appropriate food means that people have food that they are familiar and comfortable with including availability of familiar foods & varieties.

**Components or pillars of food security**

Basically, there are four components/pillars of food security. They are:

1. **Food availability**
   - Having sufficient quantity of appropriate food available.
   - It includes: domestic production, import capacity, food stocks, food aid.
2. **Accessibility**
   - Physical and economic access to food
   - Purchasing power, income of population, transport and market infrastructure
3. **Utilization**
   - It includes adequate dietary intake and ability to use nutrients in the body.
   - Food safety, hygiene and manufacturing practices applied in primary agricultural production, harvesting and storage; food processing; transportation, retail, households, diet quality and diversity: meeting needs in terms of energy, macro- and micronutrients.

## 4. Stability

- Stability of supply and access
- Weather variability, price fluctuations, political factors, economic factors

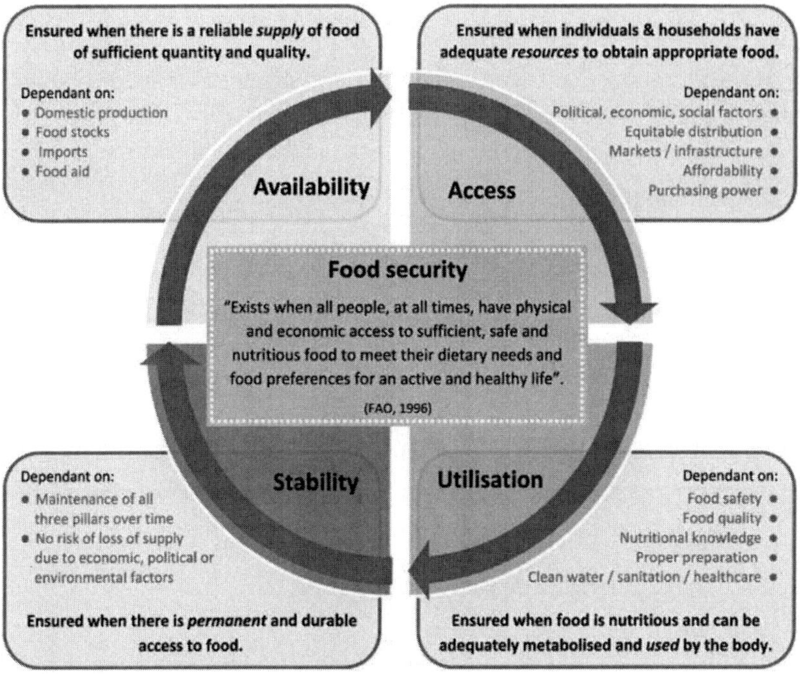


### Importance of food security in India

- "All people at all times have access to sufficient, safe, nutritious food to support a healthy and active life," said the 1996 World Food Summit.
- Food is our main source of nutrition; hence the development of the nation's food system is essential.
- Food security has long been a cause for worry at the international, national, regional, state, household, and individual levels.
- The issue of food security has become very critical in many parts of the world.

### Dimensions of food security in India

The various dimensions of food security are listed below.

- Food production domestically, food imports and the stock from the previous year kept in government granaries are all considered to be available.

- Accessibility implies that food is available to everyone.
- A person must be financially capable of purchasing enough wholesome food to satisfy their nutritional needs in order to be considered affordable.

Therefore, a nation's food security is only guaranteed if there is enough food available for everyone, everyone has the means to buy food of a sufficient standard, and there are no obstacles in the way of obtaining food.

**Dimensions of food security and its determinants**

| S/N | Availability | Accessibility | Utilization | Stability |
|---|---|---|---|---|
| 1 | Domestic production | Income, Purchasing power, Own production | Food safety and quality | Weather variability, Seasonality |
| 2 | Import capacity | Transport | Clean water | Price fluctuations |
| 3 | Food Stocks | Market infrastructure | Health and sanitization | Political factors |
| 4 | Food aid | Food distribution | Care, feeding and health seeking practices | Economic factors |

**Recent outline for food security in India**

- **Legal provision:** Though the Indian Constitution does not have any explicit provision regarding right to food, the fundamental right to life enshrined in Article 21 of the Constitution can be interpreted to include the right to live with human dignity, which may include the right to food and other basic necessities.
- **Buffer stock:** Food Corporation of India (FCI) has the prime responsibility of procuring the food grains at minimum support price (MSP) and stored in its warehouses at different locations and from there it is supplied to the state governments in terms of requirement.
- **Public distribution system:** Over the years, Public Distribution System has become an important part of Government's policy for management of the food economy in the country. PDS is supplemental in nature and is not intended to make available the entire requirement of any of the commodity. Under the PDS,

presently the commodities namely wheat, rice, sugar and kerosene are being allocated to the States/UTs for distribution. Some States/ UTs also distribute additional items of mass consumption through the PDS outlets such as pulses, edible oils, iodized salt, spices, etc.

- **National Food Security Act, 2013 (NFSA):** It marks a paradigm shift in the approach to food security from welfare to rights based approach.

  1. NFSA covers 75% of the rural population and 50% of the urban population under:

     i) Antyodaya Anna Yojana: It constitute the poorest of-the-poor, are entitled to receive 35 kg of food grains per household per month.

     ii) Priority Households (PHH): Households covered under PHH category are entitled to receive 5 kg of food grains per person per month.

  2. The eldest woman of the household of age 18 years or above is mandated to be the head of the household for the purpose of issuing ration cards.

  3. In addition, the act lays down special provisions for children between the ages of 6 months and 14 years old, which allows them to receive a nutritious meal for free through a widespread network of Integrated Child Development Services (ICDS) centres, known as Anganwadi Centres.

**Factors affecting food security**

The food security can be mainly influenced by:

1. Low per capita income
2. Low and unequal distribution of income
3. Poor and highly unstable growth performance especially in agriculture
4. Unemployment and underemployment
5. Low and declining farm size
6. Inequalities in land distribution
7. Low land utilization
8. Social discrimination
9. Population growth

10. Access to market
11. Food restrictions: certain restrictions in the food consumption due to cultural and social norms.
12. Poverty
13. Climate induced insecurity through deforestation, landslide, declining soil fertility
14. Political instability
15. Poor, marginalized, ethnic group & lower caste groups
16. High maternal and infant mortalities

**Challenges related to food security in India**

1. **Population:** Although a major part of the Indian population is engaged in agricultural activities, the availability of food for all is a challenge due to the increasing population of the country.
2. **Poverty:** This is one of the biggest challenges which need to be overcome in order to attain the desired food security in the country. The percentage of people living below the poverty line (BPL) is extremely high. Know about the Poverty Estimation in India at the linked article.
3. **Lack of awareness:** Lack of education and training on new techniques, technologies and agricultural products. Traditional farming methods are slightly more time consuming and delay the production of food grains, etc.
4. **Worsening soil health:** A key element of food production is healthy soil because nearly 95% of global food production depends on soil. Soil degradation due to excessive or inappropriate use of agrochemicals, deforestation and natural calamities is a significant challenge to sustainable food production. About one-third of the earth's soil is already degraded.
5. **Aggressive weed threats:** In the past 15 years, India has faced more than 10 major invasive pest and weed attacks. Fall Armyworm (Pest) destroyed almost the entire maize crop in the country in 2018. India had to import maize in 2019 due to the damage caused by the pest in 2018. In 2020, locust attack was reported in districts of Rajasthan and Gujarat.
6. **Lack of effective management agenda:** India lacks strict management agenda for food security. Public Distribution System faces challenges like leakages and diversion of food-grains,

inclusion/exclusion errors, fake and false ration cards and social audit mechanism.

7. **Faults in procurement:** Farmers have diverted land from producing coarse grains to the production of rice and wheat due to a minimum support price. Further, there is a tremendous wastage of around Rs.50,000 crore annually by both improper accounting and inadequate storage facilities

8. **Climate change:** The monsoon accounts for around 70% of India's annual rainfall and irrigates 60% of its net sown area. Changing precipitation patterns and growing frequency and intensity of extreme weather events such as heat waves, floods are already reducing agricultural productivity in India, posing a serious threat to food security. To increase domestic availability amid low Kharif Crop productivity this year (2022), the Government of India has banned the export of broken rice.

9. **Disruption of global supply chain:** At a time when the Covid-19 Pandemic had already impacted food supply around the world in 2020, Russia-Ukraine War in 2022 has disrupted the global supply chain and resulted in food scarcity and food inflation. Russia and Ukraine represent 27% of the world market for wheat, 26 countries, mainly in Africa, West Asia and Asia, depend on Russia and Ukraine for more than 50% of their wheat imports.

10. **Short of food distribution:** The balance between the food distributions has been varied in urban and rural areas.

11. **Corruption and fraud:** Diverting the grains to open market to get better margin, selling poor quality grains at ration shops, the irregular opening of the shops adds to the issue of food insecurity.

12. **Inadequate storage facilities:** Inadequate and improper storage facilities for grains, which are often stored outside under tarps that provide little protection from humidity and pests.

13. **Unmonitored nutrition programmes:** Emphasis must be given on introducing and enacting well-monitored nutrition programmes.

**Solutions related to food security in India**

1. **Increase in agricultural productivity:** Agricultural productivity can be increased by using various techniques mentioned below.

   i) Better techniques must be implemented in farming.

   ii) Increased irrigation facilities.

iii) Distribution of cultivable land and size of farms.

iv) Providing seeds with higher quality.

v) Providing fertilizers.

2. **Crop variation and establishing food grain banks**: Farmer community must focus on growing legumes as an alternative to rice and wheat. Also, should focus on growing non-cereal crops such as oilseeds, fruits and vegetables. With the own interests, farmers must create decentralized food banks in villages, it may help in tackling corruption.

3. **Moving towards sustainable farming:** For ensuring Food Security in India, improvement in productivity through greater use of biotechnology, intensifying watershed management, use of Nano- urea and access to micro-irrigation facilities and bridging crop yield gaps across States through collective approach should be at priority. There is also a need to look forward towards establishing Special Agriculture Zones through ICT based crop monitoring.

4. **Moving towards precision agriculture:** There is need to increase the use information technology in agriculture to ensure that crops and soil receive exactly what they need for optimum health and productivity. By adopting precision agriculture with high-tech farming practices, farmers' incomes will increase, input cost of production will be reduced, and many other issues of scale will be addressed.

5. **Refreshing Aadhar scattering of ration cards:** To speed up the process of Aadhar linking to ration cards, ground monitoring measures must be taken that will ensure no valid beneficiary is left out of their share of food grains that can give thrust to the aim of zero hunger.

6. **Food availability to below poverty line (BPL) people:** The inefficiency in Public Distribution System (PDS) due to corruption must be eliminated and should be made transparent. Try to find the exact number of people BPL. Provide subsidies to people above poverty line. Reduce food prices and increase affordability by improving inter-state movement, stocking.

7. **Direct benefit transfer (DBT) through JAM:** There is a need to rationalize food and fertilizer subsidies into direct benefit

transfers to accounts of identified beneficiaries through the JAM (Jandhan, Aadhar and Mobile) platform that will reduce enormous physical movement of food grains, provide greater self-sufficiency to beneficiaries to choose their consumption basket and promote financial inclusion.

8. **Improving food storage technologies:** Safeguard transparency in food stock assets. To improve communication with farmers can help them to get a better contract for their produce while improving storage houses with the latest technology is equally important to deal with natural disasters. Further, food grain banks can be organized at block or village level from which people may get subsidized food grains against food coupons (that can be provided to Aadhar linked beneficiaries).

9. **Increase in employment and purchasing power in rural and urban areas:** Mahatma Gandhi National Rural Employment Guarantee Act (MGNREGA) is a good program working in the direction of increasing purchasing power of people in rural areas. Government should try to come up with more innovative programs to improve purchasing power. Also, should come up with more Poverty Alleviation Programmes and should focus on workers in the informal sector.

10. **Nursing of nutritional programmes:** Health department's officials must initiate steps to monitor the implementation of various programmes. Need to make sure they have all the latest information and evaluate the effectiveness of the program implementation.

11. **Community participation and coordination:** There needs to be greater participation from the various stakeholders or sponsors like Panchayati Raj Institutions, Self Help Groups (SHG). Attention also needs to be paid to hygiene, sanitation and nutrition education.

**Food nutrition**

Food is a key outcome of agricultural activities, and, in turn, is a key input into good nutrition. Without agriculture, there is little food or nutrition, but availability of food from agriculture does not ensure good nutrition. One would assume a strong connection between the two fields of agriculture and nutrition however, in fact, there is often a significant disconnect. Extreme poverty and hunger are largely rural, with smallholder farmers and their families making up a very significant proportion of the poor and hungry.

Thus, eliminating poverty and hunger are integrally related to enhancing food production, agricultural productivity and rural incomes.

Nutrition security, by dissimilarity, exists when, in addition to having access to a healthy and balanced diet, people also have access to adequate caregiving practices and to a safe and clean environment that allows them to stay healthy and utilize the foods they eat effectively. For young children, for example, this means that they have enough of the right foods and this includes breast milk for up to two years of age along with proper quantity and quality of balancing foods starting at six months of age because breast milk can no longer fulfil all of the infant's nutrient needs after that age. In addition, young children also need caregivers who have the time, education, knowledge, physical and mental health, and nutritional well-being to care for them adequately. Adequate caregiving means that caregivers are able to attend to all their children's multiple needs, including adequate feeding, hygiene, health-seeking practices and supportive parenting. Finally, to be nutrition secure, young children must also be free of repeated (chronic) or acute infections which affect with absorption and utilization of food and nutrients for body functions.

Thus, "food and nutrition security" can be defined as, "a situation that exists when all people at all times have physical, social and economic access to food, which is consumed in sufficient quantity and quality to meet their dietary needs, requirements for growth and food preferences, and is supported by an environment of adequate sanitation, health services and caregiving" (United Nations Food and Agriculture Committee on World Food Security). This allows for appropriate utilization of food and nutrients by the body and therefore creates the conditions for a healthy and active life. Nutrition security therefore implies an optimal nutritional status.

## Linkages between food and nutrition security

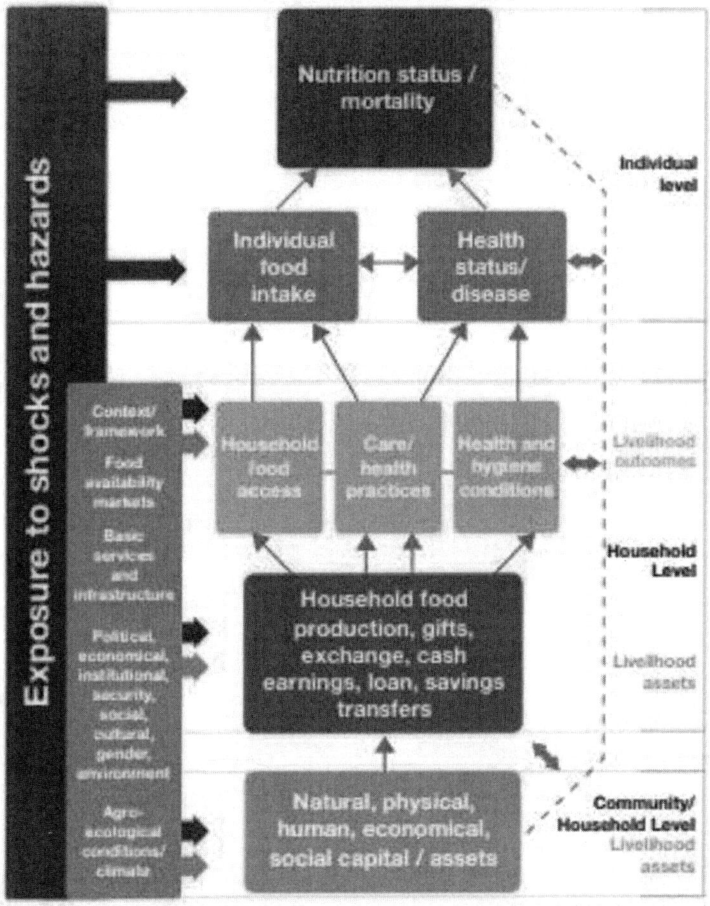


Food systems produce, market, store, trade, process, retail, and promote food for consumers to obtain and use to feed their families. Food systems also provide livelihood and income to large amounts of the population around the globe, including many of the world's poorest. The latest Food and Agriculture Organization (FAO)'s State of Food and Agriculture report, published in 2013, says that "food systems, from agricultural inputs and production, through processing, marketing and retailing to consumption can promote more nutritious and sustainable diets for everyone". The report also highlights that while the nature and causes of food insecurity and malnutrition are complex and diverse, a common denominator for both conditions is a nutritionally deficient diet. For this reason, the food system provides an opportunity to address both food insecurity and malnutrition, but in order to do so; the food system needs to be re-shaped to be more nutrition-sensitive.

The nutrition community has used extensively ever since because it provides a useful picturing of the multiple factors that affect maternal and child under nutrition. The outline is particularly useful to show the different levels of factors that affect nutrition. e. g. Factors at the national, community, household and individual levels and refers to basic, underlying and direct determinants of under nutrition. Some variations of the models also show examples of involvements that can be used to address some of the causes of under nutrition at these different levels.

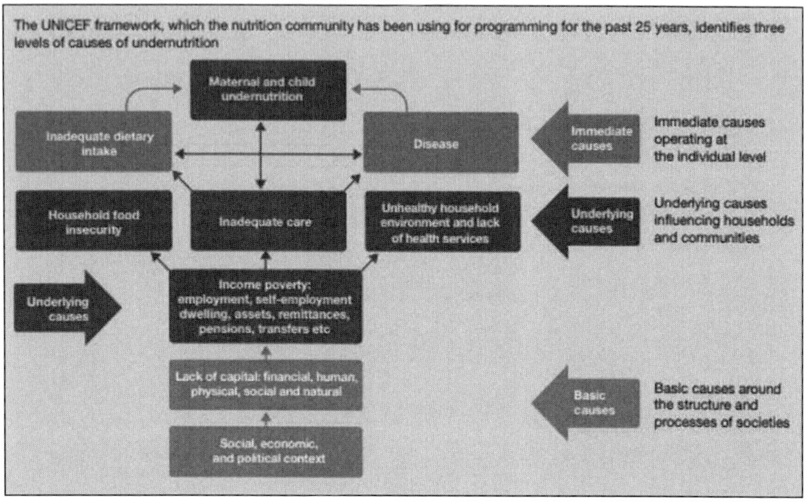


There is even stronger validation that under nutrition can trap children, families, communities and nations in an intergenerational cycle of poor nutrition, illness and poverty. More is known about the inadequate growth due to nutritional deficiencies before the age of two years with impaired brain development and following reduced performance in school. Hence, there is need to promote optimal growth during this critical period to avoid an raised risk of non-communicable diseases such as cardiovascular disease in adulthood and even in the next generation.

Indian Constitution provides the right to life to all its citizens including the Right to Food (Article 21) and directs the state to take measures to raise the level of nutrition and the standard of living of its people and improve the public health (Article 47). The Universal Declaration of Human Rights and International Convention on Economic, Social and Cultural Rights, to which India is a signatory, also cast responsibilities on all State parties to recognize the right of everyone to adequate food. Eradicating extreme poverty and hunger is one of the goals under the Millennium Development Goals of the United Nations (Anon., 2013).

In pursuance of the obligations under Indian constitution and International conventions, providing food security to all citizens has been the focus of the Government's planning and policy. National Food Security Act (NFS Act) is the latest initiative of Indian Government to achieve the food security through right based legislation. The ambitious programme of the Government, besides offering several opportunities, throws many challenges in its implementation. In this background, it is important to review the features of the NFS Act with special emphasis on its strengths and weaknesses and its implications on Indian Agriculture.

In the context to considering the climate change, India needs to upsurge investment in research, development and demonstration of technologies to improve the sustainability of food systems in the country. Building resilience of the food systems will be critical to averting large-scale future shortages and to ensuring food security and good nutrition for all. The figure below depicts the framework that helps us understand the pathways that link agriculture and nutrition as abstracted by Herforth and Harris. The pathways are interconnected rather than linear.

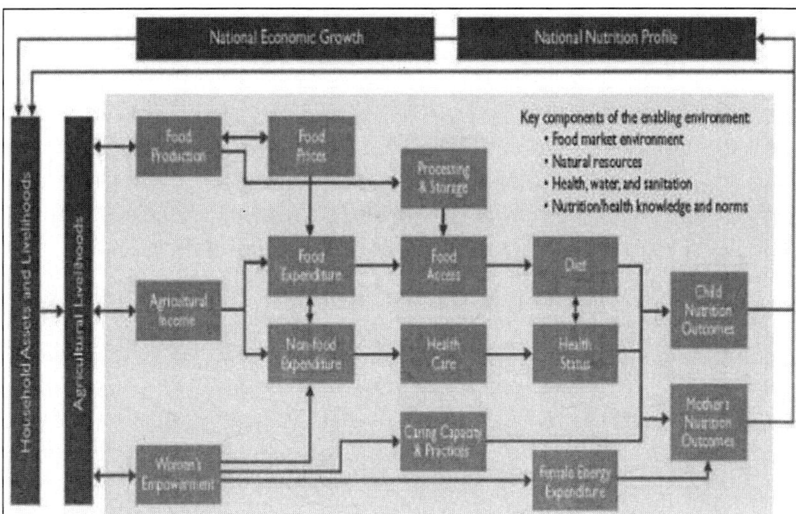


**Source:** Understanding and Applying Primary Pathways and Principles. Herforth & Harris. USAID/SPRING Project; 2014. p.

**Pathways from agriculture to nutrition**

The main pathways from agriculture to nutrition are through:

1. **Food production**

The process of converting raw ingredients into edible food is called food production. The process of food production can be segregated into different stages like cleaning, sorting, etc. Nearly all the food that we consume is received from plants and animals.

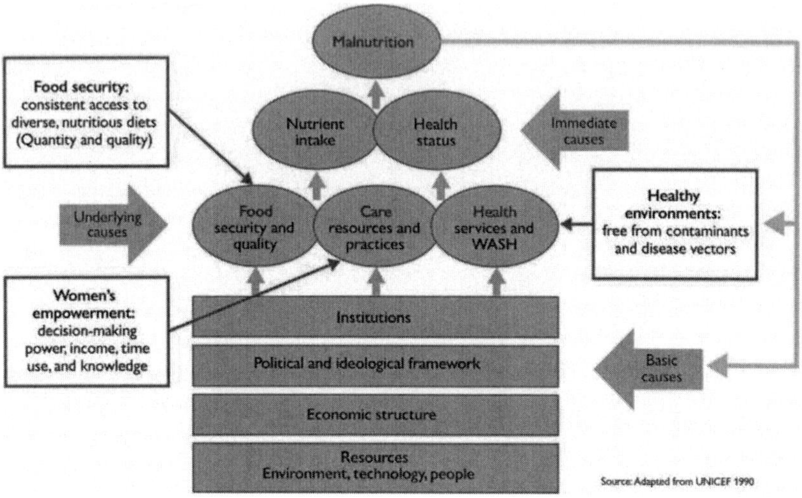


*Source*: UNICEF's Malnutrition Conceptual Framework. Source: Adapted from Understanding and Applying Primary Pathways and Principles. Herforth & Harris. USAID/SPRING Project; 2014. p. 11.

2. **Agricultural income**

Partial agricultural income is the income where the assessee is growing agricultural produce and using them as raw material for the manufacturing of products. Here income from the sale of products is partial agriculture income and partial non-agricultural income.

3. **Women's empowerment**

Women's empowerment can be defined to promoting women's sense of self-worth, their ability to determine their own choices, and their right to influence social change for themselves and others.

**Conclusion**

The day, 16[th] October supposed as a world food day on each year to highpoint the need for regular admittance to nutritious food for the millions of people all over the world who cannot meet the expense of a healthy diet. This world food day has a different theme each year and for 2022 it is "Leave no one behind". The theme assumes particular significance as the world is facing multiple challenges including, the COVID-19 pandemic, conflict, climate change, rising prices and international tensions affecting global food security. The Global Hunger Index, 2022 has ranked India 107$^{th}$ among 121 countries with a score of 29.1 putting our nation in the 'serious' category in level of hunger. All other countries in South Asia except Afghanistan are ranked higher than India. In this context, food and nutrition security are extremely important dimensions as we struggle towards higher levels of development.

Secondly, the challenges and constraints forecasted in implementing the National Food Security Act must effectively address to do well and deliver goods in the Indian economy and rural India. It is important that, India with more than 1.2 billion will get its long due position of a leader at the global level. But the hunger and malnutrition affecting more than one third of citizens will not signify well for its desires. An African quote is apt to mention in this context. "Anybody's Hunger is Everybody's Shame". Dream of India becoming a global superpower can be achieved not before all its citizens hunger is addressed. National Food Security Act though delayed, is a massive footstep in that direction.

**References**


1. Anonymous, 1996, Food and Agriculture Organization. Rome Declaration on World Food Security and World Food Summit Plan of Action, http://www.fao.org/DOCREP/003/W3613E/W3613E00.HTM. Accessed October 11, 2013. 2.

2. Anonymous, 2010, India – New Global Poverty Estimate. World Bank Report, Washington DC. 3.

3. Anonymous, 2011, Global Hunger Index Report, The challenge of hunger: taming of price spikes and excessive food price volatility, International Food Policy Research Institute, Washington DC.

4. Anonymous, 2013 a, The National Food security Act, The Gazette of India (September 10, 2013) Ministry of Law and Justice, Government of India, New Delhi.

5. Food and Nutritional Security in India: charting the way to a robust Agri-food system. Department of Economic Analysis and Research (DEAR) National Bank for Agriculture and Rural Development; November 2022

6. Gulati Ashok, Jyoti Gujral and Nandakumar T., 2012, National Food Security Bill: Challenges and Options, Discussion Paper-2, Commission for Agricultural Costs and Prices, New Delhi.

7. https://byjus.com/free-ias-prep/world-food-programme-wfp/

8. https://byjus.com/free-ias-prep/nfsa/

9. https://byjus.com/free-ias-prep/poverty-and-hunger-issues/

10. https://www.drishtiias.com/daily-updates/daily-news-editorials/food-security-in-india

11. The State of Food Security and Nutrition in the World 2019: Safeguarding Against Economic Slowdowns and Downturns. Food and Agriculture Organization (author). United Nations publications

12. Weingartner Lioba, 2005, Food and nutrition security. The Concept of Food and Nutrition Security, In WEnt 2005. Retrieved from http://www.eldis.org/vfile/upload/1/document/0803/ID22376.

# Chapter - 19

## Biological Control of Insect Pests: Harnessing Natural Enemies


**Authors**

**Trupti Rajesh Pandit**
Department of Entomology, Lovely Professional University, Punjab, India

**Dr. Sunil Kumar Dwivedi**
Department of Entomology, Lovely Professional University, Punjab, India

**Amol Pawar**
Department of Agronomy, Lovely Professional University, Punjab, India

**Mayur Patil**
Department of Agronomy, Lovely Professional University, Punjab, India

# Chapter - 19

## Biological Control of Insect Pests: Harnessing Natural Enemies

**Trupti Rajesh Pandit, Dr. Sunil Kumar Dwivedi, Amol Pawar and Mayur Patil**


### Abstract

Biological control, the utilization of natural enemies to suppress insect pests, offers a captivating and sustainable approach to pest management. In this intricate dance between predators, parasites, and their prey, we witness the marvels of nature at work. Ladybugs don their spotted armor, voraciously consuming aphids that threaten our gardens. Meanwhile, stealthy parasitic wasps inject their eggs into unsuspecting pest larvae, ensuring their demise. These natural enemies become our allies, their inherent abilities harnessed to protect crops and ecosystems. Biological control not only provides an eco-friendly alternative to chemical pesticides but also promotes biodiversity and preserves ecological balance. By cultivating the presence of natural enemies, we foster a harmonious coexistence between predator and prey. The delicate interplay of these organisms serves as a testament to the intricacy and resilience of our natural world.


**Keywords:** Predator, biological control, natural enemies, pest management.

### Introduction

Parasites, pathogens, and other predators can help reduce pest populations and the damage they cause through a process known as "biological control." For biological methods of pest bug and disease control, these "natural enemies" are crucial. As an initial illustration, consider the mite (Fig. Rangeland and wildland weeds, such as Klamath weed and St. Johnswort, can be effectively managed with the use of natural enemies for biological control. Plant diseases, nematodes, and vertebrate predators are all naturally controlled by other organisms, although it is not always easy to spot, understand, or control this biological control. These predators and parasites can be utilized through conservation efforts, artificial enhancements, and conventional biological control strategies. Parasites, pathogens, and predators are commonly used to manage insect and mite populations (Table 1). Most parasites, pathogens, and many predators only affect one type of pest. You can learn how to recognise these pests with the help of the Natural Enemies Gallery and the Natural Enemies Handbook. (Rayl *et al.*, 2019)

All or nearly all a parasite's sustenance must come from a living host, hence the name. Insect parasites can cause 2019.toms on the inside or the outside of the insect. In most cases, only the parasite's juvenile stage feeds on the host. Adult females of certain parasites (such as many wasps that attack scales and whiteflies) constitute a usually overlooked yet significant source of biological control. Although the term "parasite" is used, most parasites (including fleas and ticks) do not really result in the death of their hosts. Some of the most useful species for biological control are also the deadliest to their hosts, hence they are more properly known as "parasitoids." The orders Diptera and Hymenoptera contain most parasitic insects. Somewhere about 300 families of Hymenoptera are home to parasitic wasps. Aphididae is a family of Braconidae that feeds specifically on aphids. Trichogrammatidae are parasitoids that feed on insect eggs. The families Aphelinidae, Encyrtidae, Eulophidae, and Ichneumonidae are also home to parasitoids. Don't worry about getting stung by one of these little to medium-sized wasps because they can't harm people. Most parasitic flies belong to the hairy Tachinidae family. Adult tachinid insects often resemble common house flies. The larvae of these insects resemble maggots and feed exclusively on the host. (Rao *et al.*, 2000)

Bacteria, fungi, nematodes, protozoa, and viruses are all examples of pathogens since they are capable of infecting and ultimately killing the host. Many insect pests, such as aphids, caterpillars, and mites, flourish in environments with high humidity or in places with significant populations of other insect pests. Natural disease outbreaks (epizootics) aren't the only source of beneficial pathogens; others are sold commercially as biological or microbial pesticides. The bacteria Bacillus thuringiensis (Bt), entomopathogenic nematodes, and granulosis viruses are only a few examples. For instance, Avermectins and spinosyns are microbial metabolites found in several pesticides, yet the use of these substances is not considered biological control.

**Table 1:** Examples of insect pest and their natura enemies

| | Natural enemies | Natural enemies | Natural enemies | Natural enemies | Natural enemies | Natural enemies |
|---|---|---|---|---|---|---|
| *Pests* | Lacewings | Lady beetles | Parasitic flies | Parasitic wasps | Predatory mites | Other groups and examples |
| *Aphids* | X | X | | X | | Entomopathogenic fungi, soldier beetles, syrphid fly larvae |
| *Carpenterworm,* clearwing moth larvae | | | | X | | Entomopathogenic nematodes |
| Caterpillars (e.g., *California oakworm)* | X | | X | X | | *Bacillus thuringiensis*, birds, entomopathogenic fungi and viruses, predaceous bugs and wasps, *Trichogramma* spp. (Egg parasitic wasps), spiders |
| Cottony cushion scale | | X | X | | | *Cryptochaetum iceryae* (Parasitic fly), *vedalia beetle* |
| Elm leaf beetle | | | X | X | | *Erynniopsis antennata* (Parasitic fly), *Oomyzus* (*Tetrastichus*) spp. (Parasitic wasps) |
| Eucalyptus long horned borers | | | | X | | *Avetianella longoi* (Egg parasitic wasp) |
| Eucalyptus redgum lerp psyllid | | | | X | | *Psyllaephagus bliteus* (Parasitic wasp) |
| Giant whitefly | X | X | | X | | *Encarsia hispida*, *Encarsia noyesi*, *Entedononecremnus krauteri*, and *Idioporus affinis* (Parasitic wasp), syrphid fly larvae |
| Glassy-winged sharpshooter | X | | | X | | Assassin bugs, *Gonatocerus* spp. (egg parasitic wasps), spiders |
| Lace bugs | X | X | | X | | assassin bugs and pirate bugs, spiders |

| | | | | | | |
|---|---|---|---|---|---|---|
| Mealybugs | X | X | | X | | Mealybug destroyer lady beetle |
| Mosquitoes | | | | | | Bacillus thuringiensis spp. israelensis, mosquito-eating fish |
| Psyllids | X | X | | X | | Pirate bugs |
| Scales | X | X | | X | X | *Aphytis*, *Coccophagus*, *Encarsia*, and *Metaphycus* spp. parasitic wasps |
| Slugs, snails | | | X | | | *Rumina decollata* (Predatory snail), predaceous ground beetles, birds, snakes, toads, and other vertebrates |
| Spider mites | X | X | | | X | Bigeyed bugs and minute pirate bugs, *Feltiella* spp. (Predatory cecidomyiid fly larvae), six spotted thrips, *Stethorus picipes* (Spider mite destroyer lady beetle) |
| Thrips | X | | | X | X | Minute pirate bugs, predatory thrips |

Throughout their lives, many predators will eat the flesh of several prey. Many amphibians, birds, mammals, and reptiles depend heavily on insects as a food source. Predatory insects such as wasps and flies feast on mites and other pest insects like ants and cockroaches. Most spiders consume only other insects. Predatory mites, such as *Amblyseius* spp., *Neoseiulus spp.*, and *Galendromus occidentalis*, mostly feed on pest spider mites.

**Recognizing natural enemies**

Proper identification of pests, including being able to tell them apart from their natural enemies, is crucial for successful biological control. Keep a close eye on the insects and mites on your plants and learn their behaviors. For instance, syrphid fly larvae are frequently misidentified as caterpillars. The larvae of the syrphid fly, on the other hand, feed exclusively on aphids. Your plants have mites? Get a good hand lens and examine it closely. (Rayl *et al.,* 2019) Mites that don't feed on other insects tend to be more sedentary. Predaceous mites are larger and more solitary than their pest mite counterparts. The cited materials provide further reading for those interested in plant pests and the predators and parasites that can help you control them. If you need help determining what kind of organism you're dealing with, you can call your county agriculture commissioner's office, participate in the UC Master Gardener programme, or visit your neighborhood UC Cooperative Extension office. (Rust *et al.,* 2012).

**Conservation: PROTECT your natural enemies**

Cultural, mechanical, or chemical constraints that are not damaging to the ecosystem should be used to maintain ecological equilibrium. Keep in mind that very few insects and mites cause harm. Effective biological management relies on preserving natural predators of pests. Important measures have been taken to control ants, protect their habitat, and reduce the usage of pesticides. (Rust *et al.,* 2012).

**Pesticide management**

When employing broad-spectrum, residual insecticides (those that remain for days or weeks) fails to prevent a pest outbreak or recurrence, biological control becomes vital. Malathion and other pesticides in the carbamate and organophosphate family have been associated to unwanted side effects, such as an increase in spider mite populations. Try to stay away from utilizing pesticides with a wide range of active ingredients. Commonly used poisons such as carbamates, organophosphates, and pyrethroids can leave behind residues that can be fatal to both native predators and newcomers to the area. When honeybees and other beneficial insects translocate (migrate) inside flowers, they can be poisoned by neonicotinoids, a class of systemic insecticides (e.g., dineto furan, imidacloprid) (Rao *et al.*, 2000). If beneficials are exposed to even trace amounts of pesticide residues, they may be less successful in reproducing or eliminating pests. Use caution and forethought when dispersing insecticides. Pests can only be effectively treated by applying "spot" treatments to specific locations. Bacillus thuringiensis (Bt) is better to other pesticides since it kills just the insects it is designed to target (in this case, only caterpillars on treated leaves). Contact insecticides such azadirachtin, insecticidal soap, restricted range oil (horticultural oil), neem oil, and pyrethrins, sometimes in combination with the synergist piperonyl butoxide, can be used to manage most other species of exposed-feeding insects. When the goal is to encourage biological control, it is best to use nonpersistent pesticides, as they pose less of a threat to natural enemies that move in following treatment. Spraying your plants should start in the spring, when the number of pests is at its highest (Tooker *et al.*, 2020).. It's likely that you'll need a command suite of tools. Spinosad, which is produced by fermenting naturally existing bacteria, is an alternative to conventional contact pesticides. This insecticide has a half-life of about a week, and it is translaminar, so it diffuses readily into plant tissue. Since Spinosad's toxicity lasts for up to 14 days after application, it should not be used on blooming plants to control predatory mites, Trichogramma wasps, or syrphid fly larvae. (Brévault *et al.*, 2019)

**Table 2:** Relative toxicity of insecticide to natural enemies

| | **Toxicity to parasites and predators** | **Toxicity to parasites and predators** |
|---|---|---|
| Insecticide | Direct | Residual |
| microbial (Bacillus thuringiensis) | no | no |
| botanicals (pyrethrins) | yes/no | no |
| oil (horticultural), soap (potash soap) | yes | no |
| microbial (spinosad) | yes/no | yes/no |
| neonicotinoids (imidacloprid) | yes/no | yes |
| carbamates (carbaryl*), organophosphates (malathion), pyrethroids (bifenthrin) | yes | yes |

1. Direct contact toxicity is killing within several hours from spraying the beneficial or its habitat. Residual toxicity is killing or sublethal effects (such as reduced reproduction or ability to locate and kill pests) due to residues that

| | |
|---|---|
| | persist. |
| 2. | Toxicity depends on the specific material and how it is applied and the species and life stage of the natural enemy. See the active ingredients database for more information about specific pesticides. |

## Ant control and honeydew producers

Honeydew is produced by aphids, mealybugs, soft scales, psyllids, and whiteflies, and is devoured by Argentine ants and other ant species. Ants protect honeydew producers from other animals. Farming' is a common way to talk about ants' practice of transporting honeydew-producing insects from one plant to another. If ant populations are managed in areas with existing natural enemies, many other pest populations can be reduced over the course of several generations. The soil can be farmed near ant nests, insecticide baits can be placed near plants, and sticky ant barriers can be built around tree trunks. Additional reading is provided by cockroaches and ants. (Brévault *et al.,* 2019).

## Habitat manipulation

The best approach to ensure that beneficial insects have access to nectar, pollen, and shelter is to plant a wide variety of species that bloom at different times throughout the growing season. The larvae of many predatory insects (including green lacewings and syrphid flies) eat only on pollen and nectar, and many parasites' adult stages do the same. Many beneficials can't thrive without flowering plants that attract pollinators and provide nectar for their larvae to feed on. Plants that are native to the area and resistant to low populations of plant-feeding insects and mites should be cultivated to ensure that predators and parasites always have access to food. Cultural measures, such as limiting dust and adequately fertilizing and watering crops, can aid natural enemies. The effectiveness of predatory insects can be diminished by dust, allowing pests like spider mites to flourish. Dusty plants should be watered, and wind barriers and ground cover can be planted to lessen the problem. The fast proliferation of phloem-feeding pests like aphids can overwhelm the ability of natural enemies to keep them under control, making it vital to minimize fertilization and irrigation (Thompson *et al.,* 2022).

## Augmentation

When resident natural enemies are insufficient, their populations can sometimes be increased (optimally) by collecting and releasing wild or captive-bred animals for profit. However, there is a lack of information on how unleashing natural enemies in landscapes and gardens affects the environment. The release of an object into the environment rarely serves as a sufficient deterrent against pests. Store-bought all-natural pest control options have mixed success. Many predators have trouble thriving because they consume both potentially dangerous and beneficial species. (Tooker *et al.,* 2020). It's important to selectively promote natural enemies in outdoor spaces like gardens and landscapes. Nematodes that attack insects that feed on plants and grass are called entomopathogens. You can buy convergent lady beetles (Hippodamia convergens) in bulk online and store them in the refrigerator until you need them to control aphid numbers. While lady beetles can be an effective management tool, buying them from a store is probably not a long-term answer. The effectiveness of an improvement depends on several aspects, such as careful planning, extensive understanding of biology, the right moment of release, patience, and a place where pests and damage may be endured. When pests or damage are widespread, it's not a smart idea to amp things up (Rust *et al.,* 2012).

**Classical biological control or importation**

Importation, or classical biological control, is a common method for dealing with exotic pests that have been accidentally brought to a new region. Many creatures, even ones that cause no difficulties in their native habitats, can cause serious problems when introduced to a new one. Scientists often travel to an invasive species' native area in search of potential natural adversaries. Except in areas where they are disrupted by pesticide applications or honeydew-seeking ants, for example many insects and several plants that were historically prominent pests in California are now managed by introduced natural enemies (Rust *et al.,* 2012). Importation of predators and parasites is restricted to qualified scientists in possession of the appropriate licenses. To prevent any potential danger, scientists' study natural enemies while keeping them in a secure quarantine facility. This process is repeated until the natural enemy poses no significant threat to the developing nation. Managers of outdoor spaces must keep an eye out for and protect newly imported predators and parasites. Traditional biological control, which is carried out by authorities and institutions using tax dollars, is highly dependent on public backing if it is to continue offering long-term benefits across a large geographical area (Tooker *et al.,* 2020).

**Is biological control "Safe"?**

When compared to the widespread use of broad-spectrum insecticides, the side effects of biological management are much lower. Problems with alien species often arise from three sources: the entrance of unwanted organisms at the border, the import of pest-infested fruit by tourists, and the escape of foreign ornamentals from cultivation as weeds. These improper or illegal imports have nothing to do with the necessity of biological management. As a type of "quasi-biological control," the importation of predatory animals including frogs, mongooses, and different fish has been implemented, but the results have been unsatisfactory. Following government quarantine restrictions and focusing on host-specific natural enemies that offer minimal risk but give large advantages, scientists engaged in biological control reduce the severity of these issues (Thompson *et al.*, 2022). Because host-specific natural enemies cannot hunt or reproduce on other species, population numbers of both the pest and the biological control agent decrease as the pest is brought under biological control.

## Conclusion

When traditional chemical pesticides have failed, it may be time to adopt a biological control strategy. Through harnessing the complex network of interdependencies that governs our ecosystems, it enables humans to collaborate with rather than compete with nature. The compounds that have been fighting pests for eons without getting any recognition are the real heroes here, not the manmade chemicals. New methods of biological control are being developed as research and study continue. Potential future treatments may be more targeted and effective thanks to genetic engineering and other state-of-the-art technologies. Our expanding knowledge proves that we are the vanguard of the global anti-pest movement. We anticipate that biological methods of control will prove to be the most effective and trustworthy in the long run. We fully understand how crucial it is to lessen our impact on the environment and protect our species. Taking the necessary safety measures now will not upset the natural order of things. When it comes to biological warfare, even the tiniest of organisms can prove to be formidable foes. Natural predators like ladybirds and wasps are just two examples that help keep pests away from our crops. Their accomplishment demonstrates the power that may come from a relatively small number of enthusiastic backers.

## References

1. Dreistadt, S.H., M.L. Flint, and J.K. Clark. 2004. *Pests of Landscape Trees and Shrubs: An Integrated Pest Management Guide*. 2nd ed. Oakland: Univ. Calif. Agric. Nat. Res. Publ. 3359.
2. Flint, M.L. and S.H. Dreistadt. 1998. *Natural Enemies Handbook: The Illustrated Guide to Biological Pest Control*. Oakland: Univ. Calif. Div. Agric. Nat. Res. Publ. 3386.

3. Rust, M.K. and D.-H. Choe. 2012. *Pest Notes: Ants*. Oakland: Univ. Calif. Agric. Nat. Res. Publ. 7411.

4. Tooker, J. F., O'Neal, M. E., & Rodriguez-Saona, C. (2020). Balancing disturbance and conservation in agroecosystems to improve biological control. *Annual Review of Entomology*, *65*, 81-100.

5. Rao, M. R., Singh, M. P., & Day, R. (2000). Insect pest problems in tropical agroforestry systems: Contributory factors and strategies for management. *Agroforestry Systems*, *50*, 243-277.

6. Thompson, M. N., Medina, R. F., Helms, A. M., & Bernal, J. S. (2022). Improving Natural Enemy Selection in Biological Control through Greater Attention to Chemical Ecology and Host-Associated Differentiation of Target Arthropod Pests. *Insects*, *13*(2), 160.

7. Shanker, C., Katti, G., Padmakumari, A. P., Padmavathi, C., & Sampathkumar, M. (2011). Biological control, functional biodiversity, and ecosystem services in insect pest management. In *Crop stress and its management: perspectives and strategies* (pp. 471-495). Dordrecht: Springer Netherlands.

8. Brévault, T., & Clouvel, P. (2019). Pest management: reconciling farming practices and natural regulations. *Crop Protection*, *115*, 1-6.

9. Rayl, R. J., Shields, M. W., Tiwari, S., & Wratten, S. D. (2018). Conservation biological control of insect pests. *Sustainable Agriculture Reviews 28: Ecology for Agriculture*, 103-124.

# Chapter - 20

## The Role of Digital Technology in Agriculture

**Authors**

**Th Onchoila Maring**

Ph.D. Scholar, Department of Agricultural Extension, School of Social Sciences, College of Post Graduate Studies in Agricultural Sciences, CAU (I), Meghalaya, India

**Nang Priyanka Langkhun**

Ph.D. Scholar, Department of Genetic and Plant Breeding, School of Crop Improvement, College of Post Graduate Studies in Agricultural Sciences, CAU (I), Meghalaya, India

**Snata Kaushik**

Ph.D. Scholar, Department of Plant Pathology, School of Crop Protection, College of Post Graduate Studies in Agricultural Sciences, CAU (I), Meghalaya, India

**Pankaj Kumar**

Ph.D. Scholar, Department of Agricultural Extension, School of Social Sciences, College of Post Graduate Studies in Agricultural Sciences, CAU (I), Meghalaya, India

# Chapter - 20

## The Role of Digital Technology in Agriculture

**Th Onchoila Maring, Nang Priyanka Langkhun, Snata Kaushik and Pankaj Kumar**

### Abstract

Digital agriculture revolutionizes the farming process, providing significant benefits to developing countries by tackling rural poverty and hunger. This transformation is driven by advancements like indoor vertical farming, automation, livestock tech, precision agriculture, Artificial Intelligence (AI), and blockchain. The integration of digital technology across various agricultural areas, such as machinery, agronomy, and communication, has created widespread connectivity. The use of mobile devices has further enhanced efficiency and accessibility. Notably, artificial intelligence, the Internet of Things (IoT), and mobile solutions play pivotal roles in driving these advancements. Consequently, digital agriculture offers numerous advantages, including improved communication, monitoring, and decision-making. It also fosters time and cost savings, heightened operational efficiency through targeted applications, increased productivity, and profitability. Additionally, the sector benefits from improved marketing strategies, real-time information access, streamlined record-keeping, better risk management, and reduced regulatory burdens. Embracing these digital tools empowers agriculture to be more sustainable and efficient, surpassing conventional methods in the quest for a more productive and prosperous future.

**Keywords:** Digital agriculture, precision agriculture, AI, IoT and blockchain.

### 1. Introduction

Throughout history, the agricultural industry has undergone three significant revolutions that have transformed food production techniques and technologies. The first revolution marked the transition from a hunter-gatherer lifestyle to settled agriculture, where humans began cultivating crops and raising livestock. The second revolution introduced mechanization, with the advent of tractors and combines. These machines revolutionized farming practices by increasing efficiency in tasks like tilling, planting, and harvesting, reducing manual labor and boosting productivity. The third revolution brought about the use of agrochemicals, such as pesticides and fertilizers, along with advancements in genetic engineering. These technologies aimed to enhance crop yields, improve pest resistance, and optimize plant characteristics.

Now, the agricultural industry stands on the cusp of a fourth revolution: digital agriculture. This revolution involves integrating advanced technologies across the entire agricultural value chain to improve efficiency, productivity, and sustainability. Drones, along with other technologies like artificial intelligence, robotics, and sensors, play a crucial role in digital agriculture. Drones enable enhanced data collection, storage, and analysis, providing valuable insights for decision-making processes in agriculture. They can gather data on crop health, animal behavior, and environmental conditions, allowing farmers to make informed predictions and take appropriate actions.

In summary, digital agriculture represents the fourth revolution in the agricultural industry, and technologies like drones, robotics, and AI are shaping the future of farming by improving data-driven decision-making, automating tasks, and fostering connectivity throughout the supply chain.

## 1.1 Digital agriculture

Digital agriculture refers to the use of digital technology in agriculture to integrate the entire process from the farm to the consumer. This integration can help developing countries overcome poverty and hunger in rural areas more quickly. There are several major technological advancements in digital agriculture, including indoor vertical farming, automation and robotics, livestock technology, modern greenhouse practices, precision agriculture, artificial intelligence, and blockchain. Farm equipment is connected to software platforms that capture data on soil and climate conditions in specific locations. This data can provide farmers with guidance on seed choice and more precise application of pesticides and fertilizers.

Mobile phones have one of the highest adoption rates of any technology developed in the last century. This widespread use of mobile phones can help farmer's access information and connect with consumers more easily. Digitalisation will bring consumers and farmers closer together by providing extensive information on plants and animals, making farming more transparent to consumers. Distribution of knowledge, advisory services, and knowledge sharing, improving accessibility, efficiency, and inclusivity of extension services. Digital technology has led to increased connectivity across the world, with mobile devices being used in a smaller, faster, cheaper, and more efficient way.

This technology is assisting businesses and farmers in making more informed decisions, particularly in agriculture. Farmers are able to use digital technologies to maintain better control over their operations, including using more accurate amounts of water and fertilizers. Digital technologies also make it easier for farmers to plan farming activities, budget, report, and monitor multiple tasks and performances. Digital technology is used in various areas of agriculture, including farm machinery, livestock handling facilities, agronomy, communication, and more.

Advanced technologies such as artificial intelligence, analytics, and connected sensors have the potential to increase yields and improve the efficiency of water and other inputs. These technologies can also help in building sustainability and resilience across crop cultivation and animal husbandry.

Digital farming technologies, such as sensors, automation, and robots, are being applied in production systems to improve efficiency and productivity.

The use of these technologies can lead to more precise application of pesticides and fertilizers, which can reduce waste and improve crop yields.

Additionally, the use of digital technology can help farmers make more informed decisions about their operations, such as planning farming activities, budgeting, reporting, and monitoring multiple tasks and performances.

## 1.2 Importance of digital technology in agriculture

Technological advancements have had a profound impact on various aspects of agriculture, including fertilizers, pesticides, seed technology, mechanization, irrigation, and transportation systems. These developments have played a crucial role in increasing crop yields, improving food and fiber quality, reducing manual labor, and achieving food self-sufficiency. Digital agriculture offers numerous benefits, making farming more sustainable and efficient compared to traditional methods. By incorporating technologies such as artificial intelligence, the Internet of Things (IoT), mobile solutions, and other digital tools, the agricultural sector can experience significant advantages:

**Enhanced communication:** Digital agriculture streamlines communication by automating workflows and eliminating manual data collection and processing. Technologies like RFID tags and mobile computing enable real-time data collection in the field, improving operational efficiency and providing better visibility throughout the food supply chain.

**Effective monitoring:** Farmers can use technology to monitor the health of livestock and crops more effectively. Wearable sensors for livestock and drones for field surveys and soil assessment help farmers track temperature, plan seed planting patterns, and manage weed and pest activity.

**Informed decision-making:** Digital technology provides farmers with detailed insights to understand their performance better and make informed decisions. Big Data analysis helps interpret the vast amount of collected data, offering valuable information on planting, harvesting, pricing, and logistics to optimize operations and prevent potential issues.

**Time and cost savings:** Digital agriculture saves farmers time and money through increased efficiency and visibility. GPS technology and advanced insights help optimize field layout, leading to higher yields and faster completion of tasks, thereby reducing costs. Although initial technology investments may be required, the long-term time and cost savings outweigh the upfront expenses.

**Increased operational efficiency through targeted applications:** Digital technologies enable farmers to optimize their resource utilization by precisely applying inputs like fertilizers, water, and pesticides, leading to improved efficiency on the farm.

**Boosted productivity and profitability:** By leveraging digital tools, farmers can enhance their productivity, maximize yield potential, and ultimately increase their profitability.

**Improved marketing strategies:** Digital platforms and technologies provide farmers with the means to directly connect with consumers, expand their market reach, and effectively promote their agricultural products.

**Real-time access to information:** Digital farming provides farmers with real-time information on weather conditions, market trends, and crop performance, enabling them to make timely and informed decisions.

**Streamlined record-keeping practices:** Digital solutions simplify and streamline record-keeping processes, ensuring accurate documentation for compliance, financial analysis, and traceability purposes.

**Effective risk and uncertainty management:** Through digital tools and insights, farmers can better identify and manage risks, helping them navigate uncertainties and challenges in their agricultural operations.

**Reduction in regulatory burdens:** Digital solutions alleviate the administrative burdens associated with regulatory compliance, allowing farmers to focus more on their core activities and reducing the time and effort spent on meeting regulatory requirements.

### 1.3 Overview of key digital technology in agriculture

Digital technology plays a crucial role in transforming agriculture and driving innovation across various aspects of the industry. Here is an overview of key digital technologies in agriculture:

1. **Internet of Things (IoT):** IoT involves the connectivity of devices and sensors that collect and transmit data. In agriculture, IoT enables the monitoring of soil moisture, temperature, and other environmental factors. This data helps farmers make informed decisions regarding irrigation, pest management, and crop health.
2. **Agriculture sensors:** These wireless sensors provide farmers with data to optimize crop management based on environmental conditions. They assist in soil analysis, moisture detection, nutrient monitoring, precise location tracking, and airflow assessment. By using sensors, farmers can reduce pesticide and labor costs while efficiently applying fertilizers, maximizing yields, and conserving natural resources.
3. **Artificial Intelligence (AI):** AI technologies, such as machine learning and predictive analytics, analyze vast amounts of agricultural data to provide valuable insights. AI can assist in crop disease identification, yield prediction, optimization of resource allocation, and automated decision-making processes.
4. **Robotics and automation:** Robotic systems are increasingly used in agriculture for tasks like planting, harvesting, and weeding. These technologies improve precision and efficiency while reducing labor costs and physical strain on farmers.
5. **Remote sensing and satellite imaging:** Remote sensing and satellite imaging techniques provide real-time and accurate information on crop conditions, vegetation health, and soil moisture levels. This data helps farmers monitor crop growth, detect diseases, and optimize irrigation practices.

6. **Precision agriculture:** Precision agriculture utilizes GPS, GIS, and data analytics to optimize farm operations. Farmers can create detailed field maps, apply fertilizers and pesticides precisely, and monitor crop variability to maximize productivity and minimize resource waste.
7. **Farm management software:** Farm management software enables farmers to streamline operations, manage inventory, track machinery maintenance, and monitor financial aspects of their business. These tools enhance efficiency and enable data-driven decision-making.
8. **Blockchain technology:** Blockchain offers secure and transparent data management and transaction systems. In agriculture, blockchain can enhance traceability, supply chain management, and food safety by recording and verifying every stage of production, processing, and distribution.
9. **Big data analytics:** Big data analytics involves analyzing large and complex data sets to derive meaningful insights. In agriculture, it aids in crop modeling, weather forecasting, market trends analysis, and supply chain optimization.
10. **Drones:** Drones have gained widespread usage in agriculture, particularly in China where they are employed to survey vast areas of cotton crops covering millions of hectares. Drones provide valuable insights that may not be easily observable from the ground, aiding in determining optimal harvest timing, monitoring irrigation needs, implementing pest protection measures, and more.
11. **Cloud connectivity:** Cloud-based connectivity revolutionizes the digital solutions available to farmers by leveraging real-time internet connections. This upgrade is especially significant as many farmers still lack connectivity. Embracing cloud connectivity enables farmers to access a broader range of digital tools and services, empowering them to make real-time decisions and accomplish more efficiently. It also offers the potential for economies of scale and enhanced productivity in agricultural operations.

## 2. Precision agriculture and digital tools

Precision farming, also known as precision agriculture, is a modern approach to managing agricultural operations by utilizing specific information regarding nutrient management, remote sensing, global information systems, global positioning systems, and variable rate application. Its goal is to effectively optimize the use of production inputs by performing the appropriate actions, in the correct locations, using the right methods, at the optimal times, and following the appropriate procedures. By carefully managing inputs such as water, seeds, and fertilizers, precision farming aims to enhance crop yield, quality, and profitability while minimizing waste and promoting environmental sustainability. The objective of precision farming is to align agricultural inputs and practices with the specific requirements of crops and agro-climatic conditions, thereby improving the precision and accuracy of their application.

**Principles**

**1. Geolocating**

Geolocating is an important method for gathering and analyzing information about fields, such as soil quality, previous crop production, and nitrogen levels. There are two ways to do this: using GPS receivers in vehicles like tractors while on the ground, or by using satellite imagery to create field maps.

**2. Variables**

Variables play a crucial role and depend on factors like climate, crop production methods, soil management, cleanliness, and disease. These variables provide farmers with valuable information about the environment, helping them track crop status, implement disease control measures, and optimize crop production by monitoring factors like nitrogen and water stress.

**3. Strategies**

Strategies based on soil maps enable farmers to make informed decisions and effectively manage crop production. By analyzing big data and considering both business value and environmental impact, farmers can leverage modern technology to improve their economic situation.

**4. Implementing practices**

Implementing practices involve using advanced communication technologies and equipment in crop management processes. These practices make it easier for farmers to apply techniques like varying seed density, improving operational efficiency and effectiveness.

## 2.1 Role of sensors, drones, satellite imaginary in precision agriculture

### 2.1.1 Role of sensors in precision agriculture

A sensor is a device that detects and reacts to specific inputs like light, motion, pressure, heat, or moisture, converting them into signals readable by humans for analysis and interpretation.

**Electromagnetic sensors:** Electromagnetic sensors utilize electric circuits to measure the electrical conductivity of soil particles. Two types of EC sensors exist: contact and non-contact. Contact sensors have electrode colters that make direct contact with the soil, while non-contact sensors use electromagnetic induction without physical contact. These sensors record changes in conductivity and are widely used in precision agriculture due to their ability to cover large areas and resistance to external interference.

**Optoelectronic sensors:** Optoelectronic sensors utilize the interaction between electronic processes and light to capture specific spectral signatures. They consist of a light emitter and a detector that operate in defined wavelength ranges. Optoelectronic sensors are used for various purposes, including measuring soil organic matter (SOM), soil moisture, and cation exchange capacity (CEC). Multiple wavelength sensors provide more flexibility and can be calibrated to predict SOM over different soil types and moisture levels.

**Electrochemical sensors:** Electrochemical sensors are crucial for assessing nutrient availability and pH in the soil. They involve standardized laboratory procedures and often employ ion-selective electrodes to measure specific ions such as nitrate, potassium, or hydrogen (for pH). While not as accurate as laboratory tests, electrochemical sensors provide valuable data, especially when used with high sampling density.

**Ion selective electrode sensors:** Ion selective electrode sensors measure the potential of a specific ion in a solution and are commonly used to determine nitrogen content in the soil. The potential difference between the electrode and a stable reference electrode reflects the ion's activity, which is proportional to its concentration in the solution. This enables users to make analytical measurements of ion concentrations.

**Mechanical sensors:** Mechanical sensors assess the mechanical resistance of the soil by penetrating or cutting through it and measuring the resulting force using strain gauges or load cells. Airflow sensors measure soil air permeability, while acoustic sensors evaluate changes in noise levels caused by interactions between the sensor tool and soil particles, providing information on soil texture.

**Dielectric soil moisture sensors:** Dielectric soil moisture sensors estimate soil moisture levels based on the dielectric constant, an electrical property that changes with moisture content. These sensors are often used alongside precipitation monitoring locations on the farm.

### Agriculture Sensors IoT

Agriculture sensors connected to the Internet of Things (IoT) offer real-time information on various parameters such as temperature, rainfall, soil health, and crop status. This data helps farmers make informed decisions regarding irrigation, harvesting, and resource management.

Location sensors, such as agricultural weather stations, are positioned throughout fields to determine position, distance, and height using GPS satellites.

### 2.1.2 Role of drones in precision agriculture

**Crop surveying and mapping:** Drones have revolutionized crop monitoring and mapping by providing real-time and accurate footage, unlike the costly and less precise satellite imagery used previously. Farmers now have access to up-to-date data and can track crop progress in real-time. This enables them to make informed decisions based on current information, improving overall farm management and productivity.

**Weeds and pest control management:** Precision agriculture utilizes drones equipped with advanced cameras, sensors, and GPS technologies to accurately track and manage weeds and pests in various crops. By flying over fields and capturing photographic data, drones enable rapid identification of weed and pest infestations. This allows farmers to take prompt action, implementing targeted strategies for weed control and reducing reliance on herbicides. Drones aid in minimizing crop damage, lowering costs, and promoting biodiversity in agricultural practices.

**Soils inspection:** Drones are commonly employed to assess soil conditions, replacing the traditional manual field visits and soil sampling methods. Equipped with sensors and advanced equipment, drones provide accurate soil examinations and detailed reports.

**Irrigation and water management:** Irrigation issues pose challenges for farmers, but drones equipped with thermal cameras can help identify problems like inadequate or excessive moisture. By detecting such issues, crops can be managed to optimize drainage, prevent water pooling, and protect sensitive crops, ultimately avoiding costly water-related problems and potential yield losses. Drone surveys can proactively address irrigation concerns before they escalate.

**Spraying management:** Drones, being aerial devices, are highly effective for precision agriculture spraying. Equipped with sensors, they collect data from farmlands and apply pesticides and fertilizers precisely where needed. Drone sprayers eliminate the risks associated with manual spraying and deliver superfine spray applications, improving efficiency and reducing pesticide expenses. This technology enables targeted spraying, enhancing effectiveness while minimizing health hazards and costs for farmers and businesses.

### 2.1.3 Role of satellite imaginary in precision agriculture

Satellite imagery enables remote monitoring of crop areas, providing valuable information on water demands across farms. This allows farmers to assess and determine optimal water requirements for each plot, improving irrigation planning and management efficiency. By using satellite imagery analysis, farmers can make informed decisions regarding water usage and prevent over or under irrigation, leading to cost savings, enhanced crop quality, and increased productivity

Furthermore, satellite imagery contributes to enhanced climate resilience in agriculture. It allows farmers to monitor climatic conditions and predict weather disasters, such as floods, droughts, or untimely rains. This early knowledge enables farmers to prepare and take preventive measures to protect their crops, mitigating the impact of extreme weather events on agriculture. By providing insights into fluctuating weather patterns, satellite imagery supports the adoption of climate-resilient farming practices

Moreover, the integration of satellite imagery with Internet of Things (IoT) technology facilitates precision agriculture practices. High-resolution satellite images combined with smart devices like drones and agricultural robots enable precise and targeted applications of agrochemicals, such as fertilizers and pesticides. This minimizes chemical usage, improves pest management, and promotes sustainable cultivation practices while reducing crop failures.

Satellite imagery also empowers agricultural research and development (R&D) efforts. It provides valuable data for monitoring soil health, analyzing crop productivity, and understanding climate change impacts. This supports the development of innovative farming techniques and improved crop yields, contributing to food security and sustainability

Additionally, satellite imagery aids in farm planning through accurate yield estimations. By integrating satellite imagery with enterprise resource planning (ERP) platforms, farmers can detect vegetation, moisture, and nutrient stress issues, leading to precise yield estimates. This information enhances agricultural planning, storage, and insurance decisions, improving overall farm management

Satellite imagery plays a pivotal role in achieving sustainability in agriculture. It enables efficient water management, climate resilience, precise agrochemical usage, precision farming practices, agricultural R&D, and accurate yield estimations. By harnessing the power of satellite imagery, farmers and agribusinesses can optimize resource utilization, increase productivity, and promote sustainable agricultural practices.

## 3. Farm management system and decision support tool

Farm management systems (FMS) refer to the integration of advanced technologies, software, and practices to efficiently manage agricultural operations. These systems are designed to optimize the entire farming process, from planning and planting to harvesting and marketing. By leveraging data-driven decision-making, automation, and real-time monitoring, farm management systems offer numerous benefits to farmers, agricultural businesses, and the environment.

### Key components of farm management systems

1. Data Collection and Analysis: FMS utilizes various data sources, such as satellite imagery, sensors, drones, and weather forecasts, to collect valuable information about soil health, crop growth, pests, and other critical factors affecting farming.
2. Planning and Resource Allocation: By analyzing the data, farmers can make informed decisions on resource allocation, including the optimal use of water, fertilizers, and pesticides. This helps minimize waste and reduces production costs.
3. Precision Agriculture: FMS enables precision agriculture practices, such as variable rate application of inputs, allowing farmers to tailor treatments to specific areas of their fields based on varying needs.
4. Automation: Automation technologies, like autonomous tractors and robotic harvesters, can streamline labor-intensive tasks, leading to increased efficiency and reduced labor costs.
5. Record Keeping and Compliance: Farm management systems help farmers maintain detailed records of their operations, which is essential for compliance with regulations, certifications, and quality standards.
6. Market Analysis and Decision Support: FMS can provide valuable

insights into market trends, enabling farmers to make informed decisions about crop selection and timing for optimal profitability.

**Benefits of farm management systems**

1. Increased Productivity: With precise resource allocation, timely interventions, and optimized practices, FMS can significantly enhance farm productivity. This means higher yields and better quality produce.
2. Cost Reduction: By avoiding over-application of inputs and reducing manual labor requirements, FMS can lead to cost savings and improved profitability for farmers.
3. Sustainability: FMS promotes sustainable agriculture by optimizing resource usage, reducing environmental impacts, and promoting biodiversity.
4. Improved Decision-making: Access to real-time data and analytical tools empowers farmers to make more informed and data-driven decisions, resulting in better outcomes.
5. Risk Mitigation: Farm management systems allow farmers to identify and address potential issues early, reducing the risk of crop failure or losses due to unforeseen events.
6. Time Savings: Automation and streamlined processes free up farmers' time, allowing them to focus on other critical aspects of their operations or enjoy a better work-life balance.
7. Traceability and Quality Assurance: With better record-keeping and traceability, FMS can help ensure that agricultural products meet quality and safety standards, increasing consumer confidence.
8. Scalability: Farm management systems can be applied to farms of all sizes, making them beneficial for both small-scale and large-scale agricultural operations

## 3.1 Integration of data and technologies for farm planning and scheduling

The integration of data and technologies for farm planning and scheduling has revolutionized the way farmers manage their operations. By harnessing the power of data collection through various sources, such as sensors, satellites, drones, and weather forecasts, farmers now have access to a wealth of information about their fields' conditions and performance. This data-driven approach enables them to make well-informed decisions regarding crop selection, planting schedules, and resource allocation. With precision agriculture practices, such as variable rate application of inputs, farmers can tailor their treatments to suit specific areas of their fields, optimizing resource usage and maximizing yield potential.

Furthermore, farm management software and digital tools allow for efficient planning and scheduling of farming activities. These technologies provide real-time updates and insights, ensuring that farmers can adapt swiftly to changing conditions and unforeseen events. They can plan planting and harvesting timelines, manage irrigation schedules, and allocate labor and machinery resources more effectively, leading to increased efficiency and reduced operational costs.

The integration of data and technologies also plays a significant role in risk management. By continuously monitoring weather patterns and pest and disease outbreaks, farmers can proactively implement preventive measures and respond promptly to potential threats. This proactive approach helps mitigate risks and minimize potential crop losses, ultimately enhancing the resilience of their farming operations.

### 3.2 Decision support tool for optimizing resource allocation and improving efficacy

A decision support tool for optimizing resource allocation and improving efficacy in agriculture is a software or system that assists farmers in making data-driven decisions to maximize productivity and efficiency. Such tools typically use data from various sources, like sensors, satellite imagery, weather forecasts, and historical records, to provide insights and recommendations for resource allocation, crop management, and other farm operations. Here are some examples of decision support tools used in agriculture:

1. Crop Monitoring and Advisory Systems: These tools use satellite imagery and sensor data to monitor crop health and growth throughout the season. Farmers receive real-time insights about nutrient deficiencies, pest infestations, and other stressors affecting the crop. Based on this information, the tool provides recommendations for optimized fertilization, pest control strategies, and irrigation scheduling.
2. Variable Rate Technology (VRT) Systems: VRT tools integrate data from soil tests and crop health assessments to apply inputs, such as fertilizers and pesticides, at variable rates based on the specific needs of different areas within a field. This precision application maximizes the efficacy of inputs and minimizes waste, leading to cost savings and improved environmental sustainability.

3. Irrigation Management Systems: These decision support tools use soil moisture data from sensors placed in the field to schedule irrigation based on actual crop water needs. By avoiding under- or over-irrigation, farmers can optimize water usage, enhance crop performance, and reduce the risk of water-related stress on plants.
4. Integrated Pest Management (IPM) Software: IPM tools combine weather data, pest modeling, and historical pest occurrence records to predict pest outbreaks and provide early warning alerts. This helps farmers implement targeted and timely pest control measures, reducing the use of chemical pesticides and improving overall efficacy in pest management.
5. Farm Management Software: Comprehensive farm management software integrates various data streams, including financial records, equipment data, and yield data, to offer a holistic view of farm operations. These tools aid in planning and scheduling tasks, managing resources, and tracking performance, enabling farmers to make informed decisions for optimizing productivity and efficiency.
6. Livestock Performance Monitoring Systems: Decision support tools for livestock management collect data from wearable devices or sensors on animals to monitor health, behavior, and feed intake. Farmers can analyze this data to adjust feed rations, detect health issues early, and optimize livestock performance.
7. Market Analysis and Pricing Tools: Decision support tools that integrate market analysis and pricing data help farmers make informed decisions on crop selection and timing for selling their produce. These tools enable farmers to strategize sales to maximize profitability.

### 3.3 Real-time monitoring and reporting in a farm management system

Real-time monitoring and reporting in a farm management system have revolutionized the way farmers oversee their agricultural operations. With the continuous stream of data from sensors and other technologies, farmers gain unprecedented visibility into their fields and livestock, enabling them to monitor vital parameters at all times. For instance, soil moisture sensors provide real-time information about the moisture content, allowing farmers to adjust irrigation schedules and prevent water stress in crops. Weather monitoring ensures that farmers stay informed about weather changes, reducing the risk of weather-related losses by enabling timely actions, such as sheltering crops or livestock. Drones equipped with multispectral cameras can rapidly scan large areas, providing instant insights into crop health, identifying nutrient deficiencies, and detecting early signs of diseases or pests. Livestock tracking systems enable real-time monitoring of animal health and behavior, ensuring timely interventions in case of any abnormalities. The real-time data collected from these monitoring systems is then analyzed and presented through comprehensive reports. These reports offer data-driven insights and actionable recommendations, allowing farmers to make informed decisions swiftly. By leveraging the power of data analytics, farmers can identify trends, predict potential challenges, and optimize resource allocation. For example, if the data indicates that a particular crop is performing exceptionally well in a specific region of the field, farmers can adjust planting strategies for the following season to maximize yield potential. Similarly, by analyzing equipment performance data, farmers can identify maintenance needs and prevent costly breakdowns, thereby increasing overall operational efficiency.

Moreover, real-time monitoring and reporting enable farmers to comply with regulatory requirements and maintain quality assurance standards. By keeping detailed records of operations and data-driven practices, farmers can easily demonstrate their adherence to sustainability and environmental guidelines, promoting consumer confidence in their products.The continuous flow of real-time information and the availability of comprehensive reports also facilitate better communication and collaboration among farm teams and stakeholders. Farm managers, agronomists, and other stakeholders can access the same data and reports, fostering collaborative decision-making and ensuring that everyone is on the same page regarding the farm's status and goals.

## 4. Internet of things (IoT) and smart farming

### 4.1 Introduction to IoT in agriculture and its application

The adoption of technology in agriculture has seen remarkable growth, with various emerging technologies holding great promise for the future of farming. Among these, the Internet of Things (IoT) has transitioned from an emerging technology to a mainstream position due to its widespread adoption. In its simplest terms, IoT refers to the concept of the internet controlling things. IoT devices are "smart" devices capable of transferring data over a network. The term "Internet of Things" was coined in 1999 by Kevin Ashton, who introduced the concept while presenting radio-frequency identification (RFID) tags for supply chain management at Procter and

Gamble. The first network-connected device, a vending machine at Carnegie Mellon University in 1982, marked the beginning of IoT's journey. LG Electronics later introduced the Internet Digital DIOS refrigerator in 2000, which utilized IoT technology to track stored food items via RFID tags. Although initially met with limited success, it paved the way for more household appliances to be connected to the internet. The term "Internet of Things" gained prominence between 2008 and 2009 when the number of connected devices surpassed the world's human population, marking a significant turning point in technological advancement. Today, IoT plays a vital role in agriculture, offering opportunities for enhanced efficiency and data-driven decision-making in farm management.

### IoT technologies in agriculture

IoT advancements have given rise to smart agricultural solutions, empowering farmers to remotely oversee their crop fields with the aid of sensors and automated irrigation systems. This convenience grants farmers and relevant stakeholders the ease of monitoring field conditions from any remote location. The varied applications of IoT in agriculture encompass a wide range of solutions.

### a) Robotics in agriculture

In the wake of the industrial revolution in the 1800s, automation has continuously evolved to efficiently manage complex tasks and increase agricultural production. However, with the global surge in demand for agricultural products and a shortage of labor, a new solution has emerged - agriculture robots, commonly known as Agribots. These innovative machines are garnering attention among farmers due to their potential to address labor shortages and enhance farming practices. Leveraging recent advancements in sensors and AI technology, Agribots have become more remarkable in their ability to learn from their surroundings. The agricultural sector is now at the forefront of an ag-robotics revolution, fully embracing the potential of the Internet of Things (IoT) in agriculture. Many Agribot products are still in their early trial phases and undergoing research and development.

### Weeding robot

One significant application of Agribots in agriculture is weeding robots. These smart Agri robots utilize digital image processing techniques to compare images of weeds in their database with crops, enabling them to identify and effectively remove or spray weeds directly using their robotic